DSP-BASED ELECTROMECHANICAL MOTION CONTROL

POWER ELECTRONICS AND APPLICATIONS SERIES

Muhammad H. Rashid, Series Editor
University of West Florida

PUBLISHED TITLES

Complex Behavior of Switching Power Converters
Chi Kong Tse

DSP-Based Electromechanical Motion Control
Hamid A. Toliyat and Steven Campbell

Advanced DC/DC Converters
Fang Lin Luo and Hong Ye

FORTHCOMING TITLES

Renewable Energy Systems: Design and Analysis with Induction Generators
Marcelo Godoy Simoes and Felix Alberto Farret

DSP-BASED ELECTROMECHANICAL MOTION CONTROL

Hamid A. Toliyat
Steven Campbell

Texas A&M University
Department of Electrical Engineering
College Station, Texas

CRC Press is an imprint of the
Taylor & Francis Group, an informa business

CRC Press
Taylor & Francis Group
6000 Broken Sound Parkway NW, Suite 300
Boca Raton, FL 33487-2742

First issued in paperback 2019

ISBN-13: 978-0-8493-1918-1 (hbk)
ISBN-13: 978-0-367-39496-7 (pbk)
Library of Congress Card Number 2003058462

Library of Congress Cataloging-in-Publication Data

Toliyat, Hamid A.
DSP-Based electromechanical motion control / by Hamid A. Toliyat and Steven Campbell.
p. cm.-- (Power electronics and applications series)
Includes bibliographical references and index.
ISBN 0-8493-1918-8 (alk. paper)
1. Digital control systems. 2. Electromechanical devices. 3. Signal processing--Digital techniques. I. Campbell, Steven (Steven Gerard), 1979- II. Title. III. Series.

TJ223.M53.T64 2003
629.8—dc22 2003058462

Visit the Taylor & Francis Web site at
http://www.taylorandfrancis.com

and the CRC Press Web site at
http://www.crcpress.com

To my wife Mina, and my sons Amir and Mohammad for their love and patience while this book was being prepared.

To my parents for their continuous support and encouragement.

- H.T.

PREFACE

This book was written to provide a general application guide for students and engineers of all disciplines who want to begin utilizing a Digital Signal Processor (DSP) for the task of electromechanical motion control. While the act of learning to program and use the DSP itself is not overly difficult, utilizing the DSP in applications such as motor control can sometimes seem challenging to the first-time user.

Full mastery of all the topics and concepts presented in this text would take years of study and knowledge from many areas of engineering and science. For this reason, we will attempt to survey each topic, giving readers a basic understanding of each topic without going into great depth. We will thus leave it to the reader for in-depth study of particular topics of interest.

So why would we choose to integrate a DSP into a motion control system? Well, the advantages of such a design are numerous. DSP-based control gives us a large degree of freedom in developing computationally extensive algorithms that would otherwise be very difficult or impossible without a DSP. Advanced control algorithms can sometimes drastically increase the performance and efficiency of the electromechanical system being controlled.

For example, consider a typical Heating-Ventilation-and-Air-Conditioning (HVAC) system. A standard HVAC system contains at least three electric motors: compressor motor, condenser fan motor, and the air handler fan motor. Typically, indoor temperature is controlled by simply cycling (turning on and off) the system. This control method puts unnecessary wear on system components and is inefficient. An advanced motor drive system incorporating DSP control could continuously adjust both the air-conditioner compressor speed and indoor fan to maintain the desired temperature and optimal system performance. This control scheme would be much more energy efficient and could extend the operational lifespan of the system.

We will start by visiting the LF2407 DSP processor. Device functionality, integrated components, memory, and assembly programming will be covered. Several laboratory exercises will help the reader practice the information presented in each chapter. After several chapters are presented on the DSP, more advanced topics are presented involving several real-world applications in the area of motion control and motor drives.

ACKNOWLEDGMENTS

As most readers can imagine, creating a book is no trivial task. Besides the authors listed on the cover of each book, there are usually many others who give their time and knowledge. These contributions range from the writing of a chapter to simply proofreading the book for mistakes. This book is no exception. There are many people I would like to thank who made invaluable contributions to the creation of this book.

During the past two years that this book was in development, the many undergraduate students who took my "DSP-Based Electromechanical Motion Control Devices" course in the Department of Electrical Engineering at Texas A&M University provided invaluable feedback on the material. I am in debt to all of them.

I would also like to extend my gratitude to Texas Instruments for permitting me to use the materials in its manuals. I would also like to extend a special acknowledgment to Gene Frantz and Christina Peterson from Texas Instruments, whose help and support for materializing this book were fundamental.

Several individuals, including my past and present graduate students, have contributed to this book. They are as follows: Sebastien Gay, Dr. Masoud Hajiaghajni - Chapter 7; Dr. Lei Hao and Leila Parsa - Chapters 8, 9, and 12; Mehdi Abolhassani - Chapter 10; Nasser Qahtani - Chapter 11; Peyman Niazi - Chapter 13; Sang-shin Kwak - Chapter 15; and Baris Ozturk the Appendix.

Dr. Babak Fahimi of University of Missouri-Rolla wrote Chapter 14 and Dr. Syed Madani of the University of Puerto Rico-Mayaguez contributed to Chapter 7. Rebecca Morrison proofread several chapters.

I would also like to thank Nora Konopka, Helena Redshaw, Susan Fox of CRC Press for their patience and support while this book was being prepared.

Hamid A. Toliyat
College Station, TX

TABLE OF CONTENTS

Chapter 1

INTRODUCTION TO THE TMSLF2407 DSP CONTROLLER

1.1 Introduction

The Texas Instruments TMS320LF2407 DSP Controller (referred to as the LF2407 in this text) is a programmable digital controller with a C2xx DSP central processing unit (CPU) as the core processor. The LF2407 contains the DSP core processor and useful peripherals integrated onto a single piece of silicon. The LF2407 combines the powerful CPU with on-chip memory and peripherals. With the DSP core and control-oriented peripherals integrated into a single chip, users can design very compact and cost-effective digital control systems.

The LF2407 DSP controller offers 40 million instructions per second (MIPS) performance. This high processing speed of the C2xx CPU allows users to compute parameters in real time rather than look up approximations from tables stored in memory. This fast performance is well suited for processing control parameters in applications such as notch filters or sensorless motor control algorithms where a large amount of calculations must be computed quickly.

While the "brain" of the LF2407 DSP is the C2xx core, the LF2407 contains several control-orientated peripherals onboard (see Fig. 1.1). The peripherals on the LF2407 make virtually any digital control requirement possible. Their applications range from analog to digital conversion to pulse width modulation (PWM) generation. Communication peripherals make possible the communication with external peripherals, personal computers, or other DSP processors. Below is a brief listing of the different peripherals onboard the LF2407 followed by a graphical layout depicted in Fig. 1.1.

The LF2407 peripheral set includes:

- Two Event Managers (A and B)
- General Purpose (GP) timers
- PWM generators for digital motor control
- Analog-to-digital converter
- Controller Area Network (CAN) interface
- Serial Peripheral Interface (SPI) – synchronous serial port
- Serial Communications Interface (SCI) – asynchronous serial port
- General-Purpose bi-directional digital I/O (GPIO) pins
- Watchdog Timer ("time-out" DSP reset device for system integrity)

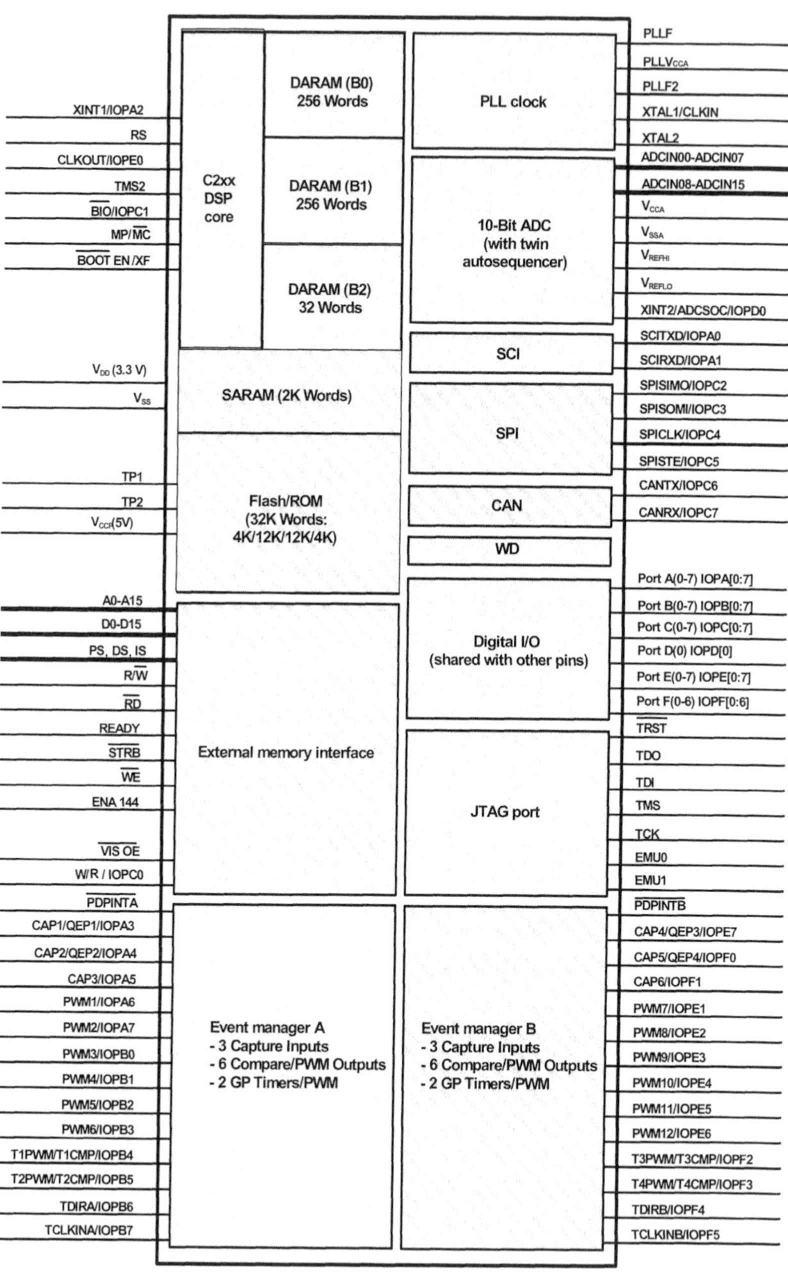

Figure 1.1 Graphical overview of DSP core and peripherals on the LF2407. (Courtesy of Texas Instruments)

1.2 Brief Introduction to Peripherals

The following peripherals are those that are integrated onto the LF2407 chip. Refer to Fig. 1.1 to view the pin-out associated with each peripheral.

Event Managers (EVA, EVB)

There are two Event Managers on the LF2407, the EVA and EVB. The Event Manager is the most important peripheral in digital motor control. It contains the necessary functions needed to control electromechanical devices. Each EV is composed of functional "blocks" including timers, comparators, capture units for triggering on an event, PWM logic circuits, quadrature-encoder–pulse (QEP) circuits, and interrupt logic.

The Analog-to-Digital Converter (ADC)

The ADC on the LF2407 is used whenever an external analog signal needs to be sampled and converted to a digital number. Examples of ADC applications range from sampling a control signal for use in a digital notch filtering algorithm or using the ADC in a control feedback loop to monitor motor performance. Additionally, the ADC is useful in motor control applications because it allows for current sensing using a shunt resistor instead of an expensive current sensor.

The Control Area Network (CAN) Module

While the CAN module will not be covered in this text, it is a useful peripheral for specific applications of the LF2407. The CAN module is used for multi-master serial communication between external hardware. The CAN bus has a high level of data integrity and is ideal for operation in noisy environments such as in an automobile, or industrial environments that require reliable communication and data integrity.

Serial Peripheral Interface (SPI) and Serial Communications Interface (SCI)

The SPI is a high-speed synchronous communication port that is mainly used for communicating between the DSP and external peripherals or another DSP device. Typical uses of the SPI include communication with external shift registers, display drivers, or ADCs.

The SCI is an asynchronous communication port that supports asynchronous serial (UART) digital communication between the CPU and other asynchronous peripherals that use the standard NRZ (non-return-to-zero) format. It is useful in communication between external devices and the DSP. Since these communication peripherals are not directly related to motion control applications, they will not be discussed further in this text.

Watchdog Timer (WD)

The Watchdog timer (WD) peripheral monitors software and hardware operations and asserts a system reset when its internal counter overflows. The WD timer (when enabled) will count for a specific amount of time. It is necessary for the user's software to reset the WD timer periodically so that an unwanted reset does not occur. If for some reason there is a CPU disruption, the watchdog will generate a system reset. For example, if the software enters an endless loop or if the CPU becomes temporarily disrupted, the WD timer will overflow and a DSP reset will occur, which will cause the DSP program to branch to its initial starting point. Most error conditions that temporarily disrupt chip operation and inhibit proper CPU function can be cleared by the WD function. In this way, the WD increases the reliability of the CPU, thus ensuring system integrity.

General Purpose Bi-Directional Digital I/O (GPIO) Pins

Since there are only a finite number of pins available on the LF2407 device, many of the pins are multiplexed to either their primary function or the secondary GPIO function. In most cases, a pin's second function will be as a general-purpose input/output pin. The GPIO capability of the LF2407 is very useful as a means of controlling the functionality of pins and also provides another method to input or output data to and from the device. Nine 16-bit control registers control all I/O and shared pins. There are two types of these registers:

- I/O MUX Control Registers (MCRx) – Used to control the multiplexer selection that chooses between the primary function of a pin or the general-purpose I/O function.
- Data and Direction Control Registers (PxDATDIR) – Used to control the data and data direction of bi-directional I/O pins.

Joint Test Action Group (JTAG) Port

The JTAG port provides a standard method of interfacing a personal computer with the DSP controller for emulation and development. The XDS510PP or equivalent emulator pod provides the connection between the JTAG module on the LF2407 and the personal computer. The JTAG module allows the PC to take full control over the DSP processor while Code Composer Studio™ is running. Figure 1.2 shows the connection scheme from computer to the DSP board.

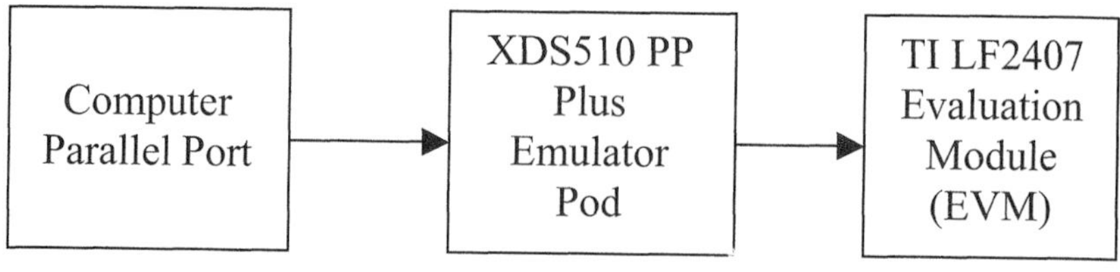

Figure 1.2 PC to DSP connection scheme.

Phase Locked Loop (PLL) Clock Module

The phase locked loop (PLL) module is basically an input clock multiplier that allows the user to control the input clocking frequency to the DSP core. External to the LF2407, a clock reference (can oscillator/crystal) is generated. This signal is fed into the LF2407 and is multiplied or divided by the PLL. This new (higher or lower frequency) clock signal is then used to clock the DSP core. The LF2407's PLL allows the user to select a multiplication factor ranging from 0.5X to 4X that of the external clock signal. The default value of the PLL is 4X.

Memory Allocation Spaces

The LF2407 DSP Controller has three different allocations of memory it can use: Data, Program, and I/O memory space. Data space is used for program calculations, look-up tables, and any other memory used by an algorithm. Data memory can be in the form of the on-chip random access memory (RAM) or external RAM. Program memory is the location of user's program code. Program memory on the LF2407 is either mapped to the off-chip RAM (MP/MC- pin =1) or to the on-chip flash memory (MP/MC- = 0), depending on the logic value of the MP/MC-pin.

I/O space is not really memory but a virtual memory address used to output data to peripherals external to the LF2407. For example, the digital-to-analog converter (DAC) on the Spectrum Digital™ evaluation module is accessed with I/O memory. If one desires to output data to the DAC, the data is simply sent to the configured address of I/O space with the "OUT" command. This process is similar to writing to data memory except that the OUT command is used and the data is copied to and outputted on the DAC instead of being stored in memory.

1.3 Types of Physical Memory

Random Access Memory (RAM)

The LF2407 has 544 words of 16 bits each in the on-chip DARAM. These 544 words are partitioned into three blocks: B0, B1, and B2. Blocks B1 and B2 are allocated for use only as data memory. Memory block B0 is different than B1 and B2. This memory block is normally configured as Data Memory, and hence primarily used to hold data, but in the case of the B0 block, it can also be configured as Program Memory. B0 memory can be configured as program or data memory depending on the value of the core level "CNF" bit.

- (CNF=0) maps B0 to data memory.
- (CNF=1) maps B0 to program memory.

The LF2407 also has 2K of single-access RAM (SARAM). The addresses associated with the SARAM can be used for both data memory and program memory, and are software configurable to the internal SARAM or external memory.

Non-Volatile Flash Memory

The LF2407 contains 32K of on-chip flash memory that can be mapped to program space if the MP/MC-pin is made logic 0 (tied to ground). The flash memory provides a permanent location to store code that is unaffected by cutting power to the device. The flash memory can be electronically programmed and erased many times to allow for code development. Usually, the external RAM on the LF2407 Evaluation Module (EVM) board is used instead of the flash for code development due to the fact that a separate "flash programming" routine must be performed to flash code into the flash memory. The on-chip flash is normally used in situations where the DSP program needs to be tested where a JTAG connection is not practical or where the DSP needs to be tested as a "stand-alone" device. For example, if a LF2407 was used to develop a DSP control solution to an automobile braking system, it would be somewhat impractical to have a DSP/JTAG/PC interface in a car that is undergoing performance testing.

1.4 Software Tools

Texas Instrument's Code Composer Studio™ (CCS) is a user-friendly Windows-based debugger for developing and debugging software for the LF2407. CCS allows users to write and debug code in C or in TI assembly language. CCS has many features that can aid in developing code. CCS features include:

- User-friendly Windows environment
- Ability to use code written in C and assembly
- Memory displays and on-the-fly editing capability
- Disassembly window for debugging
- Source level debugging, which allows stepping through and setting breakpoints in original source code
- CPU register visibility and modification
- Real-time debugging with watch windows and continuous refresh
- Various single step/step over/ step-into command icons
- Ability to display data in graph formats
- General Extension Language (GEL) capability, allows the user to create functions that extend the usefulness of CCS™

1.4.1 Becoming Aquatinted with Code Composer Studio (CCS)

This exercise will help you become familiar with the software and emulation tools of the LF2407 DSP Controller. CCS™, the current emulation and debugging software, is user-friendly and a powerful development tool.

The hardware required for this exercise and all others is the Spectrum Digital TMS320LF2407 EVM package, which includes LF2407 EVM board and the XDS510PP Plus JTAG emulator pole. You will also need a Windows-based personal computer with a parallel printer port. In this lab exercise you will learn how to:

- Open a program, build it, and load the program onto the DSP.
- View the disassembly
- View and edit memory locations
- View and edit CPU registers
- Open a Watch Window
- Reset the DSP
- Run the program in Real-time Mode
- Set breakpoints
- Single step through code
- Save and load a workspace

Since some readers may not have connected their EVM to their PC, we will start with the necessary PC to EVM connection and setup. Follow this procedure if you are first connecting the LF2407 EVM to your PC.

- First, if you have not done so, configure the parallel port of your PC and connect the emulator and target board according to the documentation that came with the LF2407 EVM.
- Before you can start using CCS™, CCS needs to be configured for the particular DSP emulator you are going to be using.

Run *CC_setup.exe*, which should be an icon under *Start/Programs/Code Composer* or at *C:\tic2xx\cc\bin\cc_setup.exe*. You should see a window appear similar to that shown in Fig. 1.3.

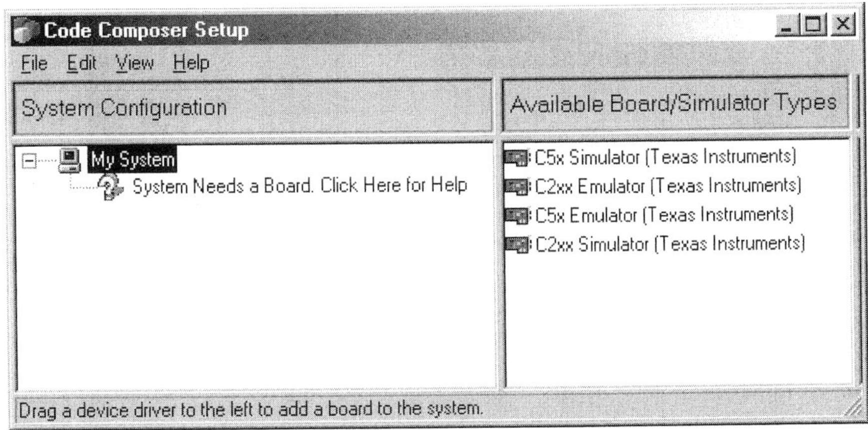

Figure 1.3 Code Composer setup window (from running Setup.exe).

Once you have entered Code Composer Setup window, the proper board/simulator needs to be added to the "System Configuration".

a. Drag the appropriate icon from the "Available Board/Simulator Types" list to the "System Configuration" list. To use the LF2407 DSP select, use the sdgo2xx icon as shown in Fig. 1.4.

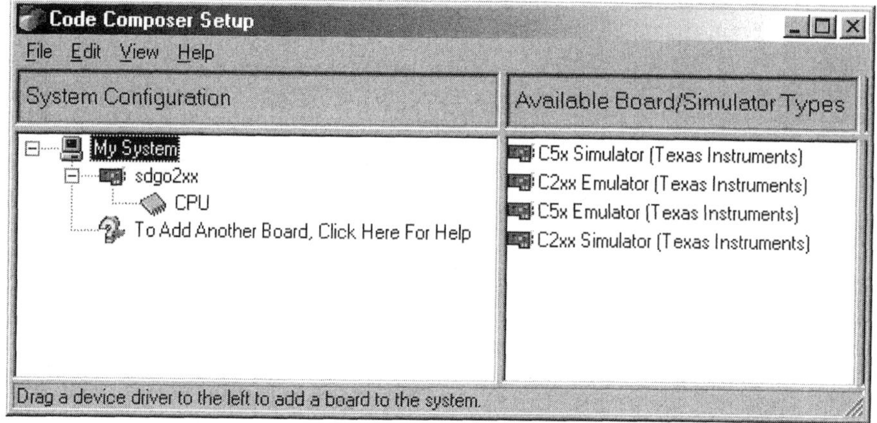

Figure 1.4 Simulator types.

b. Once you drag the sdgo2xx icon into the "System Configuration" section, a "Board Properties" box (shown in Fig. 1.5) should appear. Click on the "Board Properties" tab and set the I/O port for 378.

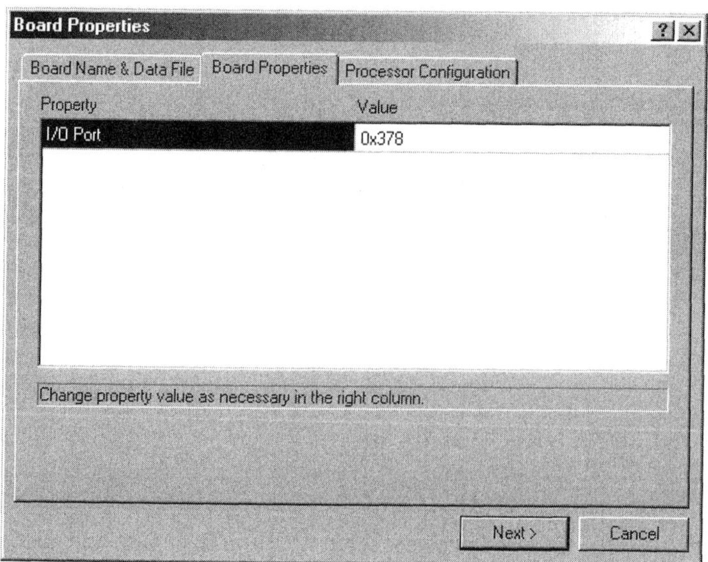

Figure 1.5 Port setting for Printer/Parallel Port.

c. Click on the "Processor Configuration" tab and select the TMS320C2400 processor. Click on the "Add Single" button. You should the see "CPU_1" under the "processors on the board" list.

d. Click on the "Finish" button located at the lower right corner of the "Board Properties" box. The setup is now complete. Go to *File/Save* to save the configuration. Close the Code Composer Setup window.

Now that everything is connected properly, we shall begin with the CCS exercise:

1. Turn on the EVM. The green LED on the top right of the board will confirm that there is power to the board.
2. Open Code Composer Studio by running *cc_app.exe* either from the desktop icon, *Start/Programs/Code Composer*, or *C:\tic2xx\cc\bin\cc_app.exe*.
3. Go to the "Project" menu, select "Open" as seen in Fig. 1.6. Open *realtime.mak*, which is found under *C:\tic2xx\c2000\tutorial\realtime*. The project file is the master file that "holds" the other files together to build a working program.

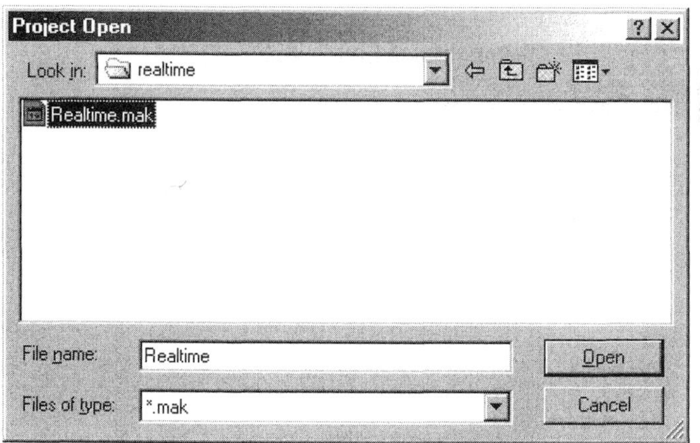

Figure 1.6 Project open window.

4. Once you have the project opened, look at the frame on the left side of the screen where "Files", "GEL files", and "Project" are listed. Expand everything in the "Project" folder. When you are done, you should see the "Include" files, "Libraries", and "Source" files as shown in Fig. 1.7. The project file (*.*mak*) is the master file that links the other necessary files together as a common filename. When you want to create a program with Code Composer, you will want to first create a new project, add a new source file(s) (*.*asm or* *.*c*) to the project, add the linker command file (*.*cmd*), and add "include" (*.*i*) or "header" (*.*h*) files.

As in other programming languages, "include" (*.*i*) and "header" (*.*h*) files are user-defined files that are common to most programs. Functionally *.*h* and *.*i* files are the same. Both types of files can define constants, macros (user defined callable functions), or variables. In this case, we want to run our program in real-time mode. Therefore, we need a real-time monitor program (*C200mntr.i* in this program). The file *X24x.h* contains variable names for data memory mapped control registers. The code that is in the header (*.*h*) or include (*.*i*) file could be written in the actual source code, but it is easier to just make general register definitions as a header file that can be used with many projects.

The linker command file (*.*cmd*) is vital to the proper building of your code. It specifies where in the program memory to place sections of the program code, defines memory blocks, contains linker options, and names input files for the linker, names the (*.out*) etc. The linker command file also specifies memory allocations. Without a proper linker command file, CCS will not build the program properly. In this case, the linker command file is named *realtime.cmd.*

Source (*.*c* or *.*asm*) files contain the actual program that is to be run on the DSP. You must have at least one source file, but may have source files that call other source files. Be sure *all* relevant source files are added to the project.

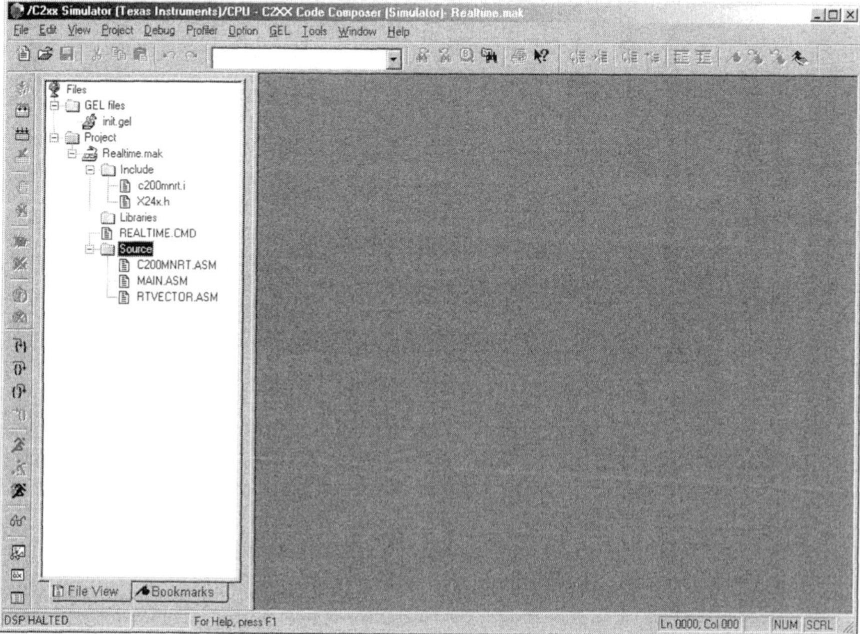

Figure 1.7 CCS window with opened project.

5. Now that you have the project opened, go to *Option/Program Load*, and check the "load program after build" box (Fig. 1.8). This will automatically load the DSP compatible version of the program (**.out*) file into the DSP after the build is complete. Building the project causes Code Composer to assemble and link your code. Basically, this creates a file that the DSP can be loaded with and run. Loading the program can also be done manually under the "File" menu.

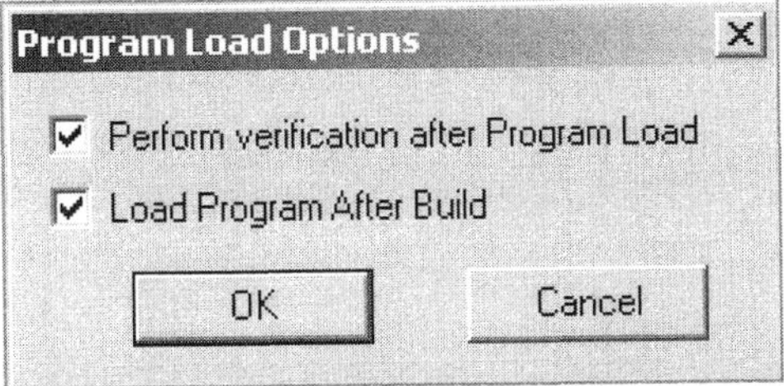

Figure 1.8 Program load options box.

6. Now go to *Project/ Rebuild* All. This will build and load the program into the DSP. If the program is being loaded onto the DSP, the disassembly window will open up automatically.

Note: *It is good practice to ALWAYS RESET THE DSP each time you build or rebuild the project. Do this by going to "Debug" menu, then "Reset DSP".*

To view the disassembly window as in Fig. 1.9 if it is not already open, go to *View/Dis-Assembly*. The disassembly window shows the assembly code that is stored in program memory. It also has a highlighted line that serves as the position marker when running the program.

```
Dis-Assembly
0125 0BAD RPT    *+,AR5
0126 0000 LAR    AR0,0h
0127 0BAD RPT    *+,AR5
0128 0000 LAR    AR0,0h
0129 0BAD RPT    *+,AR5
012A 0000 LAR    AR0,0h
012B 0BAD RPT    *+,AR5
012C 0000 LAR    AR0,0h
012D 0BAD RPT    *+,AR5
012E 0000 LAR    AR0,0h
```

Figure 1.9 Disassembly window.

Note: *When Source Level Debugging is selected (we'll get to this in a minute), a position marker also appears in the appropriate source code window (if a program is loaded into the DSP) (Fig. 1.10).*

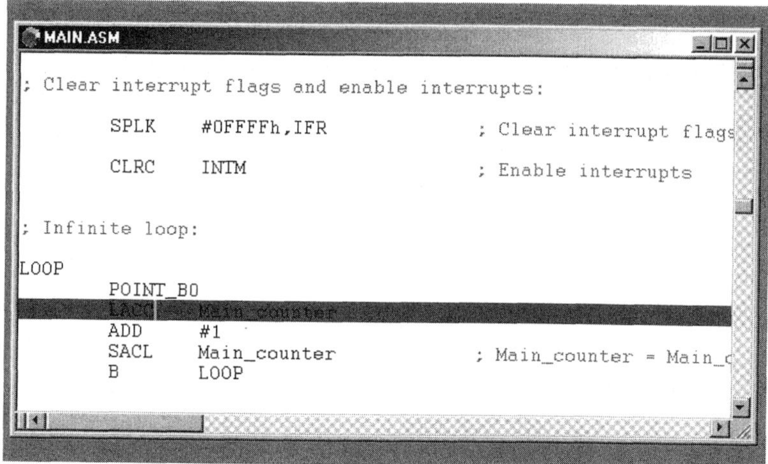

Figure 1.10 Source level debugging.

7. The CPU registers and CPU status registers are very helpful in debugging code. To view these registers, go to *View/CPU Registers* (both registers are under this menu). Open both CPU registers. You should see the registers appear in new frames on the screen.

8. The ability to view memory locations is also vital to debugging. To view memory, go to *View/Memory*. You should see a box pop up which will configure the memory window that is about to open (see Fig. 1.11). Enter 0x0300 for the start address.

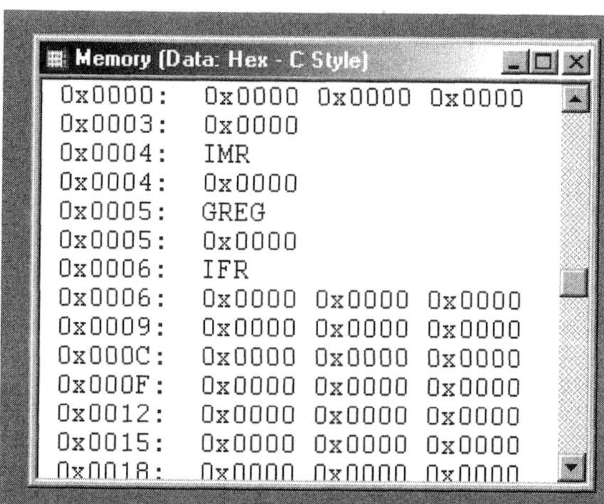

Figure 1.11 Memory viewing window.

You can also change the values for the CPU registers and memory locations by double clicking on the register or memory location. A box will pop up that will allow you to enter in a new value.

Double click on the 0x300 location in the memory window and change the value to 0x0555. The new value will appear in red signifying that the memory location has been changed.

Using the same technique, change a few registers in the CPU status and CPU register frames. Observe how the values in the registers change to the new value entered.

9. In MAIN.ASM scroll down until you see the line ".bss Main_counter,1". Highlight "Main_counter" and add that variable to the watch window.

A watch window allows us to view variables that we use in our code. Open a watch window by going to *View/Watch Window*. You can add variables to this window by highlighting the variable name in the source code and then right clicking the mouse button and selecting "add to watch window". Now, let us edit the display format of this variable in the watch window. Double click on the variable name in the watch window. When the "edit variable" box appears, add the command "*(int*)" in front of the variable name (see Fig. 1.12). This configures the variable in the watch window to be displayed as an integer, thus ensuring that a decimal value is displayed. Otherwise, a hex value will be displayed.

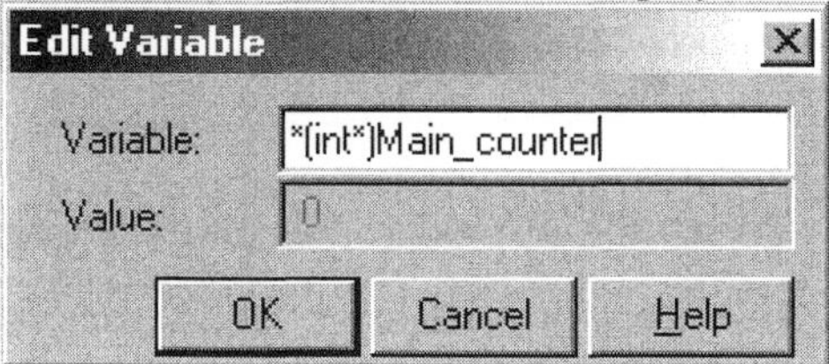

Figure 1.12 Editing a variable while in the watch window.

10. Rebuild the project (which should load the program as well) and reset the DSP by going to *Debug Menu/Reset* DSP.

Note: *If a source code window opens up as well as the disassembly window when the project is built, Source Level Debugging is enabled. If not, enable Source Level Debugging by going to Project/Options/Assembler Tab and check the "enable source level debugging" (Fig. 1.13). Source level debugging lets you see where in the source code the program is running instead of having to decipher the disassembly window information. If you have just enabled Source Level Debugging, you need to rebuild the project before it takes effect.*

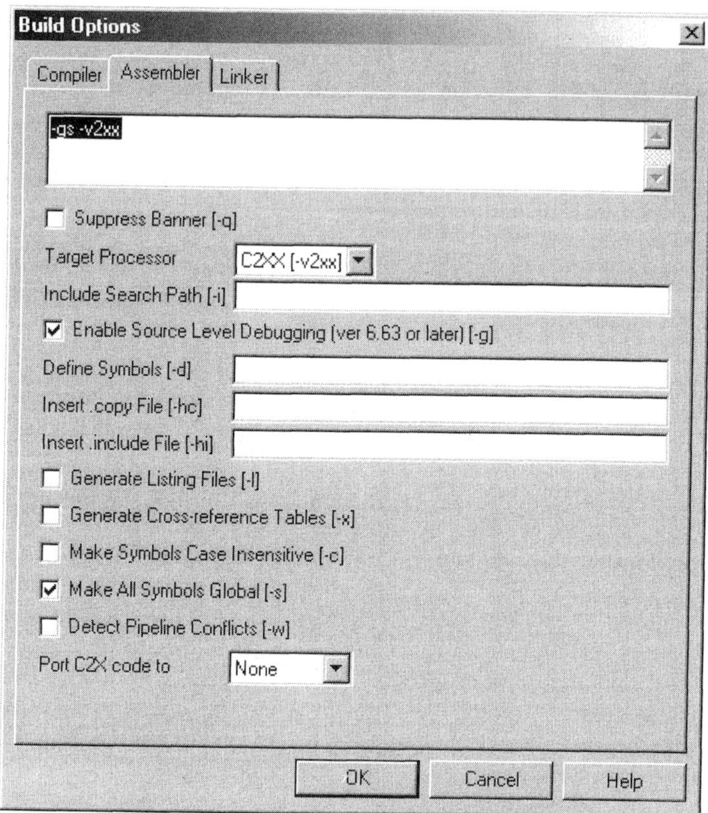

Figure 1.13 Build options menu box.

11. Enable Real Time mode by performing the following steps:

    a. The DSP *must* have the program already loaded in order to enable real-time mode. (While in real-time mode, programs cannot be loaded to the DSP.)
    b. Reset the DSP by going to *Debug Menu / Reset DSP.*
    c. Open the Command Window by going to Tools *Menu / Command Window.*
    d. Type in the Command Window "go MON_GO".
    e. Put CCS in Real-time mode by going to Debug Menu / Real-time Mode. When in real-time mode, you will se the word "REALTIME" in the bottom of the code composer screen.
    f. Reset the DSP again and the program is ready to RUN.

Note: *Real-time mode is a useful feature of CCS that allows you to see changes as they happen but is not necessary for program debugging. When CCS is not in real time mode, the values in all windows will update as soon as the program is halted or a break point occurs.*

Right click on the watch window and choose "Continuous Refresh". This will allow the values in the Watch window to change.

12. We are now ready to run the demonstration program. First make sure that no breakpoints have been set or the DSP will stop when it reaches the breakpoint.

    Run the program by going to *Debug/Run.* Running and halting the DSP can also be performed by hitting F5 to run and Shift-F5 to halt. Observe as the value of "Main_counter" in the watch window changes.

13. Halt the DSP by going to Debug/Halt. In the disassembly or source window you should see that the program is halted somewhere in the area of code entitled "Loop" (hex address 0159-015D in the disassembly window (program memory)). Left click on a line in the "Loop" section and toggle a breakpoint by right clicking the mouse and selecting "toggle breakpoint". You should see a purple line appear where the breakpoint is set (Fig. 1.14). Notice how the breakpoint appears in both the disassembly window *and* the window containing the assembly source code.

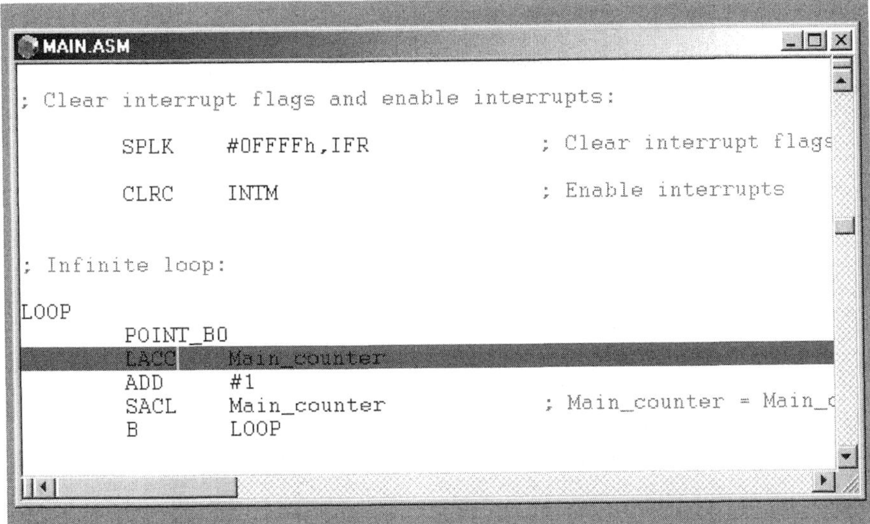

Figure 1.14 Breakpoint is located at the highlighted line (source level debug).

14. Run the program and watch as the DSP stops at the breakpoint each time it passes through the "Loop" section. (You will need to "run" the DSP each time after it hits a breakpoint because the breakpoint essentially "pauses" the DSP.) Observe as the value of Main_counter increments by 1 in the watch window each time the code is restarted after the breakpoint. Remove the breakpoint by toggling it off.

Note: *If you wish to single step through the code regardless of whether or not a breakpoint is set, you can do this by choosing Debug/Step Into or pressing F8.*

15. If you wish to save the screen configuration (position of windows, what appears on the screen, etc.) go to *File Menu/Workspace/save workspace* shown in Fig. 1.15.

    Now, when you re-open CCS in the future, you will only have to load the workspace, saving you the trouble of opening the memory, CPU, and source code windows shown in Fig. 1.16. Saving a workspace not only saves window configuration, but project configuration as well. If a previously saved workspace is opened, the project that was open at the time of the workspace save will also open. While saving a workspace saves screen configuration, it *does not* save the contents of any files or the project!

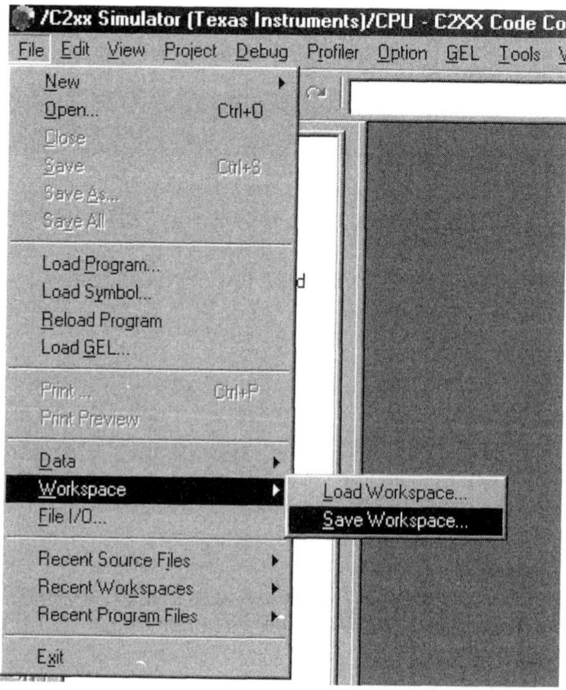

Figure 1.15 Saving a workspace.

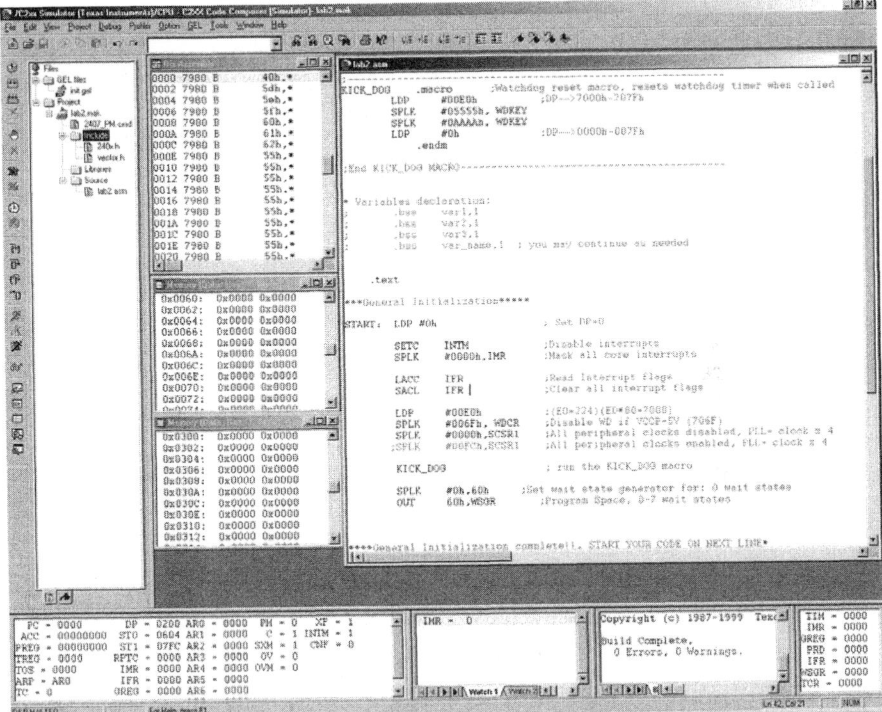

Figure 1.16 Screenshot of typical CCS™ workspace.

The screenshot shown in Fig. 1.16 displays what a typical workspace might contain. The workspace includes: several memory windows, watch window, CPU register windows, source code, and project window.

This concludes the introduction of the most common features of Code Composer Studio. There are many features not covered by this introduction that may be useful to advanced users. Consult the program Help or the Code Composer Users Guide for more information on Code Composer functions.

Chapter 2

C2xx DSP CPU AND INSTRUCTION SET

2.1 Introduction to the C2xx DSP Core and Code Generation

The heart of the LF2407 DSP Controller is the C2xx DSP core. This core is a 16-bit fixed point processor, meaning that it works with 16-bit binary numbers. One can think of the C2xx as the central processor in a personal computer. The LF2407 DSP consists of the C2xx DSP core plus many peripherals such as Event Managers, ADC, etc., all integrated onto one single chip. This chapter will discuss the C2xx DSP core, subcomponents, and instruction set.

The C2xx core has its own native instruction set of assembly mnemonics or commands. Through the use of CCS and the associated compiler, one has the freedom of writing code in both C language and the native assembly language. However, to write compact, fast executing programs, it is best to compose code in assembly language. Due to this reason, programming in assembly will be the focus of this book. However, we will also include an example of a software tool called VisSim™, by Visual Solutions. VisSim allows users to simulate algorithms and develop code in "block" form. More on VisSim will be presented in the Appendix.

2.2 The Components of the C2xx DSP Core

The DSP core (like all microprocessors) consists of several subcomponents necessary to perform arithmetic operations on 16-bit binary numbers. The following is a list of the multiple subcomponents found in the C2xx core which we will discuss further:

- A 32-bit central arithmetic logic unit (CALU)
- A 32-bit accumulator (used frequently in programs)
- Input and output data-scaling shifters for the CALU
- A (16-bit by 16-bit) multiplier
- A product-scaling shifter
- Eight auxiliary registers (AR0 – AR7) and an auxiliary register arithmetic unit (ARAU)

Each of the above components is either accessed directly by the user code or is indirectly used during the execution of an assembly command.

Central Arithmetic Logic Unit (CALU)

The C2xx performs 2s-complement arithmetic using the 32-bit CALU. The CALU uses 16-bit words taken from data memory, derived from an immediate instruction, or from the 32-bit multiplier result. In addition to arithmetic operations, the CALU can perform Boolean operations. The CALU is somewhat transparent to the user. For example, if an arithmetic command is used, the user only needs to write the command and later read the output from the appropriate register. In this sense, the CALU is "transparent" in that it is not accessed directly by the user.

Accumulator

The accumulator stores the output from the CALU and also serves as another input to the CALU (many arithmetic commands perform operations on numbers that are currently stored in the accumulator; versus other memory locations). The accumulator is 32 bits wide and is divided into two sections, each consisting of 16 bits. The high-order bits consist of bits 31 through 16, and the low-order bits are made up of bits 15 through 0. Assembly language instructions are provided for storing the high- and low-order accumulator words to data memory. In most cases, the accumulator is written to and read from directly by the user code via assembly commands. In some instances, the accumulator is also transparent to the user (similar to the CALU operation in that it is accessed "behind the scenes").

Scaling Shifters

The C2xx has three 32-bit shifters that allow for scaling, bit extraction, extended arithmetic, and overflow-prevention operations. The scaling shifters make possible commands that shift data left or right. Like the CALU, the operation of the scaling shifters is "transparent" to the user. For example, the user needs only to use a shift command, and observe the result. Any one of the three shifters could be used by the C2xx depending on the specific instruction entered. The following is a description of the three shifters:

- **Input data-scaling shifter (input shifter):** This shifter left-shifts 16-bit input data by 0 to 16 bits to align the data to the 32-bit input of the CALU. For example, when the user uses a command such as "ADD 300h, 5", the input shifter is responsible for first shifting the data in memory address "300h" to the left by five places before it is added to the contents of the accumulator.

- **Output data-scaling shifter (output shifter):** This shifter left-shifts data from the accumulator by 0 to 7 bits before the output is stored to data memory. The content of the accumulator remains unchanged. For example, when the user uses a command such as "SACL 300h, 4", the output shifter is responsible for first shifting the contents of the accumulator to the left by four places before it is stored to the memory address "300h".

- **Product-scaling shifter (product shifter)**: The product register (PREG) receives the output of the multiplier. The product shifter shifts the output of the PREG before that output is sent to the input of the CALU. The product shifter has four product shift modes (no shift, left shift by one bit, left shift by four bits, and right shift by six bits), which are useful for performing multiply/accumulate operations, fractional arithmetic, or justifying fractional products.

Multiplier

The multiplier performs 16-bit, 2s-complement multiplication and creates a 32-bit result. In conjunction with the multiplier, the C2xx uses the 16-bit temporary register (TREG) and the 32-bit product register (PREG).

The operation of the multiplier is not as "transparent" as the CALU or shifters. The TREG **always** needs to be loaded with one of the numbers that are to be multiplied. Other than this prerequisite, the multiplication commands do not require any more actions from the user code. The output of the multiply is stored in the PREG, which can later be read by the user code.

Auxiliary Register Arithmetic Unit (ARAU) and Auxiliary Registers

The ARAU generates data memory addresses when an instruction uses indirect addressing to access data memory (more on indirect addressing will be covered later along with assembly programming). Eight auxiliary registers (AR0 through AR7) support the ARAU, each of which can be loaded with a 16-bit value from data memory or directly from an instruction. Each auxiliary register value can also be stored in data memory. The auxiliary registers are mainly used as "pointers" to data memory locations to more easily facilitate looping or repeating algorithms. They are directly written to by the user code and are automatically incremented or decremented by particular assembly instructions during a looping or repeating operation. The auxiliary register pointer (ARP) embedded in status register ST0 references the auxiliary register. The status registers (ST0, ST1) are core level registers where values such as the Data Page (DP) and ARP located. More on the operation and use of auxiliary registers will be covered in subsequent chapters.

2.3 Mapping External Devices to the C2xx Core and the Peripheral Interface

Since the LF2407 contains many peripherals that need to be accessed by the C2xx core, the C2xx needs a way to read and write to the different peripherals. To make this possible, peripherals are mapped to data memory (memory will be covered shortly). Each peripheral is mapped to a corresponding block of data memory addresses. Where applicable, each corresponding block contains configuration registers, input registers, output registers, and status registers. Each peripheral is accessed by simply writing to the appropriate registers in data memory, provided the peripheral clock is enabled (see System Configuration registers).

The peripherals are linked to the internal memory interface of the CPU through the PBUS interface shown in Fig. 2.1. All on-chip peripherals are accessed through the Peripheral Bus (PBUS). All peripherals, excluding the WD timer counter, are clocked by the CPU clock (which has a selectable frequency), and must be enabled via the system configuration registers.

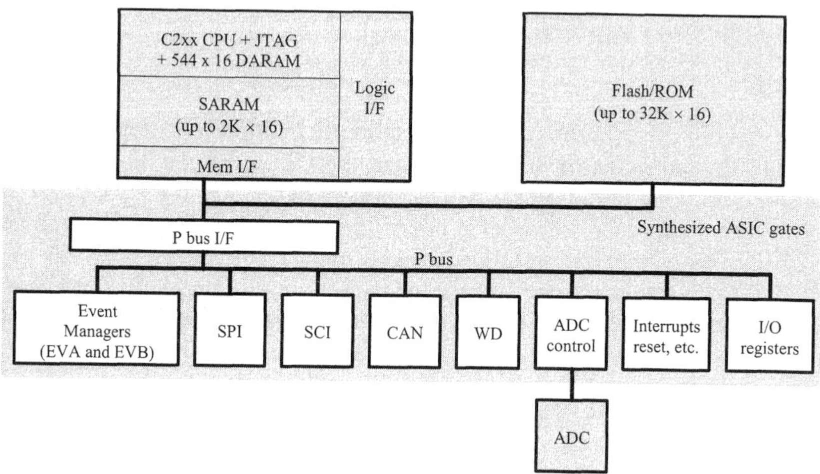

Figure 2.1 Functional block diagram of the LF2407 DSP controller.

2.4 System Configuration Registers

The System Control and Status Registers (SCSR1, SCSR2) are used to configure or display fundamental settings of the LF2407. For example, these fundamental settings include the clock speed (clock pre-scale setting) of the LF2407, which peripherals are enabled, microprocessor/microcontroller mode, etc. Bits are controlled by writing to the corresponding data memory address or the logic level on an external pin as with the microprocessor/microcontroller (*MP/MC*) select bit. The bit descriptions of these two registers (mapped to data memory) are listed below.

System Control and Status Register 1 (SCSR1) — Address 07018h

15	14	13	12	11	10	9	8
Reserved	CLKSRC	LPM1	LPM0	CLK PS2	CLK PS1	CLK PS0	Reserved
R–0	RW–0	RW–0	RW–0	RW–1	RW–1	RW–1	R–0
7	6	5	4	3	2	1	0
ADC CLKEN	SCI CLKEN	SPI CLKEN	CAN CLKEN	EVB CLKEN	EVA CLKEN	Reserved	ILLADR
RW–0	RW–0	RW–0	RW–0	RW–0	RW–0	R–0	RC–0

Note: *R = read access, W = write access, C = clear, -0 = value after reset.*

Bit 15 Reserved

Bit 14 CLKSRC. CLKOUT pin source select

0 CLKOUT pin has CPU Clock (40 MHz on a 40-MHz device) as the output

1 CLKOUT pin has Watchdog clock as the output

Bits 13–12 LPM (1:0). Low-power mode select

These bits indicate which low-power mode is entered when the CPU executes the IDLE instruction.

Description of the low-power modes:

LPM(1:0)	Low-Power mode selected
00	IDLE1 (LPM0)
01	IDLE2. (LPM1)
1x	HALT (LPM2)

Bits 11–9 PLL Clock prescale select. These bits select the PLL multiplication factor for the input clock.

CLK PS2	CLK PS1	CLK PS0	System Clock Frequency
0	0	0	4 x F_{in}
0	0	1	2 x F_{in}
0	1	0	1.33 x F_{in}
0	1	1	1 x F_{in}
1	0	0	0.8 x F_{in}
1	0	1	0.66 x F_{in}
1	1	0	0.57 x F_{in}
1	1	1	0.5 x F_{in}

Note: F_{in} is the input clock frequency.

Bit 8 Reserved

Bit 7 ADC CLKEN. ADC module clock enable control bit

0 Clock to module is disabled (i.e., shut down to conserve power)

1 Clock to module is enabled and running normally

Bit 6 SCI CLKEN. SCI module clock enable control bit

0 Clock to module is disabled (i.e., shut down to conserve power)

1 Clock to module is enabled and running normally

Bit 5 **SPI CLKEN.** SPI module clock enable control bit
0 Clock to module is disabled (i.e., shut down to conserve power)
1 Clock to module is enabled and running normally

Bit 4 **CAN CLKEN.** CAN module clock enable control bit
0 Clock to module is disabled (i.e., shut down to conserve power)
1 Clock to module is enabled and running normally

Bit 3 **EVB CLKEN.** EVB module clock enable control bit
0 Clock to module is disabled (i.e., shut down to conserve power)
1 Clock to module is enabled and running normally

Bit 2 **EVA CLKEN.** EVA module clock enable control bit
0 Clock to module is disabled (i.e., shut down to conserve power)
1 Clock to module is enabled and running normally

Note: In order to modify/read the register contents of any peripheral, the clock to that peripheral must be enabled by writing a 1 to the appropriate bit.

Bit 1 **Reserved**

Bit 0 **ILLADR.** Illegal Address detect bit
If an illegal address has occurred, this bit will be set. It is up to software to clear this bit following an illegal address detect. This bit is cleared by writing a 1 to it and should be cleared as part of the initialization sequence. Note: An illegal address will cause a Non-Maskable Interrupt (NMI).

System Control and Status Register 2 (SCSR2) — Address 07019h

15-8							
Reserved							
RW–0							
7	6	5	4	3	2	1	0
Reserved	I/P QUAL	WD OVERRIDE	XMIF HI–Z	BOOT EN	MP/MC	DON	PON
	RW–0	RC–1	RW–0	RW–BOOT EN pin	RW–MP/MC pin	RW–1	RW–1

***Note:** R = read access, W = write access, C = clear, -0 = value after reset.*

Bits 15–7 **Reserved.** Writes have no effect; reads are undefined

Bit 6 Input Qualifier Clocks.

An input-qualifier circuitry qualifies the input signal to the CAP1–6, XINT1/2, ADCSOC, and PDPINTA/B pins in the 240xA devices. The I/O functions of these pins do not use the input-qualifier circuitry. The state of the internal input signal will change only after the pin is held high/low for 6 (or 12) clock edges. This ensures that a glitch smaller than (or equal to) 5 (or 11) CLKOUT cycles wide will not change the internal pin input state. The user must hold the pin high/low for 6 (or 12) cycles to ensure that the device will see the level change. This bit determines the width of the glitches (in number of internal clock cycles) that will be blocked. Note that the internal clock is not the same as CLKOUT, although its frequency is the same as CLKOUT.

0 The input-qualifier circuitry blocks glitches up to 5 clock cycles long

1 The input-qualifier circuitry blocks glitches up to 11 clock cycles long

Note: *This bit is applicable only for the 240xA devices, not for the 240x devices because they lack an input-qualifier circuitry.*

Bit 5 Watchdog Override. (WD protect bit)

After RESET, this bit gives the user the ability to disable the WD function through software (by setting the WDDIS bit = 1 in the WDCR). This bit is a clear-only bit and defaults to a 1 after reset.

Note: *This bit is cleared by writing a 1 to it.*

0 Protects the WD from being disabled by software. This bit cannot be set to 1 by software. It is a clear-only bit, cleared by writing a 1

1 This is the default reset value and allows the user to disable the WD through the WDDIS bit in the WDCR. Once cleared, however, this bit can no longer be set to 1 by software, thereby protecting the integrity of the WD timer

Bit 4 XMIF Hi-Z Control

This bit controls the state of the external memory interface (XMIF) signals.

0 XMIF signals in normal driven mode; i.e., not Hi-Z (high impedance)

1 All XMIF signals are forced to Hi-Z state

Bit 3 Boot Enable

This bit reflects the state of the BOOT_EN / XF pin at the time of reset. After reset and device has "booted up", this bit can be changed in software to re-enable Flash memory visibility or return to active Boot ROM.

0 Enable Boot ROM — Address space 0000 — 00FF is now occupied by the on-chip Boot ROM Block. Flash memory is totally disabled in this mode. Note: There is no on-chip boot ROM in ROM devices (i.e., LC240xA)

1 Disable Boot ROM — Program address space 0000 — 7FFF is mapped to on-chip Flash memory in the case of LF2407A and LF2406A. In the case of LF2402A, addresses 0000 – 1FFF are mapped

Bit 2 Microprocessor/Microcontroller Select

This bit reflects the state of the MP/MC pin at time of reset. After reset, this bit can be changed in software to allow dynamic mapping of memory on and off chip.

0 Set to Microcontroller mode — Program Address range 0000 — 7FFF is mapped internally (i.e., Flash)

1 Set to Microprocessor mode — Program Address range 0000 — 7FFF is mapped externally (i.e., customer provides external memory device.)

Bits 1–0 SARAM Program/Data Space Select

DON	PON	SARAM status
0	0	SARAM not mapped (disabled), address space allocated to external memory
0	1	SARAM mapped internally to Program space
1	0	SARAM mapped internally to Data space
1	1	SARAM block mapped internally to both Data and Program spaces. This is the default or reset value

Note: *See memory map for location of SARAM addresses*

2.5 Memory

Memory is required to hold programs, perform operations, and execute programming instructions. There are three main blocks of memory which are present on the LF2407 chip: B0, B1, and B2. Additionally, there are two different memory "spaces" (program, data) in which blocks are used. We will discuss exactly what each memory "block" and memory "space" is, and what each is used for.

2.5.1 Memory Blocks and Types

A block of memory on the LF2407 is simply a specified range of memory addresses (each address consists of a 16-bit word of memory). There are three main memory blocks on the LF2407 that can be specified via the *Linker Command File* (we will discuss the *Linker Command File* and other files types when we cover programming).

The LF2407 has 544 16-bit words of on-chip Double Access Random Access Memory (DARAM) that are divided into three main memory blocks named B0, B1, and B2. In addition to the DARAM, there are also 2000 16-bit words of Single Access Random Access Memory (SARAM). The main difference between DARAM and SARAM is that DARAM memory can be accessed twice per clock cycle and SARAM can only be accessed once per cycle. Thus, DARAM reads and writes twice as fast as SARAM.

In addition to the RAM present on the LF2407, there is also non-volatile Flash memory. Unlike RAM, the Flash memory does not lose its contents when the LF2407 loses power. Flash memory can only be written to by "flashing" the memory, which is a process that can only be done manually by a user. Therefore, Flash memory on the LF2407 is used only to store a program that is to be run. As stated in Chapter 1, it is only necessary to use the Flash memory if the DSP is to be run independently from a PC and JTAG interface. Though we introduce Flash memory, it will not be covered in this text. However, the reader is encouraged to consult the Texas Instruments documentation on Flash memory. Flash memory can prove to be a valuable code development tool when it comes time to test a LF2407 program where having a PC connected is impractical.

2.5.2 Memory Space and Allocation

There are two ways of using the physical memory on board the LF2407: storing a program or storing data.

A program that is to be run must be stored in memory that is mapped to program space. Likewise, only memory that is in data space may be used to store data. Program memory is written to when a program is loaded into the LF2407. Data memory is normally written to during the execution of a program, where the program might use the data memory as temporary storage for calculation variables and results.

Memory blocks B1 and B2 are configured as data memory. The B0 block is primarily intended to hold data, but can be configured to act as either program *or* data memory, depending on the value of the CNF bit in Status Register ST1. CNF = 0 maps B0 in data memory, while CNF = 1 maps B0 in program memory.

The memory addresses associated with the SARAM can be configured for both data memory and program memory, and are also software configurable to either access external memory or the internal SARAM. When configured for internal, the SARAM can be used as data or program memory. However, when configured as external, these addresses are used for off-chip program memory. SARAM is useful if more memory is needed for data than the B0, B1, and B2 blocks can provide. The SARAM addresses should be configured to either program or data space via the *Linker Command File*.

The on-chip flash in the LF2407 is mapped to program memory space when the external MP/MC-pin is pulled low. When the MP/MC-pin is pulled high, the program memory is mapped to external memory addresses, access via memory that is physically external to the LF2407. In the case of the Spectrum Digital EVM, external memory is installed on the board and a jumper pulls the MP/MC pin high or low.

2.5.3 Memory Maps

Program Memory

When a program is loaded into the LF2407, the code resides in and is run from program memory space. In addition to storing the user code, the program memory can also store immediate operands and table information. Figure 2.2 shows the various program memory addresses (in hexadecimal) and how they are used.

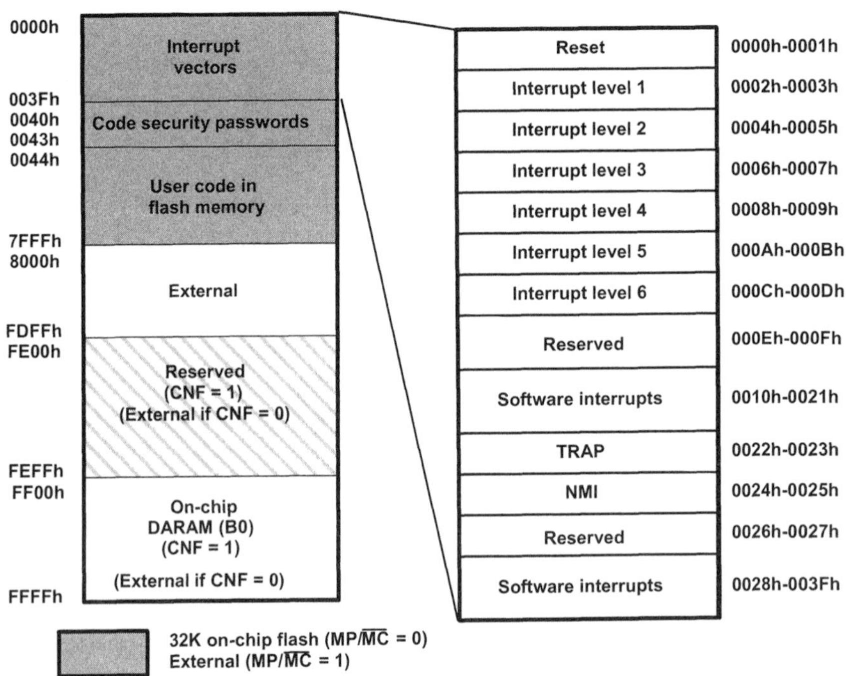

Figure 2.2 Program memory map for LF2407. (Courtesy of Texas Instruments)

Two factors determine the configuration of program memory:

CNF bit:

The CNF bit determines if B0 memory is in on-chip program space:

CNF = 0. The 256 words are mapped as external memory.

CNF = 1. The 256 words of DARAM B0 are configured for program use. At reset, B0 is mapped to data space (CNF = 0).

MP/MC pin:

The level on the MP/MC pin determines if program instructions are read from on-chip Flash/ROM or external memory:

MP/MC = 0. The device is configured in microcontroller mode. The on-chip flash EEPROM is accessible. The device fetches the reset vector from on-chip memory.

MP/MC = 1. The device is configured in microprocessor mode. Program memory is mapped to external memory.

Data Memory

For the execution of a program, it is necessary to store calculation results or look up tables in memory. The memory allocated for this function is called data memory. In order to store a value to a data memory address (*dma*), the corresponding memory block must reside in data memory space. Blocks B1 and B2 discussed earlier permanently reside in data space, while block B0 and the SARAM are configurable for either program or data.

Data memory space has the second functionality of providing an easy way to access on-chip configuration registers and peripherals. Each user configurable peripheral has associated registers in data memory addresses that may be written to or read from as needed. For example, the control registers for the analog-to-digital converter (ADC) are each located in the data memory range of **70A0**h to **70BF**h. The internal data memory includes the memory-mapped registers, DARAM blocks, and peripheral memory-mapped registers. The remaining 32K words of memory (8000h to FFFFh) form part of the external data memory.

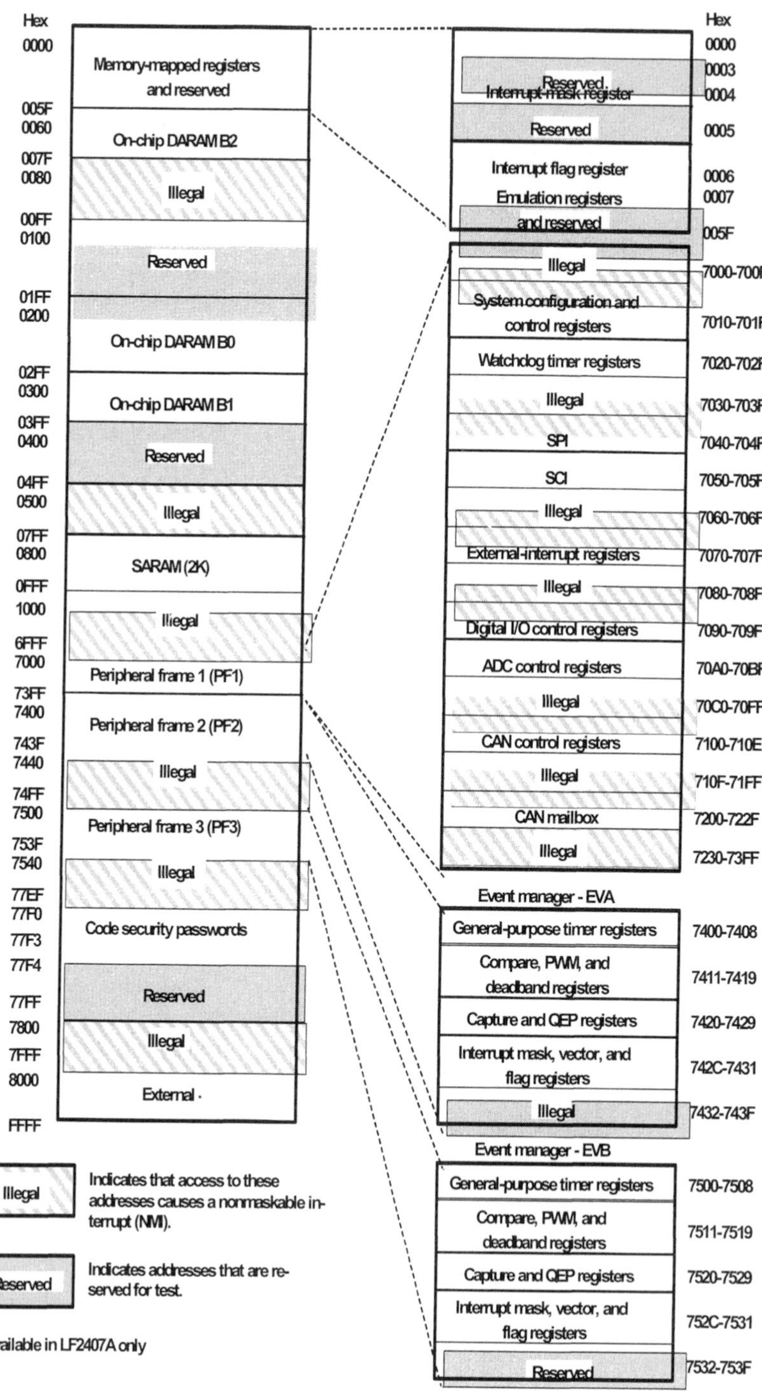

Figure 2.3 Data memory map for the LF2407. (Courtesy of Texas Instruments)

Input/Output (I/O) Space

I/O space is solely used for accessing external peripherals such as the digital-to-analog converter (DAC) on the LF2407 EVM. ***It is not to be confused with the I/O functionality of pins.*** The assembly instruction "*OUT*" is used to write to an address that is mapped to I/O space. Figure 2.4 depicts the basic memory map of the I/O space on the LF2407.

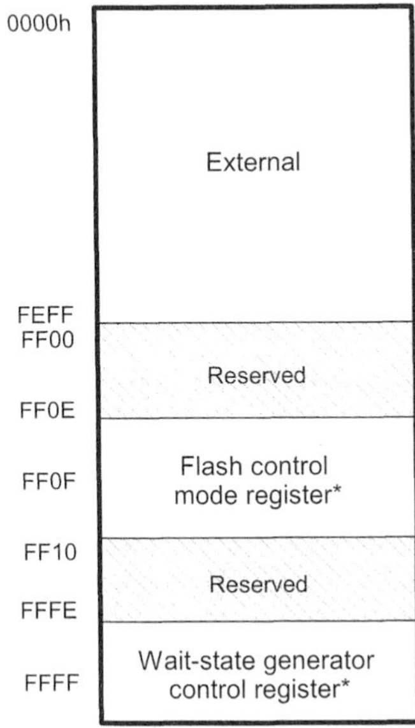

Figure 2.4 Memory map of I/O space. (Courtesy of Texas Instruments)

Within program, data, and I/O space are addresses that are reserved for system functionality and may not be written to. It is important that the user pay attention to what memory ranges are used by the program and where the program is to be loaded. It is important to make sure the *Linker Command File* is configured properly and the correct Data Page (DP) is set to avoid inadvertently writing to an undesired or reserved memory address.

Detailed information on the memory map is given in the Texas Instruments *TMS320LF/LC240xA DSP Controllers Reference Guide - System and Peripherals; Literature Number: SPRU357A.*

2.6 Memory Addressing Modes

There are three basic memory addressing modes used by the C2xx instruction set. The three modes are:

- Immediate addressing mode (does not actually access memory)
- Direct addressing mode
- Indirect addressing mode

2.6.1 *Immediate Addressing Mode*

In the immediate addressing mode, the instruction contains a constant to be manipulated by the instruction. Even though the name "immediate addressing" suggests that a memory location is accessed, immediate addressing is simply dealing with a user-specified constant which is usually included in the assembly command syntax. The "#" sign indicates that the value is an immediate address (just a constant). The two types of immediate addressing modes are:

Short-immediate addressing. The instructions that use short-immediate addressing have an 8-bit, 9-bit, or 13-bit constant as the operand.

For example, the instruction:

```
LACL    #44h        ;loads lower bits of accumulator with
                    ;eight-bit constant (44h in this case)
```

Note: *The LACL command will work only with a short 8-bit constant. If you want to load a long 16-bit constant, then use the LACC command.*

Long-immediate addressing. Instructions that use long-immediate addressing have a 16-bit constant as an operand. This 16-bit value can be used as an absolute constant or as a 2s-complement value.

For example, the instruction:

```
LACC    #4444h      ;loads accumulator with up to a 16-bit
                    ;constant (4444h in this case)
```

If you need to use registers or access locations in data memory, you must use either direct or indirect addressing.

2.6.2 *Direct Addressing Mode*

In direct addressing, data memory is first addressed in blocks of 128 words called data pages. The entire 64K of data memory consists of 512 DPs labeled 0 through 511, as shown in the Fig. 2.5. The current DP is determined by the value in the 9-bit DP pointer in status register ST0. For example, if the DP value is "0 0000 0000", the current DP is 0. If the DP value is "0 0000 0010", the current data page is 2. The DP of a particular memory address can be found easily by dividing the address (in hexadecimal) by 80h. For example:

For the data memory address 0300h, 300h/80h = 6h so the DP pointer is 6h. Likewise, the DP pointer for 200h is 4h.

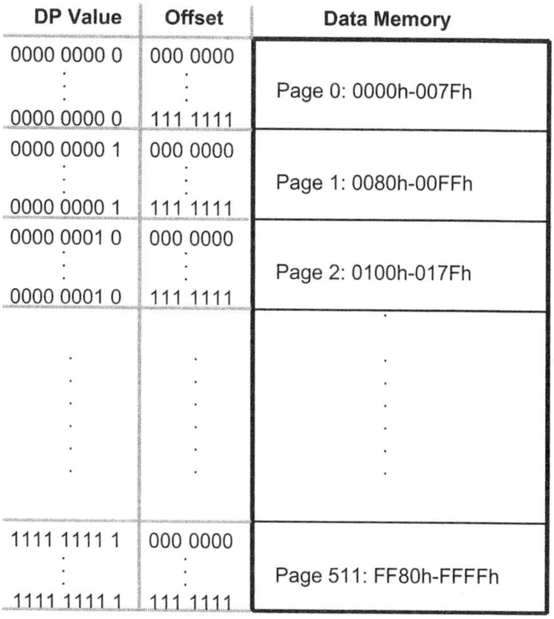

DP Value	Offset	Data Memory
0000 0000 0 ⋮ 0000 0000 0	000 0000 ⋮ 111 1111	Page 0: 0000h-007Fh
0000 0000 1 ⋮ 0000 0000 1	000 0000 ⋮ 111 1111	Page 1: 0080h-00FFh
0000 0001 0 ⋮ 0000 0001 0	000 0000 ⋮ 111 1111	Page 2: 0100h-017Fh
⋮	⋮	⋮
1111 1111 1 ⋮ 1111 1111 1	000 0000 ⋮ 111 1111	Page 511: FF80h-FFFFh

Figure 2.5 Data pages and corresponding memory ranges. (Courtesy of Texas Instruments)

In addition to the DP, the DSP must know the particular word being referenced on that page. This is determined by a 7-bit offset. The 7-bit offset is simply the 7 least significant bits (LSBs) of the memory address. The DP and the offset make up the 16-bit memory address (see Fig. 2.6).

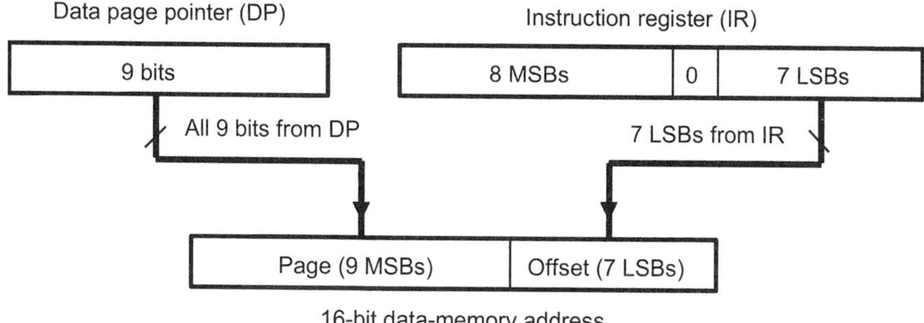

Figure 2.6 Data page and offset make up a 16-bit memory address.

When you use direct addressing, the processor uses the 9 DP bits and the 7 LSBs of the instruction to obtain the true memory address. The following steps should be followed when using direct addressing:

1. Set the DP. Load the appropriate value (from 0 to 511 in decimal or 0-1FF in hex) into the DP. The easiest way to do this is with the LDP instruction. The LDP instruction loads the DP directly to the ST0 register without affecting any other bits of the ST0.

```
LDP     #0E1h       ;sets the data page pointer to E1h
```

or

```
LDP     #225        ;sets the data page pointer to 225 decimal
                    ;which is E1 in hexadecimal
```

2. Specify the offset. For example, if you want the ADD instruction to use the value at the second address of the current data page, you would write: ADD 1h

If the data page points to 300h, then the above instruction will add the contents of 301h to the accumulator

Note: *You do not have to set the data page prior to every instruction that uses direct addressing. If all the instructions in a block of code access the same data page, you can simply load the DP before the block. However, if various data pages are being accessed throughout the block of code, be sure the DP is changed accordingly.*

2.6.3 Indirect Addressing Mode

Indirect addressing is a powerful way of addressing data memory. Indirect addressing mode is not dependent on the current data page as is direct addressing. Instead, when using indirect addressing you load the memory space that you would like to access into one of the auxiliary registers (ARx). The current auxiliary register acts as a pointer that points to a specific memory address.

The register pointed to by the ARP is referred to as the current auxiliary register or current AR. To select a specific auxiliary register, load the 3-bit auxiliary register pointer (ARP) with a value from 0 to 7. The ARP can be loaded with the MAR instruction or by the LARP instruction. An ARP value can also be loaded by using the ARx operand after any instruction that supports indirect addressing as seen below.

Example of using MAR:

```
ADD     * , AR1     ;Adds using current * , then makes AR1 the
                    ;new current AR for future uses
```

Example of using LARP

```
LARP    #2          ;this will make AR2 the current AR
```

The C2xx provides four types of indirect addressing options:

- **No increment or decrement.** The instruction uses the content of the current auxiliary register as the data memory address but neither increments nor decrements the content of the current auxiliary register.
- **Increment or decrement by 1.** The instruction uses the content of the current auxiliary register as the data memory address and then increments or decrements the content of the current auxiliary register by one.
- **Increment or decrement by an index amount.** The value in AR0 is the index amount. The instruction uses the content of the current auxiliary register as the data memory address and then increments or decrements the content of the current auxiliary register by the index amount.
- **Increment or decrement by an index amount using reverse carry.** The value in AR0 is the index amount. After the instruction uses the content of the current auxiliary register as the data memory address, that content is incremented or decremented by the index amount. The addition and subtraction process is accomplished with the carry propagation reversed and is useful in fast Fourier transforms algorithms.

Table 2.1 displays the various operands that are available for use with instructions while using indirect addressing mode.

Table 2.1 Indirect addressing operands.

Operand	Option	Example
*	No increment or decrement	**LT *** loads the temporary register TREG with the content of the data memory address referenced by the current AR.
*+	Increment by 1	**LT *+** loads the TREG with the content of the data memory address referenced by the current AR and then adds 1 to the content of the current AR.
*-	Decrement by 1	**LT *-** loads the TREG with the content of the data memory address referenced by the current AR and then subtracts 1 from the content of the current AR.
*0+	Increment by index amount	**LT *0+** loads the TREG with the content of the data memory address referenced by the current AR and then adds the content of AR0 to the content of the current AR.
*0-	Decrement by index amount	**LT *0-** loads the TREG with the content of the data memory address referenced by the current AR and then subtracts the content of AR0 from the content of the current AR.
*BR0+	Increment by index amount, adding with reverse carry	**LT *BR0+** loads the TREG with the content of the data memory address referenced by the current AR and then adds the content of AR0 to the content of the current AR, adding with reverse carry propagation.
*BR0-	Decrement by index amount, subtracting with reverse carry	**LT *BR0-** loads the TREG with the content of the data memory address referenced by the current AR and then subtracts the content of AR0 from the content of the current AR, subtracting with bit reverse carry propagation.

2.7 Assembly Programming Using the C2xx DSP Instruction Set

This section is dedicated to developing code using the C2xx assembly instruction set and Code Composer Studio (CCS). We will start by introducing the basics of using the instruction set and provide examples of different options when using an assembly instruction. Then we will cover code development in CCS including an explanation of the main file types used to create and compile a working assembly program. Finally, an exercise will be presented to allow the reader to practice the new skills presented in this chapter.

2.7.1 Using the Assembly Instruction Set

The complete detailed instruction set for the C2xx DSP core can be found in the Texas Instruments *TMS320F/C24x DSP Controllers Reference Guide: CPU and Instruction Set; Literature Number: SPRU160C.* This reference guide contains a complete descriptive listing on syntax, operands, binary opcode, instruction execution order, status bits affected by the instruction, number of memory words required to store the instruction, and clock-cycles used by the instruction. The Texas Instruments documentation on the assembly instruction set is very well written. Each assembly instruction has a complete explanation of the instruction, all optional operands, and several examples of the instructions used. Since including the instruction set and complete documentation would make this book excessively long, we will assume the reader has access to the documentation referred to above.

We will therefore focus on developing code, not the instruction set itself. Each command starts with the basic assembly instruction. Each command supports specific addressing modes and options. For example, the ADD command will work with direct, indirect, and immediate addressing. In addition to the basic command, many instructions have additional options that may be used with the instruction. For example, the ADD command supports left shifting of the data before it is added to the accumulator.

The following is the instruction syntax for the ADD command:

```
ADD    dma [, shift]            ; Direct addressing
ADD    dma, 16                  ; Direct with left shift of 16
ADD    ind [, shift [, ARn]]    ; Indirect addressing
ADD    ind, 16 [, ARn]          ; Indirect with left shift of 16
ADD    #k                       ; Short immediate addressing
ADD    #lk [, shift]            ; Long immediate addressing
```

The following is a list of the various notations used in C2xx syntax examples:

Italics Italic symbols in instruction syntax represent variables.

Example: **LACC** *dma* , you can use several way to address the *dma* (data memory address).

```
LACC    *
```

or

```
LACC    200h
```

or

```
LACC    v           ; where "v" is any variable assigned to data
                    memory
```

where *, 200h, and v are the data memory addresses

Boldface Characters Boldface characters must be included in the syntax.

Example: **LAR** *dma*, **16** ; direct addressing with left shift of 16

```
LAR     AR1, 60h, 16    ; load auxiliary AR1 register with the
                    memory contents of 60h that was left shifted
                    16 bits
```

Example: **LACC** *dma*, [shift] ; optional left shift from 0, 15 ; defaults to 0

```
LACC  main_counter,  8  ;  shifts  contents  of  the  variable
                    "main_counter"  data  8  places  to  the  left
                    before loading accumulator
```

[ ] An optional operand may be placed in the placed here.

Example: **LACC** *ind* [, shift [, **AR** n]_] Indirect addressing

```
LACC    *           ;load Accum. w/contents of the memory
                    ;location pointed to by the current AR.
LACC    * ,5        ;load Accum. with the contents of the memory
                    ;location pointed to by the current AR after
                    ;the memory contents are left shifted by 5
                    ;bits .
LACC    * ,0, AR3   ;load Accum. with the contents of the memory
                    ;location pointed to by the current AR after
                    ;the memory contents are left shifted by 5
                    ;bits .  Now you have the option of choosing
                    ;a new AR. In this case, AR3 will become the
                    ;new AR.
```

[, x1 [, x2]] Operands x1 and x2 are optional, but you cannot include x2 without also including x1.

It is optional when using indirect addressing to modify the data. Once you supply a left shift value from 0…15 (even a shift of 0), then you have the option of changing to a new current auxiliary register (AR).

\# The # sign is prefix that signifies that the number used is a constant as opposed to memory location.

Example: RPT #15 ; this syntax is using short immediate addressing. It will repeat the next instruction 15+1 times.

```
LACC    #60h        ;this will load the accumulator with the
                    ;constant 60h
LACC    60h         ;However, this instruction will load the
                    ;accumulator with the contents in the data
                    ;memory location 60h, not the constant #60h
```

We will now provide a few examples of using the instruction set. Example 2.1 performs a few arithmetic functions with the DSP core and illustrates the nature of assembly programming. Programming with the assembly instruction set is somewhat different than languages such as C. In a high-level language, to add two numbers we might just code "c = a + b". In assembly, the user must be sure to code everything that needs to happen in order for a task to be executed. Take the following example:

Example 2.1 - Add the two numbers "2" and "3":

```
LDP     #6h         ;loads the proper DP for dma 300h
SPLK    #2, 300h    ;store the number "2" in memory address 300h
LACL    #3          ;load the accumulator with  the number "3"
ADD     300h        ;adds contents of 300h ("2") to the contents
                    ;of the accumulator("3");  accumulator = 5
```

Another way:

```
LDP     #6h         ;loads the proper DP for dma 300h
SPLK    #2h, 300h   ;store the number "2h" in memory address
                    ;300h
SPLK    #3h, 301h   ;stores the number "3h" into memory address
                    ;301h
LACL    300h        ;load the accumulator with the contents in
                    ;memory location 300h
ADD     301h        ;adds contents of memory address 301h ("3h")
                    ;to the contents of the accumulator ("2h")
                    ;accumulator = 5h
```

Looping algorithms are very common in all programming languages. In high-level languages, the "For" and "While" loops can be used. However, in assembly, we need a slightly different approach to perform a repeating algorithm. The following example is an algorithm that stores the value "1" to memory locations 300h, 301h, 302h, 303h, and 304h.

Example 2.2- Looping Algorithm Using the Auxiliary Register

```
        LAR     AR0, #4     ;load auxiliary register 0 with #4
        LAR     AR1, #300h  ;this AR will be used as a memory pointer
        LACL    #1h         ;loads "1" into the accumulator


LOOPER  MAR       *, AR1     ;makes AR1 the next current AR
        SACL     *+, AR0     ;writes contents of accumulator to address
                             ;pointed to by AR1, the "+" increments AR1
                             ;by 1, next current AR is AR0
        BANZ     LOOPER      ;branch to LOOPER while current AR is not 0
                             ;decrements current AR by 1 and branches
                             ;back to LOOPER
```

One might wonder if assembly language is so tedious to use, why not just program in a high-level language all the time. When code written in a high level language is compiled into assembly, the length of the code increases substantially. For example, if an assembly program takes up 50 lines, the same program written in C might take 150 lines after it is compiled. For this reason, code written in assembly almost always executed faster and uses less memory than high-level language code.

2.7.2 Code Generation in Code Composer Studio (CCS)

In order to develop a working program in CCS, one needs to understand the main file types and structure of the code composer project file. The project file (**.mak*) is the main file that links the other necessary sub-files together. The sub-files mentioned include source files (**.asm* for assembly), header (**.h*) files, include (**.i*) files, and linker command files (**.cmd*).

During the "building" of a project, CCS "assembles" and "links" the source file(s) and creates a DSP compatible (*.out) file that will ultimately be loaded onto the DSP. The out file contains the user program and also information as to where in program memory it will be placed. We will start the explanation of files with the assembly source file.

The assembly source file contains the code that will be executed when the DSP program runs. While there may be many source files in large projects, it is really only necessary to have one source file. If many source files exist, the linking order must be specified in CCS in order to ensure that the code will be operational. The file that follows is *template.asm*. This template source file encompasses the basics of a source file. The file *template.asm* contains many lines of assembly instructions which include comments to explain their function.

The Assembly Source File

Important Note: *Only a comment ";" or " *" or label such as "KICK_DOG" may be written starting on the first space of a line. ALL assembly commands must start at least from the second space on a line!*

```
;* Source File "template.asm"  for 2407 DSP programs         *
;* Target Hardware: 2407 DSP EVM      *
        . include 240x.h   ;this is the register definition file, MUST
                           ;INCLUDE!!
        . include vector.h ;this is the interrupt vector file that you
                           ;must include ;if you are using interrupts,
                           ;if you are not using any interrupts, it
                           ;doesn't hurt to include it anyway
 ;-------------------------------------------------------------
 ; M A C R O - Definition
 ;-------------------------------------------------------------
 KICK_DOG       .macro            ;Watchdog reset macro, resets watchdog
        LDP     #00E0h            ;DP-->7000h-707Fh
        SPLK    #05555h, WDKEY
        SPLK    #0AAAAh, WDKEY
        LDP     #0h               ;DP-->0000h-007Fh
.endm
 ;*******END KICK_DOG MACRO***
continued.......
continued from above ...........
;* Variables declaration: these are commented out, and are not needed
unless you want to define a variable
 ;    .bss     var1,1
 ;    .bss     var2,1
 ;    .bss     var3,1
 ;    .bss     var_name,1        ;you may continue as needed

    .text        ; this is the start of the actual program
;***General Initialization*****
START:  LDP     #0h               ;Set DP=0
        SETC    INTM              ;Disable interrupts
        SPLK    #0000h, IMR       ;Mask all core interrupts
        LACC    IFR               ;Read Interrupt flags
        SACL    IFR               ;Clear all interrupt flags
        LDP     #00E0h            ;(E0=224)(E0*80=7000)
        SPLK    #006Fh, WDCR      ;Disable WD if VCCP=5V (706F)
        SPLK    #0000h, SCSR1     ;All peripheral clocks disabled,
                                  ;PLL= clock x 4
        SPLK    #00FCh, SCSR1     ;All peripheral clocks enabled, PLL
                                  ;= clock x 4
        KICK_DOG                  ; run the KICK_DOG macro
        SPLK    #0h, 60h    ;Set wait state generator for: 0 wait states
```

```
        OUT     60h, WSGR  ;Program Space, 0-7 wait states

;***General Initialization complete!!, START YOUR CODE ON NEXT LINE***
;****END YOUR CODE HERE, if interrupts are used, add interrupt service
;routine under GISRx, if no interrupts are used, leave following code
;as is ******

PHANTOM         KICK_DOG    ;This dummy loop is just in case a wild
        B   PHANTOM         ;interrupt happens, all non-used interrupts
                            ;are set to branch to PHANTOM, this
                            ;routine just resets the watchdog timer.
GISR1:  RET                 ;Interrupt #1 service routine
GISR2:  RET                 ;Interrupt #2 service routine
GISR3:  RET                 ;Interrupt #3 service routine
GISR4:  RET                 ;Interrupt #4 service routine
GISR5:  RET                 ;Interrupt #5 service routine
GISR6:  RET                 ;Interrupt #6 service routine
   .end                     ;this tells the assembler that this is the
                            ;end of the program, YOU MUST INCLUDE the
                            ;".end"
```

Starting at the top of the file, we can see the comments indicated by the "*" or ";"s. Some versions of CCS might not recognize "*" as a comment indicator. Further down, the ".include" command tells CCS what header files will be included in the assembling of the project. We will discuss the purpose of header files shortly. After the header files, we see the KICK_DOG macro. The KICK_DOG macro is a subroutine that will run every time the line "KICK_DOG" is written. Basically, KICK_DOG resets the watchdog timer on the LF2407. The watchdog timer, fed by the system clock, counts up and will generate a general system reset if it reaches an overflow. This ensures that if some event (software error) causes an inadvertent system lock-up, then the DSP will be reset automatically. The basic idea is that after the lock-up subsequent reset by the watchdog, the DSP will reinitialize itself and start to function normally, thereby increasing system integrity (if the same event does not occur again!). For our learning purposes, we will disable the watchdog timer.

After the KICK_DOG macro declaration, we see an optional variable declaration section. The "***.bss***" command can be used to define variables, of which the values will be stored in sequential order in the memory locations specified by the "***.bss***" section in the *linker command file*. It is still necessary to set the data page for the corresponding memory address when using a variable.

The ".text" line signifies the start of the program. This is followed by the "START" label. Labels are the only syntax that may be placed on the first space of any line. The next lines initialize the LF2407, disable the watchdog timer, and set for zero wait states. Wait states introduce a delay into the external memory interface for accessing slow external memory, but are not needed for our purposes.

After these instructions are written, the functional part of the user code can be considered.

Finally, we come to the interrupt vectors GISR1, GISR2, etc. The operation of interrupts will be explained in Chapter 5. Because no interrupts are currently being used, the interrupt vectors simply are set so that any hypothetical random interrupt will cause the DSP to return to the program and do nothing. We must include the (GISR1, GISR2,…) labels because they are referenced in the *vector.h* header file.

Header Files

Header files (**.h*) like *240x.h* and *vector.h* (both below) discussed previously serve the purpose of providing definitions or other information that would otherwise add a substantial amount of code lines to the source file. Although it is not *absolutely* necessary to include a header file in the source file, header files allow the user to avoid re-writing commonly used definitions from program to program. The information contained in the file *240x.h* consists of setting variables in the name of actual registers to their respective data memory addresses. This allows the programmer to simply type the name of the intended control register rather than having to constantly look up the actual memory.

For example, the SCSR1 register is located at 7018h in data memory. In the file *240x.h* you can see how register names are set as variables with the memory address as the value. The purpose for this is to provide for more user-friendly programming. Instead of coding:

```
    SPLK      #00FCh, 7018h  ;which would work just fine for writing
                             ;FCh to 7018h (the SCSR1 register)
```

we may write:

```
    SPLK      #00FCh, SCSR1  ;not only is this easier, but it aids in
                             ;documentation as well
```

The file "240x.h" contains the control register definitions. A section of the file is included below:

```
;**********************************************************************
; File name:  240x.h
; Description:240x register definitions, Bit codes for BIT instruction
;**********************************************************************
; 240x CPU core registers

IMR     .set     0004h       ; Interrupt Mask Register
IFR     .set     0006h       ; Interrupt Flag Register

; System configuration and interrupt registers
SCSR1   .set 7018h           ; System Control &  Status register. 1
SCSR2   .set 7019h           ; System Control &  Status register. 2
```

```
DINR    .set 701Ch          ; Device Identification Number register.
PIVR    .set 701Eh          ; Peripheral Interrupt Vector register.
PIRQR0  .set 7010h          ; Peripheral Interrupt Request register 0
PIRQR1  .set 7011h          ; Peripheral Interrupt Request register 1
PIRQR2  .set 7012h          ; Peripheral Interrupt Request register 2
.
.
```

The header file *vector.h* contains interrupt vector information. This file specifies what section of source code the processor will branch to when an interrupt occurs. By looking at the file, we can see that we have many branch statements. When an interrupt occurs, the processor first branches to **0h** in **program memory**. It then sequentially checks each program memory address for an identifiable interrupt label (INTx) that corresponds to the pending interrupt. When it finds a match, it executes the instruction on that line, which in this case is a "B GISRx" instruction. The branch instruction causes the DSP to branch to the GISRx label (which is in the source code) under which the interrupt service routine is written. The processor then starts executing the code under that section.

INT1 through INT6 are the corresponding labels for these interrupts in the vector file. Notice the "B GISRx" command after each "INTx". More on interrupts will be covered later in Chapter 5.

The file "vector.h" contains the interrupt vectors for the LF2407. A section of this file is shown below:

```
;*********************************************************************
; File name:  vector.h
; Interrupt Vector declarations
; This section contains the vectors for various interrupts in the
; 240x. Unused interrupts are shown to branch to a "phantom" interrupt
; service routine which loops to itself. Users should replace the
; label PHANTOM with the label of their interrupt subroutines in case
; these interrupts are used.
;*********************************************************************
            .sect "vectors"
RSVECT          B           START       ; Reset Vector
INT1            B           GISR1       ; Interrupt Level 1
INT2            B           GISR2       ; Interrupt Level 2
INT3            B           GISR3       ; Interrupt Level 3
INT4            B           GISR4       ; Interrupt Level 4
INT5            B           GISR5       ; Interrupt Level 5
INT6            B           GISR6       ; Interrupt Level 6
RESERVED        B           PHANTOM     ; Reserved
SW_INT8         B           PHANTOM     ; Software Interrupt
SW_INT9         B           PHANTOM     ; Software Interrupt
SW_INT10        B           PHANTOM     ; Software Interrupt
```

```
SW_INT11        B         PHANTOM    ; Software Interrupt
SW_INT12        B         PHANTOM    ; Software Interrupt
SW_INT13        B         PHANTOM    ; Software Interrupt
SW_INT14        B         PHANTOM    ; Software Interrupt
SW_INT15        B         PHANTOM    ; Software Interrupt
SW_INT16        B         PHANTOM    ; Software Interrupt
TRAP            B         PHANTOM    ; Trap vector
NMI             B         NMI        ; Non-maskable Interrupt
EMU_TRAP    B             PHANTOM    ; Emulator Trap
SW_INT20        B         PHANTOM    ; Software Interrupt
SW_INT21        B         PHANTOM    ; Software Interrupt
SW_INT22        B         PHANTOM    ; Software Interrupt
.
.
```

The *linker command file* (**.cmd*) specifies to CCS where valid memory exists in both program and data memory. It also specifies where the ***.text***, ***.sect***, ***.bss***, and other sections will be placed in memory.

Looking at the linker command file, it is broken into three "pages". Page 0 refers to program memory (memory where the user code actually resides). Page 1 refers to data memory that contains control registers and memory. Page 2 defines memory that is reserved for I/O using external peripherals (you do not normally use the I/O memory except for disabling the watchdog timer or writing to the DAC). In each page, we can see the declarations of different memory ranges, their start address, and their length.

NOTE: Memory that is not defined in the linker command file will not be recognized by the program and cannot be used even if the memory physically exists.

The SECTIONS title lists assembler directives that are used to assign particular parts of code or variables to certain sections in memory. For example:

The ***.text*** directive tells the linker to put this section of the code in program memory starting at 0000h. The ***.bss*** directive tells the linker that all code written under this title should be placed in the location defined in the linker command file (in this case, the location ***BLK_B2*** which starts at 60h in data memory and is 20h in length).

Each section is defined in either the source file or the header file. Near the top of the *vector.h* file, the line *".sect" and ".vectors"* relates to the *2407.cmd* file where the section "vectors" is defined. This tells CCS to place the following vector code in the memory defined in the linker command file as "vectors". In this case, "vectors" is defined to start at 0h in program memory.

*NOTE: Both **.text** and **.vectors** sections are listed to start at 0h in program memory in "2407_PM.cmd". **.vectors** will be placed before the **.text** section in program memory.*

You are not necessarily required to use the above linker command file as shown. For example, if you wanted to start BLK_B1 at 301h instead of 300h, you can modify the *ORIGIN* of BLK_B1 to do so, but you would have to modify the *LENGTH* definition to account for the new memory range.

The following file is the linker command file "2407_PM.cmd".

```
MEMORY
{
PAGE 0:         /* PROGRAM MEMORY    */
PM              :ORIGIN=0h, LENGTH=08000h    /*On chip flash */
SARAM_P         :ORIGIN=08000h,  LENGTH=0800h    /*SARAM*/      /*program
                            External RAM*/
EX1_PM          :ORIGIN=08800h,  LENGTH=07600h
B0_PM           :ORIGIN=0FF00h,  LENGTH=0100h    /*On-chip*/ /*DARAM if
                            CNF=1, else external*/
                                          /*B0_PM = FF00 to FFFF */

PAGE 1:         /*DATA MEMORY       */
REGS            :ORIGIN=0h,      LENGTH=60h      /*Memory mapped*/
                            /*regs & reserved address */
BLK_B2          :ORIGIN=60h,     LENGTH=20h            /*Block B2*/
BLK_B0          :ORIGIN=200h ,   LENGTH=100h           /*Block B0*/
                            /*On chip DARAM if CNF=0*/
BLK_B1          :ORIGIN=300h ,   LENGTH=100h           /*Block B1*/
SARAM_D         :ORIGIN=0800h ,  LENGTH=0800h    /*2K SARAM*/
                            /*in data*/
PERIPH          :ORIGIN=7000h ,  LENGTH=1000h    /*Peripheral regs*/
EX2_DM          :ORIGIN=8000h ,  LENGTH=8000h    /*External RAM*/

PAGE 2:         /* I/O MEMORY    */
IO_EX           :ORIGIN=0000h ,   LENGTH=0FFF0h /*External         I/O*/
                            /*mapped peripherals   */
IO_IN           :ORIGIN=0FFF0h,   LENGTH=0Fh    /* On chip I/O */
                            /*mapped peripherals      */
}
SECTIONS
{
        vectors      :{}  > PM    PAGE 0
        .text        :{}  > PM    PAGE 0
```

```
        .bss        :{}  > BLK_B2   PAGE 1
        .data       :{}  > BLK_B1   PAGE 1
}
```

2.7.3 Code Generation Exercise Using Code Composer

This exercise will help you to become familiar with using the instruction set of the C2xx and the different modes of memory addressing. This exercise is intended to be an introduction to programming the C2xx core.

1. Start up CCS.
2. Open a new project by going to *Project/New*. Enter "lab2" for the project name.
3. You are now ready to start adding files to your project.
4. Create a new source file called "lab2.asm". Do this by going to *File/New/Source* in CCS. Be sure to use the *.include* directive to include the *240x.h* and *vector.h* header files as in *template.asm.*
5. Next add "lab2.asm" to the project. Go to *Project/Add files to Project*, and add "lab2.asm" to the project. See Fig. 2.7.

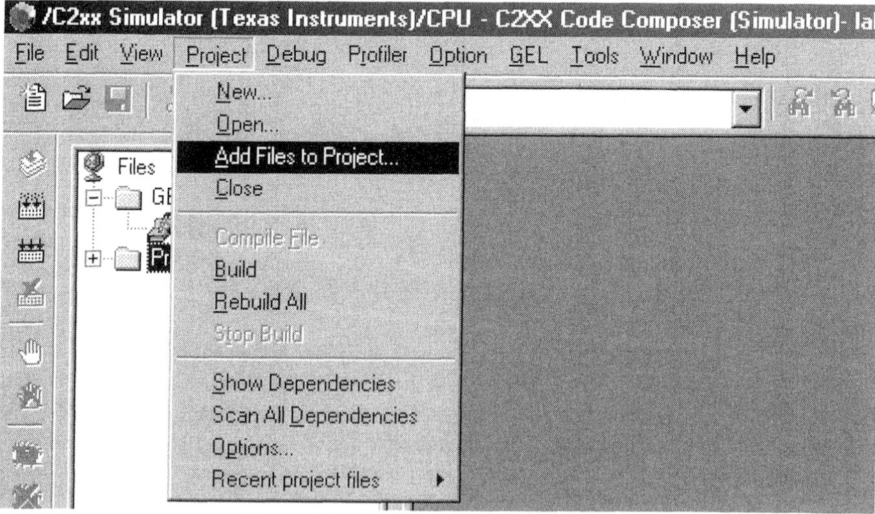

Figure 2.7 Adding files to a project.

6. Find the files "*240x.h*", "*vector.h*", and "*2407_PM.cmd*". Copy the files to the same directory in which your project is stored.
7. Add "*2407_PM.cmd*" to the project in the same way as "*lab2.asm*".
8. Now that the source file and linker command file are added to the project, go to *Project/Scan All Dependencies*" (see Fig. 2.8). Notice how the files now appear under the "Include" folder in the project window.

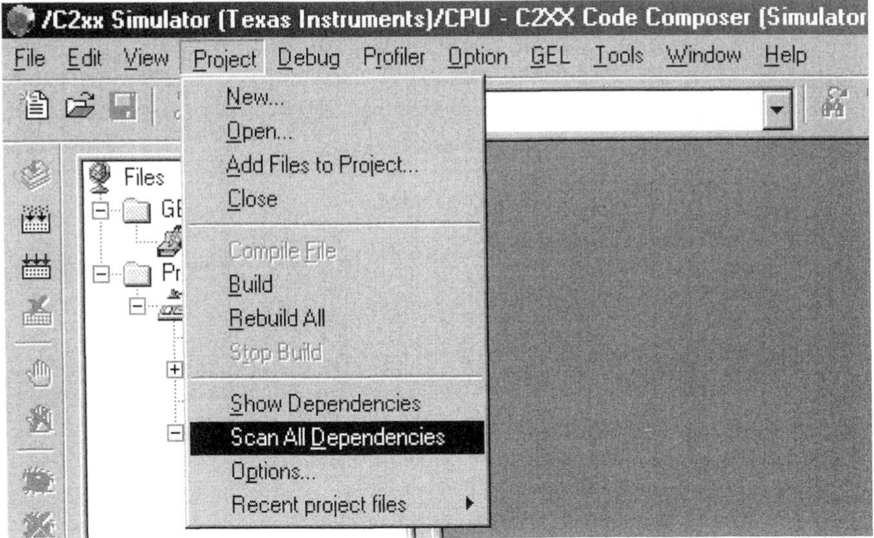

Figure 2.8 Scanning files for dependent sub-files.

9. Open "*lab2.asm*" by double-clicking on the file from the project menu.

10. In the "*lab2.asm*" source file, write a simple program that stores the number "35" into data memory location "305h" and adds it to the number "10" stored in data memory location "306h". Store the result in the data memory location "60h".

Hint: The following commands might be useful: SPLK, LACL, LACC, SACL, and ADD. Refer to the C2xx Instruction Set for more information about assembly commands.

Always be sure to set the appropriate data page pointer for the memory addresses. (Remember, this can be done by simply dividing the memory address (in hexadecimal) by "80h".)

If you get the build error ">> Warning: entry point symbol _c_int0 undefined" try changing the C Initialization from "ROM Auto initialization" to "No Auto initialization" by going to Project/Options/Linker tab/C Initialization.

11. Using indirect addressing and a looping routine, add another algorithm which writes #0h to 300h, writes #1h to 301h, writes #3h to 303h etc... all the way to 30Fh.
12. Add another routine to check memory 300h through 30Fh for the proper data (0h-Fh). If all the registers contain the proper data, your program

should write "Ah" in memory location 310h. If even one memory location has the incorrect data in it, write "DEADh" to 310h.

13. Write another algorithm to multiply the hex numbers in memory locations 300h through 30Fh by "#5h" and stores them in memory locations 320h through 32Fh.

This laboratory exercise has now concluded.

Chapter 3

GENERAL PURPOSE INPUT/OUTPUT (GPIO) FUNCTIONALITY

3.1 Pin Multiplexing (MUX) and General Purpose I/O Overview

Due to the limited number of physical pins on the LF2407 DSP, it is necessary to multiplex two functions onto most of the pins. That is, each pin can be programmed for either a primary or secondary (GPI/O) function (see Fig. 3.1). Once the pins on the LF2407 are multiplexed, the effective pin-out of the device is doubled. This provides enough effective pin-out for six General Purpose Input Output (GPIO) ports to be configured as the secondary function on most pins. Each Input/Output Port (IOP) consists of eight pins when they are configured to their secondary function.

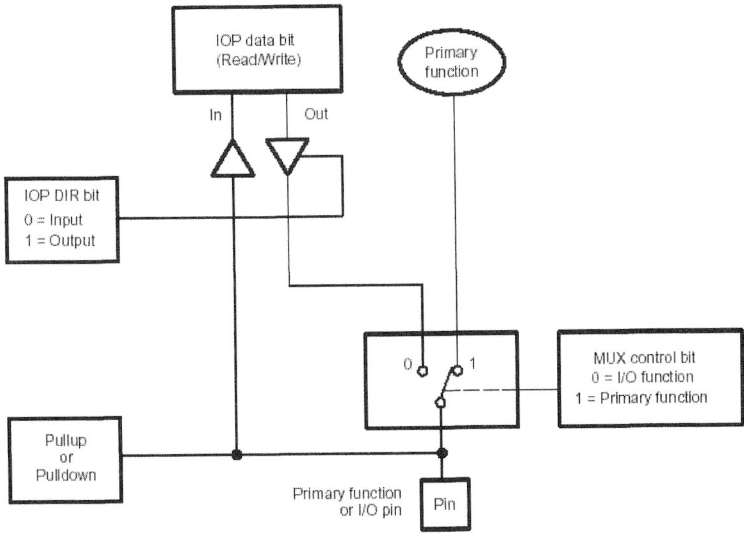

Figure 3.1 Block diagram of the multiplexing of a single pin. (Courtesy of Texas Instruments)

GPIO pins are grouped in sets of eight pins called ports. There are six ports total, ports A through F. Even though the pins are grouped in ports, each pin can be individually configured as primary or secondary (GPIO) functionality; and if GPIO, then either input or output. The multiplexing of primary pin functions with secondary GPIO functions provides a flexible method of controlling both the dedicated and secondary pin functions.

Each multiplexed pin's primary/secondary functionality is controlled by a corresponding bit in the appropriate MUX control register. Additionally, when the pin is in GPIO mode, there are port data and direction (PxDATDIR) control registers which control the direction (input or output) and data of the port/pin. If the pin is configured as an output, then the data (voltage) on the pin is determined by what value is written to the pin's data bit. Inversely, if the pin is configured as an input, then the voltage level applied to the pin determines the value of the pin's corresponding data bit.

If the pin is configured as an output pin, it can either be set to a logic high "1" (3.3 Volts) or a logic low "0" (0 Volts) by writing to its corresponding data bit in the corresponding PxDATDIR register. If the pin is configured as an input, the pin's corresponding bit in the appropriate PxDATDIR register will be "1" if 3.3 Volts or "0" if 0 Volts is applied to the pin. The data bits in the PxDATDIR can then be read by the user code and the values used in the program. The input and output ports provide a convenient way to input or output binary data (each pin = 1 bit). For example, a seven-segment display could be controlled by a GPIO port configured as output.

Note*: There is no relationship between the GPIO pins and the I/O space of the LF2407.*

3.2 Multiplexing and General Purpose I/O Control Registers

The three MUX control registers and six data/direction control registers are all mapped to data memory (see Table 3.1). They control all dedicated and shared pin functions:

- I/O MUX Control Registers (MCRA, MCRB, MCRC): These 16-bit registers determine whether a pin will operate in its primary function or secondary GPIO function. Two ports are assigned to each MUX control register. For example, the MCRA register controls ports A and B.

- Data and Direction Control registers (PxDATDIR): Once a pin is configured in I/O mode by the appropriate MUX control register, the appropriate PxDATDIR register is used to configure each pin as input or output; and if output, whether the pin is high (3.3 Volts) or low (0 Volts).

Table 3.1 GPIO Control Register Summary

Data Memory Address	**Register Name**	**Description**
7090h	MCRA	I/O MUX Control Register A
7092h	MCRB	I/O MUX Control Register B
7094h	MCRC	I/O MUX Control Register C
7098h	PADATDIR	I/O Port A Data and Direction Register
709Ah	PBDATDIR	I/O Port B Data and Direction Register
709Ch	PCDATDIR	I/O Port C Data and Direction Register
709Eh	PDDATDIR	I/O Port D Data and Direction Register
7095h	PEDATDIR	I/O Port E Data and Direction Register
7096h	PFDATDIR	I/O Port F Data and Direction Register

3.2.1 I/O Multiplexing (MUX) Control Registers

I/O MUX Control Register A (MCRA) Configuration

15	14	13	12	11	10	9	8
MCRA.15	MCRA.14	MCRA.13	MCRA.12	MCRA.11	MCRA.10	MCRA.9	MCRA.8
RW–0	RW–0	RW–0	RW–0	RW–0	RW–0	RW–0	RW–0
7	6	5	4	3	2	1	0
MCRA.7	MCRA.6	MCRA.5	MCRA.4	MCRA.3	MCRA.2	MCRA.1	MCRA.0
RW–0	RW–0	RW–0	RW–0	RW–0	RW–0	RW–0	RW–0

Note: *R = read access, W = write access, -0 = value after reset.*

Bit #	Name.bit #	Pin Function Selected (MCA.n = 1) (Primary)	Pin Function Selected (MCA.n = 0) (Secondary)
0	MCRA.0	SCITXD	IOPA0
1	MCRA.1	SCIRXD	IOPA1
2	MCRA.2	XINT1	IOPA2
3	MCRA.3	CAP1/QEP1	IOPA3
4	MCRA.4	CAP2/QEP2	IOPA4
5	MCRA.5	CAP3	IOPA5
6	MCRA.6	PWM1	IOPA6
7	MCRA.7	PWM2	IOPA7
8	MCRA.8	PWM3	IOPB0
9	MCRA.9	PWM4	IOPB1
10	MCRA.10	PWM5	IOPB2
11	MCRA.11	PWM6	IOPB3
12	MCRA.12	T1PWM/T1CMP	IOPB4
13	MCRA.13	T2PWM/T2CMP	IOPB5
14	MCRA.14	TDIRA	IOPB6
15	MCRA.15	TCLKINA	IOPB7

I/O MUX Control Register B (MCRB) Configuration

15	14	13	12	11	10	9	8
MCRB.15	MCRB.14	MCRB.13	MCRB.12	MCRB.11	MCRB.10	MCRB.9	MCRB.8
RW-1	RW-1	RW-1	RW-1	RW-1	RW-1	RW-1	RW-0

7	6	5	4	3	2	1	0
MCRB.7	MCRB.6	MCRB.5	MCRB.4	MCRB.3	MCRB.2	MCRB.1	MCRB.0
RW-0	RW-0	RW-0	RW-0	RW-0	RW-0	RW-1	RW-1

Note: *R = read access, W = write access, -0 = value after reset.*

Bit #	Name.bit #	Pin Function Selected	
		(MCB.n = 1) (Primary)	(MCB.n = 0) (Secondary)
0	MCRB.0	W/R	IOPC0
1	MCRB.1	BIO	IOPC1
2	MCRB.2	SPISIMO	IOPC2
3	MCRB.3	SPISOMI	IOPC3
4	MCRB.4	SPICLK	IOPC4
5	MCRB.5	SPISTE	IOPC5
6	MCRB.6	CANTX	IOPC6
7	MCRB.7	CANRX	IOPC7
8	MCRB.8	XINT2/ADCSOC	IOPD0
9	MCRB.9	EMU0	Reserved
10	MCRB.10	EMU1	Reserved
11	MCRB.11	TCK	Reserved
12	MCRB.12	TDI	Reserved
13	MCRB.13	TDO	Reserved
14	MCRB.14	TMS	Reserved
15	MCRB.15	TMS2	Reserved

I/O MUX Control Register C (MCRC) Configuration

15	14	13	12	11	10	9	8
Reserved	Reserved	MCRC.13	MCRC.12	MCRC.11	MCRC.10	MCRC.9	MCRC.8
		RW–0	RW–0	RW–0	RW–0	RW–0	RW–0
7	6	5	4	3	2	1	0
MCRC.7	MCRC.6	MCRC.5	MCRC.4	MCRC.3	MCRC.2	MCRC.1	MCRC.0
RW–0	RW–0	RW–0	RW–0	RW–0	RW–0	RW–0	RW–1

Note: *R = read access, W = write access, -0 = value after reset.*

		Pin Function Selected	
it #	**Name.bit #**	**(MCC.n = 1) (Primary)**	**(MCC.n = 0) (Secondary)**
0	MCRC.0	CLKOUT	IOPE0
1	MCRC.1	PWM7	IOPE1
2	MCRC.2	PWM8	IOPE2
3	MCRC.3	PWM9	IOPE3
4	MCRC.4	PWM10	IOPE4
5	MCRC.5	PWM11	IOPE5
6	MCRC.6	PWM12	IOPE6
7	MCRC.7	CAP4/QEP3	IOPE7
8	MCRC.8	CAP5/QEP4	IOPF0
9	MCRC.9	CAP6	IOPF1
10	MCRC.10	T3PWM/T3CMP	IOPF2
11	MCRC.11	T4PWM/T4CMP	IOPF3
12	MCRC.12	TDIRB	IOPF4
13	MCRC.13	TCLKINB	IOPF5
14	MCRC.14	Reserved	IOPF6
15	MCRC.15	Reserved	Reserved

3.2.2 Port Data and Direction Control Registers

Port A Data and Direction Control Register (PADATDIR)

15	14	13	12	11	10	9	8
A7DIR	A6DIR	A5DIR	A4DIR	A3DIR	A2DIR	A1DIR	A0DIR
RW–0	RW–0	RW–0	RW–0	RW–0	RW–0	RW–0	RW–0
7	6	5	4	3	2	1	0
IOPA7	IOPA6	IOPA5	IOPA4	IOPA3	IOPA2	IOPA1	IOPA0
RW–†	RW–†	RW–†	RW–†	RW–†	RW–†	RW–†	RW–†

† The reset value of these bits depends upon the state of the respective pins.

Note: *R = read access, W = write access, -0 = value after reset.*

Bits 15–8 AnDIR – Direction Bits

0 Configure corresponding pin as an input

1 Configure corresponding pin as an output

Bits 7–0 IOPAn – Data Bits

If AnDIR = 0, then:

0 Corresponding I/O pin is read as a low

1 Corresponding I/O pin is read as a high

If AnDIR = 1, then:

0 Set corresponding I/O pin low

1 Set corresponding I/O pin high

Port B Data and Direction Control Register (PADATDIR)

15	14	13	12	11	10	9	8
B7DIR	B6DIR	B5DIR	B4DIR	B3DIR	B2DIR	B1DIR	B0DIR
RW–0	RW–0	RW–0	RW–0	RW–0	RW–0	RW–0	RW–0
7	6	5	4	3	2	1	0
IOPB7	IOPB6	IOPB5	IOPB4	IOPB3	IOPB2	IOPB1	IOPB0
RW–†	RW–†	RW–†	RW–†	RW–†	RW–†	RW–†	RW–†

† The reset value of these bits depends upon the state of the respective pins.

Note: *R = read access, W = write access, -0 = value after reset.*

Bits 15–8 BnDIR – Direction Bits

0 Configure corresponding pin as an input

1 Configure corresponding pin as an output

Bits 7–0 IOPBn – Data Bits
If BnDIR = 0, then:

0	Corresponding I/O pin is read as a low
1	Corresponding I/O pin is read as a high

If BnDIR = 1, then:

0	Set corresponding I/O pin low
1	Set corresponding I/O pin high

Port C Data and Direction Control Register (PCDATDIR)

15	14	13	12	11	10	9	8
C7DIR	C6DIR	C5DIR	C4DIR	C3DIR	C2DIR	C1DIR	C0DIR
RW–0	RW–0	RW–0	RW–0	RW–0	RW–0	RW–0	RW–0
7	6	5	4	3	2	1	0
IOPC7	IOPC6	IOPC5	IOPC4	IOPC3	IOPC2	IOPC1	IOPC0
RW–†	RW–†	RW–†	RW–†	RW–†	RW–†	RW–†	RW–x

† The reset value of these bits depends upon the state of the respective pins.

Note: *R = read access, W = write access, -0 = value after reset, x = undefined.*

Bits 15–8 CnDIR – Direction Bits

0	Configure corresponding pin as an input
1	Configure corresponding pin as an output

Bits 7–0 IOPCn – Data Bits
If CnDIR = 0, then:

0	Corresponding I/O pin is read as a low
1	Corresponding I/O pin is read as a high

If CnDIR = 1, then:

0	Set corresponding I/O pin low
1	Set corresponding I/O pin high

Port D Data and Direction Control Register (PDDATDIR)

15-9	8
Reserved	D0DIR
	RW–0
7-1	0
Reserved	IOPD0
	RW–†

† The reset value of this bit depends upon the state of the respective pins.

Note: *R = read access, W = write access, -0 = value after reset.*

Bits 15–9 Reserved

Bit 8 D0DIR – Direction Bits

0 Configure corresponding pin as an input
1 Configure corresponding pin as an output

Bits 7–1 Reserved

Bit 0 IOPD0 – Data Bit

If D0DIR = 0, then:
0 Corresponding I/O pin is read as a low
1 Corresponding I/O pin is read as a high
If D0DIR = 1, then:
0 Set corresponding I/O pin low
1 Set corresponding I/O pin high

Port E Data and Direction Control Register (PEDATDIR)

15	14	13	12	11	10	9	8
E7DIR	E6DIR	E5DIR	E4DIR	E3DIR	E2DIR	E1DIR	E0DIR
RW–0	RW–0	RW–0	RW–0	RW–0	RW–0	RW–0	RW–0
7	6	5	4	3	2	1	0
IOPE7	IOPE6	IOPE5	IOPE4	IOPE3	IOPE2	IOPE1	IOPE0
RW–†	RW–†	RW–†	RW–†	RW–†	RW–†	RW–†	RW–x

† The reset value of these bits depends upon the state of the respective pins.

Note: *R = read access, W = write access, -0 = value after reset, x = undefined.*

Bits 15–8 EnDIR – Direction Bits

0 Configure corresponding pin as an input
1 Configure corresponding pin as an output

Bits 7–0 IOPEn – Data Bits

If EnDIR = 0, then:
0 Corresponding I/O pin is read as a low
1 Corresponding I/O pin is read as a high
If EnDIR = 1, then:
0 Set corresponding I/O pin low
1 Set corresponding I/O pin high

Port F Data and Direction Control Register (PFDATDIR)

15	14	13	12	11	10	9	8
Reserved	F6DIR	F5DIR	F4DIR	F3DIR	F2DIR	F1DIR	F0DIR
	RW–0	RW–0	RW–0	RW–0	RW–0	RW–0	RW–0
7	6	5	4	3	2	1	0
Reserved	IOPF6	IOPF5	IOPF4	IOPF3	IOPF2	IOPF1	IOPF0
	RW–†	RW–†	RW–†	RW–†	RW–†	RW–†	RW–†

† The reset value of these bits depends upon the state of the respective pins.

Note: *R = read access, W = write access, -0 = value after reset.*

Bit 15 Reserved

Bits 14–8 FnDIR – Direction Bits

0 Configure corresponding pin as an input

1 Configure corresponding pin as an output

Bit 7 Reserved

Bits 6–0 IOPFn – Data Bits

If FnDIR = 0, then:

0 Corresponding I/O pin is read as a low

1 Corresponding I/O pin is read as a high

If FnDIR = 1, then:

0 Set corresponding I/O pin low

1 Set corresponding I/O pin high

3.3 Using the General Purpose I/O Ports

The GPIO functionality is relatively simple to use and provides a valuable way of imputing and outputting data to and from the DSP. To use the GPIO functionality of a particular pin or groups of pins, the following steps must be followed to configure the DSP:

1. Set the bits in the appropriate MUX control register to configure the desired pins for GPIO function. This can be done by writing a "0" to the corresponding bits in the appropriate MUX. It may not be absolutely necessary to do this due to the fact that upon a reset (power on) the pins in the LF2407 are by default in their GPIO functionally. However, configuring the MUX register anyway is good programming practice.

2. Now that the desired pins are configured as GPIO, set the Port Data and Direction (PxDATDIR) register(s) that corresponds to the desired pins. When configuring the PxDATDIR, the most significant bits control the direction (input or output) and the lower bits determine (output) or display

(input) the pin data. If an input pin is desired, only the direction bit needs to be set since when the direction bit is set to input, writing to the data bit has no effect. The corresponding data bit will be used to display the logic value applied to that pin. If an output pin is required, both the direction and data bits need to be configured because the data bit will determine what logic value the pin will be set to.

3. The selected pins are now configured. The input data on pins can be obtained by reading the entire PxDATDIR register and obtaining the data for desired bits. For output, new values can be written to the pins by writing to the corresponding entire PxDATDIR register.

***Note**: When a pin is configured as input, it is important to note whether the pin has either a pull-up or pull-down resistor. If the input pin is not connected to anything, the pin's data bit will read "1" if a pull-up or "0" if a pull-down resistor exists. The pullup/down resistor comes into play only when the pin is an input and not connected. When the input pin is connected to either a logic "1" or logic "0" voltage, the pull-up/down resistor is overridden and has no effect. The reason behind the pullup/down resistor is that a digital input pin should never be completely floating.*

Example 3.1 illustrates configuring all the pins in a port for output and writing "1" to each of the eight pins in the port.

Example 3.1 Display the binary number "00100010"b with the eight pins on port A:

1. Configure the bits in MCRA so port A is I/O ("XXXXXXXX00000000"). The most significant bits in MCRA control port B; therefore, in this example, we do not care what we write to them.

2. Set pins to output in PADATDIR by setting bits 15-8 as "11111111", with "00100010"b as the data (bits 7-0). "X" designates "don't care" bits.

3.4 General Purpose I/O Exercise

This exercise allows the reader to become familiar with using the GPIO functionality on the LF2407 DSP controller. Practical applications using the GPIO functionality are very similar to the algorithms presented in this exercise. In addition, this exercise helps the reader practice writing assembly programs.

The XF pin is introduced during this exercise. The XF pin on the LF2407 is a general purpose output pin which is controlled by the XF bit in the C2xx DSP core. Because it is core controlled, the XF bit can be set and cleared without having to write to a register. The XF is easily set (made 3.3V) by the "SETC XF" command or cleared (made 0V) by the CLRC XF command. This pin can be useful in testing code to see if your code ran to a certain point, where the code would set or clear the XF bit. The level of the XF pin may be checked by either an oscilloscope or volt meter. In order to assure a correct signal is read, be sure the ground lead of the tester is connected to the digital ground of the LF2407 EVM.

1. Startup CCS and create a new project titled "lab3" in the same manner as the exercise in the previous chapter. Name the source file "lab3.asm" and include the same header files as before.

2. Write a program that first stores a certain set of values into data memory locations starting at memory address 300h. The values should be such that they control the "up counting" of a seven-segment display from "0" to "F". The program should then read the memory locations and send the values out on port A; displaying each value for a second or so.

3. Place a seven-segment LED display on a breadboard and connect the port A pins to the display in the appropriate positions.

4. Run the program and watch as the seven-segment display "counts" from 0 to F.

5. When the display has finished counting a 1000-hertz square wave should be produced by toggling the XF pin.

6. Connect the oscilloscope to the XF pin and digital ground of the LF2407 EVM. Use the oscilloscope to view the 1000-hertz waveform.

This exercise is now concluded.

Chapter 4

INTERRUPTS ON THE TMS320LF2407

4.1 Introduction to Interrupts

The interrupts on the LF2407 allow the device hardware to trigger the CPU of the LF2407 (CPU=C2xx DSP core) to break from the current task, branch to a new section of code and start a new task, then return back to the initial task. The "new task" referred to in the previous sentence is known as the Interrupt Service Routine (ISR). The ISR is simply a separate user-written subroutine, which the core will branch to every time a certain interrupt occurs.

For example, say the ADC is being used and we want the program to load the conversion value into the accumulator every time the ADC finishes a conversion. The ADC can be configured to generate an interrupt whenever a conversion is finished. When the ADC generates its interrupt, the interrupt signal makes its way through the interrupt hierarchy to the core and the core then branches to the appropriate ISR.

In a more general sense, when an interrupt occurs, the core branches to the ISR (GISR1, GISR2 etc… depending on the interrupt) where an interrupt service routine is located. In the ISR, after the instructions are executed, the interrupt hierarchy is "reset" to allow for future interrupts. This usually entails clearing the peripheral level interrupt flag bit and clearing the INTM bit. These steps ensure that future interrupts of the same origin will be able to pass through to the core. The final instruction in the ISR is the RET command, which instructs the core to return to where it was before the interrupt occurred.

4.2 Interrupt Hierarchy

This section will explain the different hierarchical levels and how an interrupt request signal propagates through them. The different control registers and their operations will be reviewed.

4.2.1 Interrupt Request Sequence

There are two levels of interrupt hierarchy in the LF2407 as seen in Fig. 4.1 below. There is an interrupt flag bit and an interrupt enable bit located in each peripheral configuration register for each event that can generate an interrupt. The peripheral interrupt flag bit is the first bit to be set when an interrupt generating event occurs. The interrupt enable bit acts as a "gate". If the interrupt enable bit is not set, then the setting of the peripheral flag bit will not be able to generate an interrupt signal. If the enable bit is set, then the peripheral flag bit will generate an interrupt signal. That interrupt signal will then leave the peripheral level and go to the next hierarchal level.

Once an interrupt signal leaves the peripheral level, it is then multiplexed through the Peripheral Interrupt Expansion (PIE) module. The PIE module takes the many individual interrupts and groups them into six priority levels (INT1 through INT6). Once an interrupt reaches the PIE, a code identifying the individual interrupt is loaded into the Peripheral Interrupt Vector Register (PIVR). This allows the ISR to determine which interrupt was actually asserted when multiple interrupts from the same level occur. After passing through the PIE module, the interrupt request signal has now entered the upper level of hierarchy or the "CPU level".

The six interrupt groupings from the PIE module feed into the CPU level. The final stage of the CPU level is the CPU itself (C2xx core). From Fig. 4.1, we can see the six interrupt levels and the many individual peripheral interrupts assigned to priority level. Each of the six levels has a corresponding flag bit in the Interrupt Flag Register (IFR). Additionally there is an Interrupt Mask Register (IMR) which acts similar to the interrupt enable bits at the peripheral level. Each of the six bits in the IMR behaves as a "gate" to each of the corresponding six bits in the IFR. If the corresponding bits in both the IFR and IMR are both set, then the interrupt request signal can continue through to the C2xx core itself.

Once the interrupt request signal has entered the CPU level and has passed through the IFR/IMR, there is one more gateway the signal must pass through in order to cause the core to service the interrupt. The Interrupt Mask (INTM) bit must be cleared for the interrupt signal to reach the core. When the core acknowledges a pending interrupt, the INTM bit is automatically set, thereby not allowing any more interrupts from reaching the core while a current interrupt is being serviced.

When the core is finished with the current interrupt, only the flag bit in the IFR is cleared automatically. The INTM bit and the peripheral level flag bit must be cleared "manually" via software. When this is done, the core will acknowledge the highest priority pending interrupt request signal.

Additionally, if an interrupt request signal occurs, but the signal never reaches the core, all flag bits "downstream" of the point where the signal was halted will still remain set until cleared by software. The IFR bits will be cleared if: (1) the interrupt path to the core is opened, and the interrupt is acknowledged normally or (2) the bit is cleared "manually" by software. If no interrupt request has occurred but the peripheral level IF bit is set and the peripheral IE bit is later set without clearing the IF bit, then an interrupt request signal will be asserted and the corresponding IFR bit will be set.

Furthermore, in the event that two interrupts of different priority groupings (INTx) occur at the same time, the highest priority interrupt will be acknowledged first by the core.

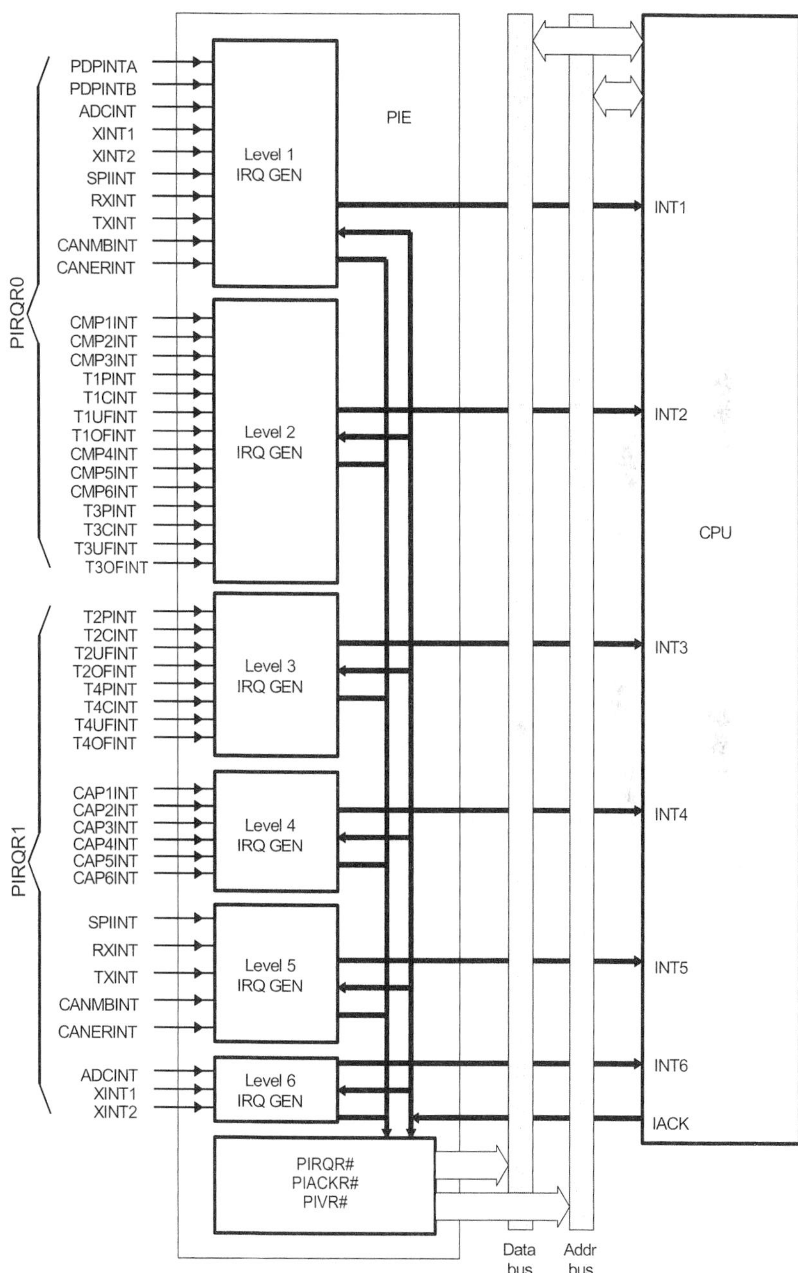

Figure 4.1 Interrupt hierarchy in the LF2407. (Courtesy of Texas Instruments)

4.2.2 Reset and Non-Maskable Interrupts

There are two special interrupts on the LF2407 which have not been covered thus far; the Reset (RS) and the Non-Maskable Interrupt (NMI). Both of these interrupts bypass the usual interrupt hierarchy and feed straight to the DSP core. A reset causes the core to branch to address 0000h in program memory. Resets are activated during power on, when the external RESET pin is brought to logic "0" (O Volts), or by the Watchdog Timer. If the Watchdog is not disabled, it will pull the reset pin to "0" if not periodically reset.

When an illegal memory space is written to, the illegal address flag (ILLADR) in System Control and Status Register 1 (SCSR1) will be set. When this flag is set, a non-maskable interrupt (NMI) will be generated, causing the core to branch to address 0024h in program memory. The illegal address flag (ILLADR) will remain set following an illegal address condition until it is cleared by software or a DSP reset.

4.3 Interrupt Control Registers

This section will review the interrupt control registers. The IFR, IMR, and PIVR registers as well as the INTM bit discussed in the previous section will be presented in more detail. We will not discuss peripheral level interrupt bits in this chapter, as they will be discussed in each section dealing with the specific peripherals.

There are three registers used at the CPU level, the Interrupt Flag Register (IFR), the Interrupt Mask Register (IMR), and the Peripheral Interrupt Vector Register (PIVR). The IFR and IMR control the interrupt signal at the beginning of the CPU level. The PIVR register, while actually loaded in the PIE, provides information about the specific interrupt that occurred at the peripheral level. This information can be used by the ISR in determining the source of the interrupt signal. In addition to these registers, the INTM bit at the CPU level provides the final "gateway" that the interrupt signal must pass through to reach the core itself.

In addition to the peripheral interrupts, there are two External Interrupts (XINT1, XINT2). Their interrupt request operation is exactly like the peripheral interrupts. However, external interrupts are triggered by a logic edge transition on their external pin. The external interrupt control registers will also be discussed.

4.3.1 Interrupt Flag Register (IFR)

The IFR is a 16-bit (only 6 bits are really used) register mapped to address 0006h in data memory. The IFR is used to identify and clear pending interrupts at the CPU level and contains the interrupt flag bits for the maskable interrupt priorities INT1–INT6.

A flag bit in the IFR is set to "1" when an individual interrupt request signal makes its way out of the peripheral level and into the CPU level. The particular flag bit set depends on what priority the individual interrupt is grouped under. After the interrupt is serviced, the IFR bit corresponding to the interrupt is automatically cleared (to "0") by the DSP.

In addition to triggering the CPU level during the standard interrupt process, the IFR can also be read by software. If a desired situation occurred where the INTM bit was set (meaning no interrupt signals make it to the core) and an interrupt signal was generated at the below levels, the corresponding bit in the IFR would still be set. In this situation, the IFR could be read by software to identify pending interrupt requests.

If desired, to "manually" clear a bit in the IFR, software needs to write a "1" to the appropriate bit (see IFR bit descriptions). The flag bits can be thought of as "toggling" when a "1" is written to them. Loading the IFR into the accumulator, then storing the contents of the IFR back into itself clears all bits in the IFR. However, if the peripheral level interrupt flag bit is still set, the corresponding bit in the IFR will immediately become set right after it is cleared.

Notes:

1. *To clear an IFR bit, we must write a one to it, not a zero.*
2. *When an interrupt is acknowledged,* ***only the IFR bit is cleared automatically****. The flag bit in the corresponding peripheral control register is* ***not*** *automatically cleared. If an application requires that the control register flag be cleared, the bit must be cleared by software.*
3. *IFR registers pertain to interrupts at the CPU level only. All peripherals have their own interrupt mask and flag bits in their respective control/configuration registers.*
4. *When an interrupt is requested by the INTR assembly instruction and the corresponding IFR bit is set, the CPU does not clear the bit automatically. If an application then requires that the IFR bit needs to be cleared, the bit must be cleared by software.*

Interrupt Flag Register (IFR) — Address 0006h

15-6	5	4	3	2	1	0
Reserved	INT6 flag	INT5 flag	INT4 flag	INT3 flag	INT2 flag	INT1 flag
0	RW1C–0	RW1C–0	RW1C–0	RW1C–0	RW1C–0	RW1C–0

Note: *0 = always read as zeros, R = read access, W1C = write 1 to this bit to clear it, -0 = value after reset.*

Bits 15–6 **Reserved.** These bits are always read as zeros.

Bit 5 **INT6.** Interrupt 6 flag. This bit is the flag for interrupts connected to interrupt level INT6.

0 No INT6 interrupt is pending

1 At least one INT6 interrupt is pending. Write a 1 to this bit to clear it to 0 and clear the interrupt request

Bit 4 **INT5.** Interrupt 5 flag. This bit is the flag for interrupts connected to interrupt level INT5.
- 0 No INT5 interrupt is pending
- 1 At least one INT5 interrupt is pending. Write a 1 to this bit to clear it to 0 and clear the interrupt request

Bit 3 **INT4.** Interrupt 4 flag. This bit is the flag for interrupts connected to interrupt level INT4.
- 0 No INT4 interrupt is pending
- 1 At least one INT4 interrupt is pending. Write a 1 to this bit to clear it to 0 and clear the interrupt request

Bit 2 **INT3.** Interrupt 3 flag. This bit is the flag for interrupts connected to interrupt level INT3.
- 0 No INT3 interrupt is pending
- 1 At least one INT3 interrupt is pending. Write a 1 to this bit to clear it to 0 and clear the interrupt request

Bit 1 **INT2.** Interrupt 2 flag. This bit is the flag for interrupts connected to interrupt level INT2.
- 0 No INT2 interrupt is pending
- 1 At least one INT2 interrupt is pending. Write a 1 to this bit to clear it to 0 and clear the interrupt request

Bit 0 **INT1.** Interrupt 1 flag. This bit is the flag for interrupts connected to interrupt level INT1.
- 0 No INT1 interrupt is pending
- 1 At least one INT1 interrupt is pending. Write a 1 to this bit to clear it to 0 and clear the interrupt request

4.3.2 Interrupt Mask Register (IMR)

The Interrupt Mask Register (IMR) is a 16-bit (only 6 bits are used) register located at address 0004h in data memory. It contains a mask bits for each of the six interrupt priority levels INT1–INT6. When an IMR bit is "0", the corresponding interrupt is "masked". When an interrupt is masked, the interrupt will be halted at the CPU level; the core will not be able to receive the interrupt request signal, regardless of the INTM bit status. When the interrupt's IMR bit is set to "1", the interrupt will be acknowledged if the corresponding IFR bit is "1" and the INTM bit is "0". The IMR may also be read to identify which interrupts are masked or unmasked.

Interrupt Mask Register (IMR) — Address 0004h

15-6	5	4	3	2	1	0
Reserved	INT6 mask	INT5 mask	INT4 mask	INT3 mask	INT2 mask	INT1 mask
0	RW	RW	RW	RW	RW	RW

Note: *0 = always read as zeros, R = read access, W = write access, bit values are not affected by a device reset.*

Bits 15–6 **Reserved**. These bits are always read as zeros.

Bit 5 **INT6.** Interrupt 6 mask. This bit masks or unmasks interrupt level INT6.
0 Level INT6 is masked
1 Level INT6 is unmasked

Bit 4 **INT5.** Interrupt 5 mask. This bit masks or unmasks interrupt level INT5.
0 Level INT5 is masked
1 Level INT5 is unmasked

Bit 3 **INT4.** Interrupt 4 mask. This bit masks or unmasks interrupt level INT4.
0 Level INT4 is masked
1 Level INT4 is unmasked

Bit 2 **INT3.** Interrupt 3 mask. This bit masks or unmasks interrupt level INT3.
0 Level INT3 is masked
1 Level INT3 is unmasked

Bit 1 **INT2.** Interrupt 2 mask. This bit masks or unmasks interrupt level INT2.
0 Level INT2 is masked
1 Level INT2 is unmasked

Bit 0 **INT1.** Interrupt 1 mask. This bit masks or unmasks interrupt level INT1.
0 Level INT1 is masked
1 Level INT1 is unmasked

Note: *A device reset does not affect The IMR bits.*

4.3.3 Peripheral Interrupt Vector Register (PIVR)

The Peripheral Interrupt Vector Register (PIVR) is a 16-bit read-only register located at address 701Eh in data memory. Each interrupt has a unique code which is loaded into the PIVR when in the PIE module. When a peripheral interrupt signal is passed through the PIE module, the PIVR is loaded with the vector of the pending interrupt which has the highest priority level. This assures that if two interrupts of different priorities happen simultaneously, the higher priority interrupt will be serviced first.

Peripheral Interrupt Vector Register (PIVR) — Address 701Eh

15	14	13	12	11	10	9	8
V15	V14	V13	V12	V11	V10	V9	V8
R-0	R-0	R-0	R-0	R-0	R-0	R-0	R-0

7	6	5	4	3	2	1	0
V7	V6	V5	V4	V3	V2	V1	V0
R-0	R-0	R-0	R-0	R-0	R-0	R-0	R-0

***Note:** R = read access, -0 = value after reset.*

Bits 15–0 V15–V0. Interrupt vector. This register contains the peripheral interrupt vector of the most recently acknowledged peripheral interrupt.

External Interrupt Control Registers

The external interrupts (XINT1, XINT2) are controlled by the XINT1CR and XINT2CR control registers, respectively. If these interrupts are enabled in their control registers, an interrupt will be generated when the XINT1 or XINT2 logic transition occurs for at least 12 CPU clock cycles.

For example, if XINT1 was configured for generating an interrupt on a low (0 Volts) to high (3.3 Volts) transition and the XINT1 pin only went high for 6 clock cycles, then back down to low, an interrupt request would not occur. However, if the pin was brought high for 12 or more cycles, an interrupt request signal would be generated.

External Interrupt 1 Control Register (XINT1CR) – Address 7070h

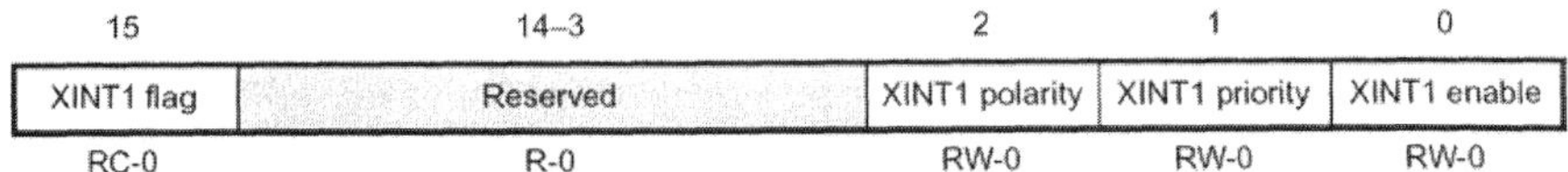

15	14–3	2	1	0
XINT1 flag	Reserved	XINT1 polarity	XINT1 priority	XINT1 enable
RC-0	R-0	RW-0	RW-0	RW-0

***Note:** R = read access, W = write access, C = clear by writing a 1, -0 = value after reset.*

Bit 15 XINT1 Flag

This bit indicates if the selected transition has been detected on the XINT1 pin and is set whether or not the interrupt is enabled. This bit is cleared by software writing a 1 (writing a 0 has no effect), or by a device reset.

0 No transition detected

1 Transition detected

Note: *the description in the TI user guide can be misleading: this bit is* ***not*** *cleared automatically during the interrupt acknowledge sequence.*

Bits 14–3 Reserved. Reads return zero; writes have no effect.

Bit 2 XINT1 Polarity

This read/write bit determines if interrupts are generated on the rising edge or the falling edge of a signal on the pin.

0 Interrupt generated on a falling edge (high-to-low transition)

1 Interrupt generated on a rising edge (low-to-high transition)

Bit 1 XINT1 Priority

This read/write bit determines which interrupt priority is requested. The CPU interrupt priority levels corresponding to low and high priority are coded into the peripheral interrupt expansion controller. These priority levels are shown in Table 2–2, *240xA Interrupt Source Priority and Vectors,* in Chapter 2 on page 2-9.

0 High priority

1 Low priority

Bit 0 XINT1 Enable

This read/write bit enables or disables external interrupt XINT1.

0 Disable interrupt

1 Enable interrupt

External Interrupt 2 Control Register (XINT2CR) – Address 7071h

15	14–3	2	1	0
XINT2 flag	Reserved	XINT2 polarity	XINT2 priority	XINT2 enable
RC-0	R-0	RW-0	RW-0	RW-0

Note: *R = read access, W = write access, C = Clear by writing a 1, -0 = value after reset.*

Bit 15 XINT2 Flag

This bit indicates if the selected transition has been detected on the XINT2 pin and is set whether or not the interrupt is enabled. This bit is cleared by software writing a 1 (writing a 0 has no effect), or by a device reset.

0 No transition detected

1 Transition detected

Note: *the description in the TI user guide can be misleading: this bit is* ***not*** *cleared automatically during the interrupt acknowledge sequence.*

Bits 14–3 **Reserved.** Reads return zero; writes have no effect.

Bit 2 **XINT2 Polarity**

This read/write bit determines if interrupts are generated on the rising edge or the falling edge of a signal on the pin.

0 Interrupt generated on a falling edge (high-to-low transition)

1 Interrupt generated on a rising edge (low-to-high transition)

Bit 1 **XINT2 Priority**

This read/write bit determines which interrupt priority is requested. The CPU interrupt priority levels corresponding to low and high priority are coded into the peripheral interrupt expansion controller. These priority levels are shown in Table 2–2, *240xA Interrupt Source Priority and Vectors,* in Chapter 2 on page 2-9.

0 High priority

1 Low priority

Bit 0 **XINT2 Enable**

This read/write bit enables or disables the external interrupt XINT2.

0 Disable interrupt

1 Enable interrupt

4.4 Initializing and Servicing Interrupts in Software

In order to utilize the interrupt functions of the LF2407, several steps should be taken to initialize the DSP and interrupt related registers. This will assure that no false interrupts are asserted. While it is unlikely that a false interrupt would be generated, writing code that would ignore a false interrupt is good practice.

Servicing the interrupt requires that a few steps also be taken to "reset" the interrupt so that future interrupts of the same origin can also occur.

4.4.1 Configuring the LF2407 for Interrupt Operation

Several steps should be performed via software to prepare the DSP and interrupt system for use before any sort of algorithm is entered. The following provides for a general procedure for initializing the DSP interrupts and peripherals:

1. The first instruction after the START label should be to set the INTM bit. This assures that no interrupts can occur during initialization.
2. Once the INTM bit is set, then the second step is to mask each of the six CPU level interrupts by writing "0" to the IMR.
3. Once all bits in the IMR are "0", the IFR value should be loaded into the accumulator, and then the accumulator should be written to the IFR. This writes the IFR back into itself, thereby clearing all flag bits.

4. Now is the time to disable the Watchdog timer by writing "6Fh" to the WDCR (watchdog control register). Also, the DSP should be configured by setting the System Control Registers (SCSR1,SCSR2) for desired operation and the enabling the clock to desired peripherals.
5. If applicable, set the reset bit in the selected peripheral control registers. Configure peripheral for desired operation.
6. Configure the IMR to unmask only those interrupt levels which correspond to the selected peripheral.
7. Clear the INTM bit to allow future interrupts to reach the CPU.
8. If applicable, bring the selected peripherals out of reset/enable operation via peripheral control register.

Example 4.1 - The following block is a segment of code which provides an example of interrupt initialization.

```
START:
        LDP     #0h             ;set DP=0
        SETC    INTM            ;Disable interrupts
        SPLK    #0000h,IMR      ;Mask all core interrupts
        LACC    IFR             ;Read Interrupt flags
        SACL    IFR             ;Clear all interrupt flags
        LDP     #WDKEY >> 7h    ;Peripheral page
        SPLK    #006Fh, WDCR    ;Disable WD if VCCP=5V
        SPLK    #0000h,  SCSR1  ;
KICK_DOG
        SPLK    #0h,GPR0        ;Set wait state generator for:
        OUT     GPR0,WSGR       ;Program Space, 0-7 wait states
        LDP     #0E1h
        SPLK    #00004h, MCRA   ;Configure XINT pin for primary
        LDP     #0E0h
        SPLK    #5h, XINT1CR    ;Configures XINT1 pin for
                                ;polarity(low to high) priority(high),
                                ;and enable bit
        LDP     # 0h
        SPLK    #1, IMR         ;XINT is INT1 so set IMR to "1"
        CLRC    INTM            ;Enables interrupts to core
LOOP            B    LOOP       ;loops here until interrupt occurs
```

The above code will enable the XINT1 pin to generate an interrupt of INT1 when a "low to high" transition is detected on the pin.

4.4.2 Servicing Interrupts

Each of the interrupt priority levels INT1 through INT6 has a corresponding memory address 0001h through 0006h in program memory to which the core will branch upon receiving the interrupt. The header file *vector.h* assigns the labels "INT1, INT2, ...INT6" to addresses 0001h through 0006h. This header file also instructs the core to branch to the corresponding General Interrupt Service Routines (GISR1 through GISR6) labels which are located in the assembly source file.

It is under the appropriate "GISRx" label in the source file where the interrupt service routine (ISR) is written. In the ISR, a variety of algorithms may be used. The ISR is simply an algorithm to which the core will execute whenever it encounters an interrupt. The first action in the ISR should be to perform a "context save" by saving the value of the accumulator, status registers, and anything else that could change as a result of the ISR, so that when the core exits from the interrupt, it is essentially in the same state as when it entered.

If multiple peripheral interrupts in the same priority level are enabled, then each of these interrupts would cause the core to branch to the same GISRx. In this case, it would be necessary to first run a PIVR reading and selection algorithm under the GISR which would determine what specific interrupt actually occurred. Then the algorithm would then branch to a Specific Interrupt Service Routine (SISR). Example 4.2 is pseudo-code which is an example of the selection algorithm discussed previously.

Example 4.2 – Two peripheral interrupts (RXINT and TXINT) are both assigned to priority level INT1. The following pseudo-code is a sample algorithm to determine which interrupt occurred and service the interrupt.

```
-------------------------------------------------------------------------------------------
GISR1        - GISR1 corresponds to ONLY INT1 interrupts
        Read the PIVR
                Does the PIVR contain the vector for RXINT ?
                        Yes – Branch to R_ISR
                        No – Continue to next instruction
                Does the PIVR contain the vector for TXINT?
                        Yes – Branch to T_ISR
                        No – Branch to ERROR
R_ISR
This would be the first SISR, the name of the SISR does not matter
User defined algorithm plus reset interrupt for next occurrence and exit ISR

T_ISR
This would be the second SISR, the name of the SISR does not matter
User defined algorithm plus reset interrupt for next occurrence and exit ISR
ERROR
User defined algorithm plus reset interrupt for next occurrence and exit ISR
-------------------------------------------------------------------------------------------
```

Interrupt Vectors

Information on the different peripheral interrupts and their corresponding PIVR codes can be found in Table 4.1, which lists the peripheral interrupt vector codes that load into the PIVR. The vector is essentially an identification number for each interrupt. Note that an interrupt may have a different overall priority and grouping based on the (low or high) priority level that the interrupt is set to in its corresponding peripheral control register. For example, XINT1 (high priority) is assigned vector 0001h and is grouped INT1 with overall priority 7. XINT1 (low priority) is still assigned vector 0001h but is grouped INT6 with overall priority 33.

Table 4.1 Interrupt vectors. (Courtesy of Texas Instruments)

Overall Priority	Interrupt Name	CPU Interrupt Vector	Peripheral Interrupt Vector	Maskable?	Source Peripheral	Description
1	Reset	RSN 0000h	N/A	N	RS Pin, Watchdog	Reset from pin, watchdog time out
2	Reserved	- 0026h	N/A	N	CPU	Emulator trap
3	NMI	NMI 0024h	N/A	N	Nonmaskable interrupt	Nonmaskable interrupt

(a) **INT1** *(level 1)*

Overall Priority	Interrupt Name	CPU Interrupt Vector	Peripheral Interrupt Vector	Maskable?	Source Peripheral	Description
4	PDPINTA	INT1 0002h	0020h	Y	EVA	Power drive protection interrupt pin
5	PDPINTB	INT1 0002h	0019h	Y	EVB	Power drive protection interrupt pin
6	ADCINT	INT1 0002h	0004h	Y	ADC	ADC interrupt in high–priority mode
7	XINT1	INT1 0002h	0001h	Y	External interrupt logic	External interrupt pin in high–priority mode
8	XINT2	INT1 0002h	0011h	Y	External interrupt logic	External interrupt pin in high–priority mode
9	SPIINT	INT1 0002h	0005h	Y	SPI	SPI interrupt in high–priority mode
10	RXINT	INT1 0002h	0006h	Y	SCI	SCI receiver interrupt in high–priority mode
11	TXINT	INT1 0002h	0007h	Y	SCI	SCI transmitter interrupt in high–priority mode
12	CANMBINT	INT1 0002h	0040h	Y	CAN	CAN mailbox interrupt (high–priority mode)
13	CANERINT	INT1 0002h	0041h	Y	CAN	CAN error interrupt (high–priority mode)

(b) **INT2** *(level 2)*

Overall Priority	Interrupt Name	CPU Interrupt Vector	Peripheral Interrupt Vector	Maskable?	Source Peripheral	Description
14	CMP1INT	INT2 0004h	0021h	Y	EVA	Compare 1 interrupt
15	CMP2INT	INT2 0004h	0022h	Y	EVA	Compare 2 interrupt
16	CMP3INT	INT2 0004h	0023h	Y	EVA	Compare 3 interrupt
17	T1PINT	INT2 0004h	0027h	Y	EVA	Timer 1 period interrupt
18	T1CINT	INT2 0004h	0028h	Y	EVA	Timer 1 compare interrupt
19	T1UFINT	INT2 0004h	0029h	Y	EVA	Timer 1 underflow interrupt
20	T1OFINT	INT2 0004h	002Ah	Y	EVA	Timer 1 overflow interrupt
21	CMP4INT	INT2 0004h	0024h	Y	EVB	Compare 4 interrupt
22	CMP5INT	INT2 0004h	0025h	Y	EVB	Compare 5 interrupt
23	CMP6INT	INT2 0004h	0026h	Y	EVB	Compare 6 interrupt
24	T3PINT	INT2 0004h	002Fh	Y	EVB	Timer 3 period interrupt
25	T3CINT	INT2 0004h	0030h	Y	EVB	Timer 3 compare interrupt
26	T3UFINT	INT2 0004h	0031h	Y	EVB	Timer 3 underflow interrupt
27	T3OFINT	INT2 0004h	0032h	Y	EVB	Timer 3 overflow interrupt

(c) **INT3** *(level 3)*

Overall Priority	Interrupt Name	CPU Interrupt Vector	Peripheral Interrupt Vector	Maskable?	Source Peripheral	Description
28	T2PINT	INT3 0006h	002Bh	Y	EVA	Timer 2 period interrupt
29	T2CINT	INT3 0006h	002Ch	Y	EVA	Timer 2 compare interrupt
30	T2UFINT	INT3 0006h	002Dh	Y	EVA	Timer 2 underflow interrupt
31	T2OFINT	INT3 0006h	002Eh	Y	EVA	Timer 2 overflow interrupt
32	T4PINT	INT3 0006h	0039h	Y	EVB	Timer 4 period interrupt
33	T4CINT	INT3 0006h	003Ah	Y	EVB	Timer 4 compare interrupt
34	T4UFINT	INT3 0006h	003Bh	Y	EVB	Timer 4 undeflow interrupt
35	T4OFINT	INT3 0006h	003Ch	Y	EVB	Timer 4 overflow interrupt

(d) **INT4** *(level 4)*

Overall Priority	Interrupt Name	CPU Interrupt Vector	Peripheral Interrupt Vector	Maskable?	Source Peripheral	Description
36	CAP1INT	INT4 0008h	0033h	Y	EVA	Capture 1 interrupt
37	CAP2INT	INT4 0008h	0034h	Y	EVA	Capture 2 interrupt
38	CAP3INT	INT4 0008h	0035h	Y	EVA	Capture 3 interrupt
39	CAP4INT	INT4 0008h	0036h	Y	EVB	Capture 4 interrupt
40	CAP5INT	INT4 0008h	0037h	Y	EVB	Capture 5 interrupt
41	CAP6INT	INT4 0008h	0038h	Y	EVB	Capture 6 interrupt

(e) **INT5** *(level 5)*

Overall Priority	Interrupt Name	CPU Interrupt Vector	Peripheral Interrupt Vector	Maskable?	Source Peripheral	Description
42	SPIINT	INT5 000Ah	0005h	Y	SPI	SPI interrupt (low priority)
43	RXINT	INT5 000Ah	0006h	Y	SCI	SCI receiver interrupt (low–priority mode)
44	TXINT	INT5 000Ah	0007h	Y	SCI	SCI transmitter interrupt (low–priority mode)
45	CANMBINT	INT5 000Ah	0040h	Y	CAN	CAN mailbox interrupt (low–priority mode)
46	CANERINT	INT5 000Ah	0041h	Y	CAN	CAN error interrupt (low–priority mode)

(f) **INT6** *(level 6)*

Overall Priority	Interrupt Name	CPU Interrupt Vector	Peripheral Interrupt Vector	Maskable?	Source Peripheral	Description
47	ADCINT	INT6 000Ch	0004h	Y	ADC	ADC interrupt (low priority)
48	XINT1	INT6 000Ch	0001h	Y	External interrupt logic	External interrupt pins (low–priority mode)
49	XINT2	INT6 000Ch	0011h	Y	External interrupt logic	External interrupt pins (low–priority mode)
	Reserved	000Eh	N/A	Y	CPU	Analysis interrupt
N/A	TRAP	0022h	N/A	N/A	CPU	TRAP instruction
N/A	Phantom Interrupt Vector	N/A	0000h	N/A	CPU	Phantom interrupt vector

4.5 Interrupt Usage Exercise

This exercise will help the reader become familiar with interrupt operation and writing interrupt service routines in software. The skills practiced in this exercise are extremely relevant in sequential chapters where interrupts must be understood for peripheral use.

1. Create a new project and source file each named "lab4". Add the same header files as in previous exercises.
2. Create a program which first properly configures the LF2407 and XINT1 interrupt registers for operation on a low to high clock edge. On the LF2407 EVM module, jumper the XF pin the XINT1. Configure the XF pin to initially output logic "0". The program should utilize a looping algorithm and the XINT1 interrupt to perform the following tasks:

a. Start with the value "0h" in the accumulator. Store the number in the accumulator to the data memory address 300h. When this operation is complete, set the XF pin to be "1" (logic high). This should trigger an XINT1 interrupt.
b. In the ISR, keep count of the number of interrupts generated in the address "030Fh". Start counting at "0" for the first interrupt generated. Reset the XF pin to logic "0". Re-enable the interrupt.
c. Keep repeating steps (a) and (b), but use the numbers "0001h" through "000Ah" instead and store them to memory address 301h through 30Ah.

The program should store a total of 11 numbers (0h to Ah) to memory addresses 300h through 30Ah. There should be exactly 11 interrupts counted with the number "Ah" stored in memory address 30Fh.

d. After steps (a) through (c) are complete, perform the calculations "Ah" multiplied by "3h" , "1h" multiplied by "5h", and "11h" multiplied by "7h" one after the other. Create a transition on the XF pin so an XINT1 interrupt will be generated after each calculation is complete. Count the number of these interrupts and store them in data memory address 310h.

When the program is finished with the task, have it loop infinitely until halted by the user. This exercise is now concluded.

Chapter 5

THE ANALOG-TO-DIGITAL CONVERTER (ADC)

5.1 ADC Overview

The Analog-to-Digital Converter (ADC) on the LF2407 allows the DSP to sample analog or "real-world" voltage signals. The output of the ADC is an integer number which represents the voltage level sampled. The integer number may be used for calculations in an algorithm. The resolution of the ADC is 10 bits, meaning that the ADC will generate a 10-bit number for every conversion it performs. However, the ADC stores the conversion results in registers that are 16 bits wide. The 10 most significant bits are the ADC result, while the least significant bits (LSBs) are filled with "0"s. We usually want to truncate the useless zeros, so the value in the result register is simply right shifted by six places.

If the ADC performs a conversion on a 3.3V signal, it will theoretically generate "1111111111000000b" (or "FFC0h") in the appropriate result register and "0h" if a 0V signal is sampled. In actuality, the least significant of the 10 bits will vary slightly; this is the result of random noise picked up by the ADC.

There are a total of 16 input channels to the single input ADC. The control logic of the ADC consists of auto-sequencers, which control the sampling of the 16 input channels to the ADC. The auto-sequencers not only control which channels (input channels) will be sampled by the ADC, but also the order of the channels that the ADC performs conversions on. The two 8-conversion auto-sequencers can operate independently or cascade together as a "virtual" 16-conversion ADC.

5.1.1 Summary of the LF2407 ADC

- 10-bit ADC with built-in Sample and Hold (S/H)
- Fast conversion time of 500 ns
- Sixteen (16) multiplexed analog inputs (ADCIN0 – ADCIN15)
- Auto-sequencing capability – up to 16 "auto-conversions" in a single session. Each conversion session can be programmed to select any one of the 16 input channels
- Two independent 8-state sequencers (SEQ1 and SEQ2) that can be operated individually in dual-sequencer mode or cascaded into one large 16-state sequencer (SEQ) in cascaded mode
- Four Sequencing Control Registers (CHSELSEQ1..4) that determine the sequence of analog channels that are taken up for conversion in a given sequencing mode
- Sixteen (individually addressable) result registers to store the converted values (RESULT0 – RESULT15)
- Multiple trigger sources for start-of-conversion (SOC)
    a. Software: Software start (using SOC SEQn bit)
    b. EVA: Event manager A (multiple event sources within EVA)

   c. EVB: Event manager B (multiple event sources within EVB)
   d. External: ADCSOC pin
- Interrupt control allows interrupt generation on every end-of-sequence (EOS) or every other EOS
- Sequencer can operate in start/stop mode, allowing multiple time-sequenced triggers to synchronize conversions
- EVA and EVB can independently trigger SEQ1 and SEQ2, respectively (this is applicable for dual-sequencer mode only)
- Sample-and-hold acquisition time window has separate prescale control
- Built-in calibration mode and built-in self-test mode

5.2 Operation of the ADC

Using the ADC on the LF2407 is relatively simple. The user first needs to configure the ADC for the desired operation. Like all peripherals, all registers relating to ADC operation have addresses in data memory space. The first step in configuring the ADC should be to reset the ADC. After the ADC is reset, the next step is to configure the main ADC control registers (ADCTRL1, ADCTRL2) for desired ADC operation. Then, load the MAXCONV register with the desired number of automatic conversions minus 1. For example, if seven auto-conversions are desired, MAXCONV would be loaded with "6". The desired input channels and their order of conversion need to be specified in the CHSELSEQn registers. Finally, a SOC trigger will start the sampling process. A short example of the assembly code performing the above listed steps is provided in Example 5.1.

Example 5.1- The following code gives an example of initializing the ADC, setting up the CHSELSEQn registers and starting the conversion sequence:

```
LDP     #0E1h
SPLK    #0100000000000000b,ADCTRL1
NOP
SPLK    #0011000000010000b,ADCTRL1
        ; the following explains bits in ADCCTRL1:
        ; 15 - RSVD | 14 - Reset(1) | 13,12 - Soft & Free
        ; 11,10,9,8 - Acq. prescalers | 7 - Clock prescaler
        ; 6 - Cont. run (1) | 5 - Int. priority (Hi.0)
        ; 4 - Seq. casc (0-dual)
SPLK    #15, MAXCONV            ;Setup for 16 conversions
SPLK    #03210h, CHSELSEQ1      ;Conv Ch 0,1,2,3
SPLK    #07654h, CHSELSEQ2      ;Conv Ch 4,5,6,7
SPLK    #0BA98h, CHSELSEQ3      ;Conv Ch 8,9,10,11
SPLK    #0FEDCh, CHSELSEQ4      ;Conv Ch 12,13,14,15
SPLK    #2000b, ADCTRL2         ;Start the conversions by bit 13
```

After the conversion process is complete, each 10-bit result can be read from the result registers RESULTn. The conversion results are stored sequentially in result registers RESULT0 to RESULT15. The first result is stored in RESULT0, the second result in RESULT1, and so on. For example, if ADC channel 1 is selected for four consecutive conversions, the results will appear in registers RESULT0 through RESULT3. *There is no correlation between ADC Channel 1 and the RESULT register 1 or Channel 2 and RESULT 2 etc.*

We will discuss each of these steps in detail in the following sections, starting with the different operating modes of the ADC. This will aid the reader in configuring the ADC control registers by helping to determine what operating mode is needed. After the reader is familiar with the ADC operating modes, we will cover the MAXCONV and CHSELSEQn registers. Various SOC trigger methods will then be discussed. Finally, the ADC conversion result register reading will be discussed.

5.2.1 Sequencer Configurations of the ADC

The first operating parameter the user needs to select is to configure the ADC to operate as either one 16-conversion sequencer or two 8-conversion sequencers. The ADC sequencer consists of two independent 8-conversion sequencers (SEQ1 and SEQ2) that can be cascaded together to form one 16-conversion sequencer (SEQ).

When the ADC is configured to operate as one cascaded 16-convesion sequencer, it may perform up to 16 conversions on any combination of the 16 input channels. For example, it could be programmed to perform 14 conversions on channel 1, or in another instance, 10 total conversions on a combination of channels depending on what the CHSELSEQn registers are set for. The diagram Fig. 5.1 shows the configuration of the cascaded 16-conversion sequencer. When in cascaded mode there is only one sequencer (SEQ) and the MAXCONV register is programmed for the maximum number of conversions. The results are stored in RESULT0 through RESULT 15 depending on the number of conversions performed.

If the ADC is configured as two 8-conversion sequencers, then each sequencer operates independently. When the two sequencers are used independently, the current active sequencer has priority over the inactive one. The start of conversion request from the "inactive" sequencer will be taken as soon as the sequence initiated by the "active" sequencer is completed. For example, if Sequencer 1 (SEQ1) is currently performing a conversion and Sequencer 2 (SEQ2) requests a start of conversion, the ADC will finish the conversion from SEQ1, and then start the SEQ2 conversion. See Fig. 5.2 for a diagram of the dual-sequencer configuration. In dual-sequencer operation, the MAXCONV register is "split" up so that the same register contains data for the maximum number of conversions for both SEQ1 and SEQ2. The 16 result registers are also split up so SEQ1 uses RESULT0 through RESULT7 and SEQ2 uses RESULT8 through RESULT15. A summary of the cascaded and dual-sequencer configurations is listed in Table 5.1.

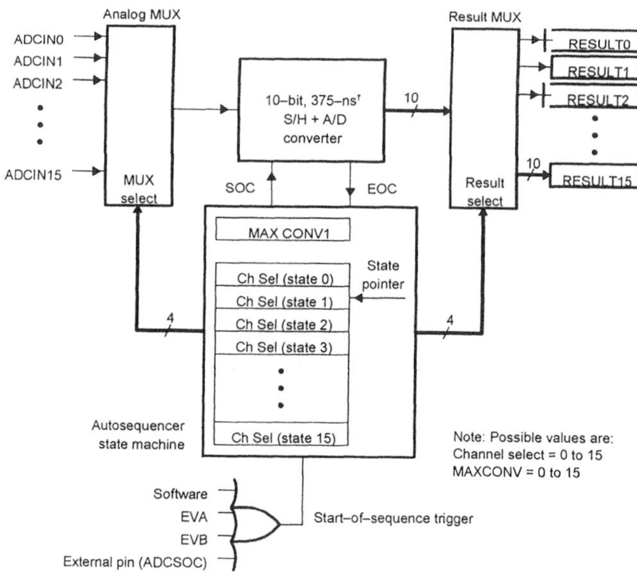

Figure 5.1 Block diagram of ADC in cascaded sequencer mode. (Courtesy of Texas Instruments)

Table 5.1 Comparasion table of dual (SEQ1 and SEQ2) versus cascaded sequencer configuration

Feature	Single 8-state sequencer #1 (SEQ1)	Single 8-state sequencer #2 (SEQ2)	Cascaded 16-state sequencer (SEQ)
Start-of-conversion triggers	EVA, software, external pin	EVB, software	EVA, EVB, software, external pin
Maximum number of autoconversions (i.e., sequence length)	8	8	16
Autostop at end-of-sequence (EOS)	Yes	Yes	Yes
Arbitration priority	High	Low	Not applicable
ADC conversion result register locations	0 to 7	8 to 15	0 to 15
CHSELSEQn bit field assignment	CONV00 to CONV07	CONV08 to CONV15	CONV00 to CONV15

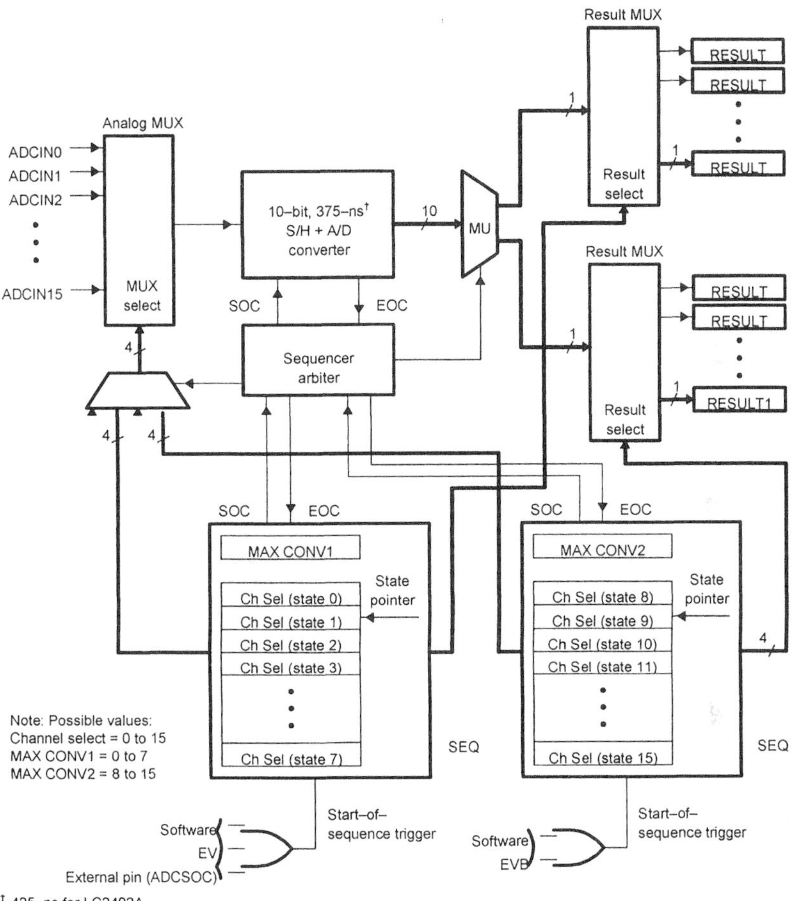

Figure 5.2 Block diagram of ADC in dual sequencer mode. (Courtesy of Texas Instruments)

5.2.2 Sequencer Operating Modes

Once the sequencer configuration has been chosen, it is necessary to determine in what mode each sequencer will operate. The sequencer operation mode depends on the continuous-run mode bit (CONT RUN) in ADCCTRL1. The ADC's interrupt flag is always set when the ADC completes the number of conversions specified by (MAXCONV + 1) regardless of the CONT RUN bit. The two ADC operation modes which apply to both dual (SEQ1, SEQ2) and cascaded (SEQ) sequencer modes are:

- Start/Stop Auto-Sequencer Mode
- Continuous Auto-Sequencer Mode

Start/Stop Auto-Sequencer Mode

If the CONT RUN bit is not set, upon receiving a trigger, the ADC performs all conversions and halts at the last conversion state (CONVxx) in the corresponding CHSELSEQn. To perform another batch of conversions, the ADC is normally reset to its initial state via the RST SEQn bit in the ADCTRL2 register and reinitialized. After being reinitialized, another trigger is given and the whole process starts over again. Figure 5.3 is a flowchart of the operation of the ADC under start/stop mode.

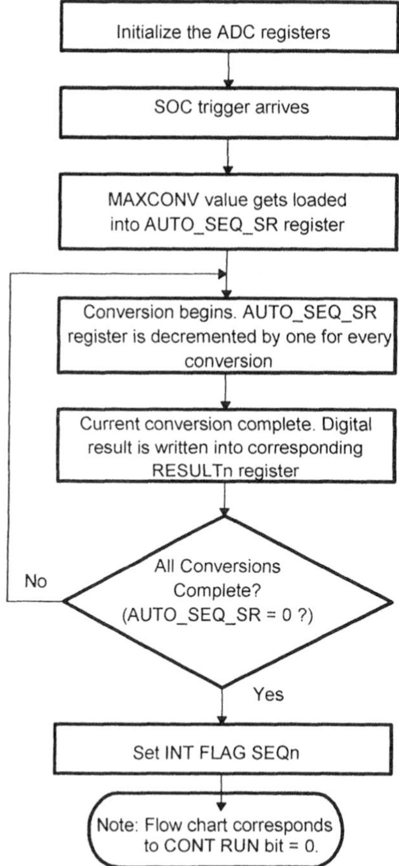

Figure 5.3 Flowchart for Start/Stop Auto-Sequencer Mode (CONT RUN=0).

In the case when another trigger signal is given and the ADC has not been reset, the ADC performs another specified number of conversions (MAXCONV + 1) from the current conversion state and then halts. Another trigger signal will simply restart the sequencer from the point where it halted. When the ADC is given multiple triggers without being reset in between, this operation is referred to as multiple time-sequenced trigger operation. Example 5.2 illustrates a situation where a multiple time-sequenced trigger operation might be used.

Example 5.2 – The following is a situation where a multiple time-sequenced trigger operation might be used.

An application requires conversions on all 16 channels, but not all at once. The application requires conversions on channels 0 through 3, perform a few calculations, convert channels 4 through 7, do a few more calculations, convert channels 8 through 11, etc. until conversions are performed on all 16 input channels. The four CHSELSEQ registers would be loaded only once with all 16 channels in the desired order. The MAXCONV register would be loaded with the number "3", which configures the auto-sequencer for four conversions. Each time the sequencer pauses, the algorithm would branch to the section of code that performs calculations and retrigger the ADC. This "branching" could either occur as a result of an interrupt or bit polling algorithm.

Continuous Auto-Sequencer Mode

The continuous-run mode bit is set to "1" for this mode of operation. When in this mode, the ADC completes the number of conversions specified, resets itself to the first conversion state (CONV00), and then performs the whole operation over again. This operation is similar to the start/stop mode except that the ADC is put in a continuous "looping" operation.

Note: *If the CONT RUN (continuous run) mode is selected, the user must be sure that the result registers are read before the next conversion sequence begins. This is because every time the ADC runs, the result registers will be overwritten with the most current results.*

5.2.3 Triggering Sources for the LF2407 ADC

In order to start the conversion sequence on the ADC, the sequencer must be triggered. There are several different trigger sources on the LF2407. Triggers may come from a SOC signal from EVA: external pin or software. A software trigger is the trigger thus far used as an example. The software trigger is generated by setting the SOC SEQ1 bit (cascaded mode) or SOC SEQ1,2 bits (dual mode) in the ADCTRL2 register. Other than software triggers, hardware in the form of an external pin or on-chip peripheral can also trigger the ADC. Table 5.2 lists the possible triggering sources which generate a SOC for the ADC. Each trigger input can be enabled /disabled.

Table 5.2 SOC Trigger Sources for the ADC

SEQ1 (sequencer 1)	SEQ2 (sequencer 2)	Cascaded SEQ
Software trigger (software SOC)	Software trigger (software SOC)	Software trigger (software SOC)
Event manager A (EVA SOC)	Event manager B (EVB SOC)	Event manager A (EVA SOC)
External SOC pin (ADC SOC)		Event manager B (EVB SOC)
		External SOC pin (ADC SOC)

The following conditions apply to trigger operation:

a. A SOC trigger can initiate an auto-conversion sequence whenever a sequencer is in an idle state. An idle state is either just after reset (CONV00), or any state where the sequencer has just finished a conversion sequence, i.e., when SEQ CNTR has reached zero.
b. If a SOC trigger occurs while a current conversion sequence is underway, it sets the SOC SEQn bit. If yet another SOC trigger occurs, that trigger is ignored. This basically operates as a SOC trigger "buffer" that will catch a trigger even though the ADC might be currently performing a conversion.
c. Once triggered, the sequencer cannot be stopped/halted in mid sequence. The program must either wait until an End-of-Sequence (EOS) or initiate a sequencer reset, which brings the sequencer immediately back to the idle start state (CONV00 for SEQ1 and cascaded cases; CONV08 for SEQ2).
d. When SEQ1 and SEQ2 are used in **cascaded mode, triggers going to SEQ2 are ignored**, while **SEQ1 triggers are active**. Cascaded mode can be viewed as SEQ1 with 16 conversion states instead of 8.

5.2.4 The ADCTRL1 and ADCTRL2 Control Registers

ADC Control Register 1 (ADCTRL1) — Address 70A0h

15	14	13	12	11	10	9	8
Reserved	RESET	SOFT	FREE	ACQ PS3	ACQ PS2	ACQ PS1	ACQ PS0
	RS–0	RW–0	RW–0	RW–0	RW–0	RW–0	RW–0
7	**6**	**5**	**4**	**3**	**2**	**1**	**0**
CPS	CONT RUN	INT PRI	SEQ CASC	CAL ENA	BRG ENA	HI/LO	STEST ENA
RW–0	RW–0	RW–0	RW–0	RW-0	RW-0	RW-0	RW-0

***Note:** R = read access, W = write access, S = set only, -0 = value after reset.*

Bit 15 Reserved

Bit 14 **RESET**. ADC module software reset. This bit causes a master reset on the entire ADC module. All register bits and sequencer state machines are reset to the initial state as occurs when the device reset pin is pulled low (or after a power-on reset).

0 No effect

1 Resets entire ADC module (bit is then set back to 0 by ADC logic)

Note: *Using the RESET Bit in the ADCTRL1 Register*

The ADC module is reset during a system reset. If an ADC module reset is desired at any other time, you can do so by writing a 1 to this bit. After a NOP, you can then write the appropriate values to the ADCTRL1 register bits:

```
SPLK #01xxxxxxxxxxxxxxb,ADCTRL1 ;Resets the ADC (RESET = 1)
NOP                             ;Provides the required delay between
                                ; two writes to the ADCTRL1
SPLK #00xxxxxxxxxxxxxxb,ADCTRL1 ;Takes ADC out of Reset(RESET= 0)
```

Note: *The second SPLK is not required if the default/power-on configuration of the ADC is sufficient.*

Bits 13, 12 **SOFT and FREE**. Soft and Free bits. These bits determine what happens with the ADC when an emulation-suspend occurs (due to the debugger hitting a breakpoint, for example). In free-run mode, the peripheral can continue with whatever it is doing. In stop mode, the peripheral can either stop immediately or stop when the current operation (i.e., the current conversion) is complete.

Soft	Free	
0	0	Immediate stop on suspend
1	0	Complete current conversion before stopping
X	1	Free run, continue operation regardless of suspend

Bits 11–8 **ACQ PS3 – ACQ PS0**. Acquisition time window – pre-scale bits 3–0 These bits define the ADC clock pre-scale factor applied to the acquisition portion of the conversion and determine over what time period each ADc sample will take place. The pre-scale values are defined in the following table.

#	ACQ PS3	ACQ PS2	ACQ PS1	ACQ PS0	PRE-SCALER (div. by)	Acquisition Time Window	Source Z (CPS=0)	Source Z (CPS=1)
0	0	0	0	0	1	2 x Tclk	67	385
1	0	0	0	1	2	4 x Tclk	385	1020
2	0	0	1	0	3	6 x Tclk	702	1655
3	0	0	1	1	4	8 x Tclk	1020	2290
4	0	1	0	0	5	10 x Tclk	1337	2925
5	0	1	0	1	6	12 x Tclk	1655	3560
6	0	1	1	0	7	14 x Tclk	1972	4194
7	0	1	1	1	8	16 x Tclk	2290	4829
8	1	0	0	0	9	18 x Tclk	2607	5464
9	1	0	0	1	10	20 x Tclk	2925	6099
A	1	0	1	0	11	22 x Tclk	3242	6734
B	1	0	1	1	12	24 x Tclk	3560	7369
C	1	1	0	0	13	26 x Tclk	3877	8004
D	1	1	0	1	14	28 x Tclk	4194	8639
E	1	1	1	0	15	30 x Tclk	4512	9274
F	1	1	1	1	16	32 x Tclk	4829	9909

Notes:

1) Period of Tclk is dependent on the "Conversion Clock Prescale" bit (Bit 7); i.e.,
CPS = 0: Tclk = 1/CLK (example, for CLK = 30 MHz, Tclk = 33 ns)
CPS = 1: Tclk = 2(1/CLK) (example, for CLK = 30 MHz, Tclk = 66 ns)
2) Source impedance Z is a design estimate only.

Bit 7 **CPS**. Conversion clock prescale. This bit defines the ADC conversion logic clock prescale

0 ^{F}clk = CLK/1

1 ^{F}clk = CLK/2

CLK = CPU clock frequency

Bit 6 **CONT RUN**. Continuous run

This bit determines whether the sequencer operates in continuous conversion mode or start-stop mode. This bit can be written while a current conversion sequence is active. This bit will take effect at the end of the current conversion sequence, i.e., software can set/clear this bit until EOS has occurred for valid action to be taken. In the continuous conversion mode, there is no need to reset the sequencer; however, the sequencer must be reset in the start-stop mode to put the converter in state CONV00.

0 Start-stop mode. Sequencer stops after reaching EOS. This is used for multiple time-sequenced triggers.

1 Continuous conversion mode. After reaching EOS, the sequencer starts all over again from state CONV00 (for SEQ1 and cascaded) or CONV08 (for SEQ2).

Bit 5 **INT PRI**. ADC interrupt request priority

0 High priority

1 Low priority

Bit 4 **SEQ CASC**. Cascaded sequencer operation. This bit determines whether SEQ1 and SEQ2 operate as two 8-state sequencers or as a single 16-state sequencer (SEQ).

0 Dual-sequencer mode. SEQ1 and SEQ2 operate as two 8-state sequencers.

1 Cascaded mode. SEQ1 and SEQ2 operate as a single 16-state sequencer (SEQ).

Bit 3 **CAL ENA**. Offset calibration enable

When set to 1, CAL ENA disables the input channel multiplexer, and connects the calibration reference selected by the bits HI/LO and BRG ENA to the ADC core inputs. The calibration conversion can then be started by setting bit 14 of ADCTRL2 register (STRT CAL) to 1. Note that CAL ENA should be set to 1 first before the STRT CAL bit can be used.

Note: *This bit should not be set to 1 if STEST ENA = 1*

0 Calibration mode disabled

1 Calibration mode enabled

Bit 2 **BRG ENA**. Bridge enable

Together with the HI/LO bit, BRG ENA allows a reference voltage to be converted in calibration mode. See the description of the HI/LO bit for reference voltage selections during calibration.

0 Full reference voltage is applied to the ADC input

1 A reference midpoint voltage is applied to the ADC input

Bit 1 **HI/LO**· ^{V}REFHI/^{V}REFLO selection

When the fail self-test mode is enabled (STEST ENA = 1), HI/LO defines the test voltage to be connected. In calibration mode, HI/LO defines the reference source polarity; see Table 7–5. In normal operating mode, HI/LO has no effect.

0 ^{V}REFLO is used as precharge value at ADC input

1 ^{V}REFHI is used as precharge value at ADC input

Reference Bit Voltage Selection

BRG ENA	HI/LO	CAL ENA = 1 Reference voltage (V)	STEST ENA = 1 Reference voltage (V)
0	0	V_{REFLO}	V_{REFLO}
0	1	V_{REFHI}	V_{REFHI}
1	0	$\|(V_{REFHI} - V_{REFLO}) / 2\|$	V_{REFLO}
1	1	$\|(V_{REFLO} - V_{REFHI}) / 2\|$	V_{REFHI}

Bit 0 **STEST ENA**. Self-test function enable
- 0 Self-test mode disabled
- 1 Self-test mode enabled

ADC Control Register 2 (ADCTRL2) — Address 70A1h

15	14	13	12	11	10	9	8
EVB SOC SEQ	RST SEQ1/ STRT CAL	SOC SEQ1	SEQ1 BSY	INT ENA SEQ1 (Mode 1)	INT ENA SEQ1 (Mode 0)	INT FLAG SEQ1	EVA SOC SEQ1
RW–0	RS–0	RW–0	R–0	RW–0	RW–0	RC–0	RW–0
7	6	5	4	3	2	1	0
EXT SOC SEQ1	RST SEQ2	SOC SEQ2	SEQ2 BSY	INT ENA SEQ2 (Mode 1)	INT ENA SEQ2 (Mode 0)	INT FLAG SEQ2	EVB SOC SEQ2
RW–0	RS–0	RW–0	R–0	RW–0	RW–0	RC–0	RW–0

Note: *R = read access, W = write access, S = set only, C = clear, -0 = value after reset.*

Bit 15 **EVB SOC SEQ**. EVB SOC enable for cascaded sequencer *(Note: This bit is active only in cascaded mode.)*
- 0 No action
- 1 Setting this bit allows the cascaded sequencer to be started by an Event Manager B signal. The Event Manager can be programmed to start a conversion on various events. See Chapter 6 for details.

Bit 14 **RST SEQ1 / STRT CAL**. Reset Sequencer1/Start Calibration

Case: Calibration Disabled (Bit 3 of ADCTRL1) = 0

Writing a 1 to this bit will reset the sequencer immediately to an initial "pre-triggered" state, i.e., waiting for a trigger at CONV00. A currently active conversion sequence will be aborted.
- 0 No action
- 1 Immediately reset sequencer to state CONV00

Case: Calibration Enabled (Bit 3 of ADCTRL1) = 1

Writing a 1 to this bit will begin the converter calibration process.
- 0 No action

1 Immediately start calibration process

Bit 13 SOC SEQ1. SOC trigger for Sequencer 1 (SEQ1). This bit can be set by the following triggers:

S/W – Software writing a 1 to this bit
EVA – Event Manager A
EVB – Event Manager B (only in cascaded mode)
EXT – External pin (i.e., the ADCSOC pin)

When a trigger occurs, there are three possibilities:

Case 1: SEQ1 idle and SOC bit clear. SEQ1 starts immediately (under arbiter control). This bit is set and cleared, allowing for any "pending" trigger requests.
Case 2: SEQ1 busy and SOC bit clear. Bit is set signifying a trigger request is pending. When SEQ1 finally starts after completing current conversion, this bit will be cleared.
Case 3: SEQ1 busy and SOC bit set. Any trigger occurring in this case will be ignored (lost).

0 Clears a pending SOC trigger.

Note: *If the sequencer has already started, this bit will automatically be cleared, and, hence, writing a zero will have no effect; i.e., an already started sequencer cannot be stopped by clearing this bit.*

1 Software trigger – Start SEQ1 from currently stopped position (i.e., idle mode)

Note: *The RST SEQ1 (ADCTRL2.14) and the SOC SEQ1 (ADCTRL2.13) bits should not be set in the same instruction. This will reset the sequencer, but will not start the sequence. The correct sequence of operation is to set the RST SEQ1 bit first, and the SOC SEQ1 bit in the following instruction. This ensures that the sequencer is reset and a new sequence started. This sequence applies to the RST SEQ2 (ADCTRL2.6) and SOC SEQ2 (ADCTRL2.5) bits also.*

Bit 12 SEQ1 BSY. SEQ1 Busy
This bit is set to a 1 while the ADC auto-conversion sequence is in progress. It is cleared when the conversion sequence is complete.
0 Sequencer is idle (i.e., waiting for trigger)
1 Conversion sequence is in progress

Bits 11–10 INT ENA SEQ1. Interrupt-mode-enable control for SEQ1

Bit 11	Bit 10	Operation Description
0	0	Interrupt is disabled
0	1	Interrupt **Mode 1.** Interrupt requested immediately when INT FLAG SEQ1 flag is set
1	0	Interrupt **Mode 2** Interrupt requested only if INT FLAG SEQ1 flag is already set. If clear,† INT FLAG SEQ1 flag is set and INT request is suppressed. (This mode allows interrupt requests to be generated for every other EOS.)
1	1	Reserved

† This means that the last completed sequence is the first of the two sequences needed to assert an interrupt.

Bit 9 **INT FLAG SEQ1.** ADC interrupt flag bit for SEQ1

This bit indicates whether an interrupt event has occurred or not. This bit must be cleared by the user writing a 1 to it.

0 No interrupt event

1 An interrupt event has occurred

Checking for ADC Peripheral Interrupt Flag

After a SOC is initiated, we can check the INT FLAG SEQn bit to see if the results are in the result registers.

Example code:

```
ADC_LOOP1:
    LDP    #0E1h                            ;data page - ADCTRL2
    SPLK   #0100000000000000b,ADCTRL2       ;Reset for SEQ1
    NOP
    NOP
    NOP
    NOP
    SPLK   #0010000000000000b,ADCTRL2       ;SOC for SEQ1
CHK_INTFLAG:
    BIT    ADCTRL2, 6              ;Wait for INT Flag to set
    BCND   CHK_INTFLAG, NTC        ;If TC=0, keep looping.
```

Bit 8 **EVA SOC SEQ1.** Event Manager A SOC mask bit for SEQ1

0 SEQ1 cannot be started by EVA trigger.

1 Allows SEQ1/SEQ to be started by Event Manager A trigger. The Event Manager can be programmed to start a conversion on various events. See Chapter 6 for details.

Bit 7 **EXT SOC SEQ1.** External signal SOC bit for SEQ1

0 No action

1 Setting this bit enables an ADC auto-conversion sequence to be started by a signal from the ADCSOC device pin.

Bit 6 **RST SEQ2.** Reset SEQ2

0 No action

1 Immediately resets SEQ2 to an initial "pre-triggered" state, i.e., waiting for a trigger at CONV08. A currently active conversion sequence will be aborted.

Bit 5 **SOC SEQ2.** SOC trigger for Sequencer 2 (SEQ2)

(Only applicable in dual-sequencer mode; ignored in cascaded mode.)

This bit can be set by the following triggers:

S/W – Software writing of 1 to this bit

EVB – Event Manager B

When a trigger occurs, there are three possibilities:

Case 1: SEQ2 idle and SOC bit clear

SEQ2 starts immediately (under arbiter control) and the bit is cleared, allowing for any pending trigger requests.

Case 2: SEQ2 busy and SOC bit clear

Bit is set signifying a trigger request is pending. When SEQ2 finally starts after completing current conversion, this bit will be cleared.

Case 3: SEQ2 busy and SOC bit set

Any trigger occurring in this case will be ignored (lost).

0 Clears a pending SOC trigger.

Note: If the sequencer has already started, this bit will automatically be cleared, and hence, writing a zero will have no effect; i.e., an already started sequencer cannot be stopped by clearing this bit.

1 Software trigger – Start SEQ2 from currently stopped position (i.e., idle mode)

Bit 4 **SEQ2 BSY**. SEQ2 Busy

This bit is set to a 1 while the ADC auto-conversion sequence is in progress. It is cleared when the conversion sequence is complete.

0 Sequencer is idle (i.e., waiting for trigger).
1 Conversion sequence is in progress.

Bits 3–2 INT ENA SEQ2. Interrupt-mode-enable control for SEQ2

Bit 3	Bit 2	Operation Description
0	0	Interrupt is disabled
0	1	Interrupt **Mode 1** Interrupt requested immediate on INT FLAG SEQ2 flag set
1	0	Interrupt **Mode 2** Interrupt requested only if INT FLAG SEQ2 flag is already set. If clear,† INT FLAG SEQ2 flag is set and INT request is suppressed. (This mode allows interrupt requests to be generated for every other EOS.)
1	1	Reserved

†This means that the last completed sequence is the first of the two sequences needed to assert an interrupt.

Bit 1 **INT FLAG SEQ2.** ADC interrupt flag bit for SEQ2
This bit indicates whether an interrupt event has occurred or not. This bit must be cleared by the user writing a 1 to it.
0 No interrupt event.
1 An interrupt event has occurred.

Note: *The bit polling algorithm discussed after the bit 9 description is also valid for the INT FLAG SEQ2 bit.*

Bit 0 **EVB SOC SEQ2.** Event Manager B SOC mask bit for SEQ2
0 SEQ2 cannot be started by EVB trigger.
1 Allows SEQ2 to be started by Event Manager B trigger. The Event Manager can be programmed to start a conversion on various events. See Chapter 6 for details.

This concludes the main operating modes of the ADC sequencers. Now that the reader has a general idea of the basic modes of operation (necessary for the initialization of registers ADCTRL1 and ADCTRL2), we will now discuss the configuration of the other ADC registers.

5.2.5 *Specifying the Maximum Number of Auto-Conversions*

The MAXCONV register is used to specify the maximum number of conversions that the ADC will automatically perform once triggered. The MAXCONV register should be loaded with the maximum number of desired auto-conversions minus 1. In this case, since 16 is the maximum number of conversions that the ADC can perform, the maximum value that should be loaded in the MAXCONV register is "0Fh".

When the ADC is in dual sequencer mode, the MAXCONV register is "split" and serves both SEQ1 and SEQ2. The lower half of the register serves SEQ1, while the upper half serves SEQ2. See the bit description of MAXCONV below.

Maximum Conversion Channels Register (MAXCONV) — Address 70A2h

15-8							
Reserved							
R–x							
7	6	5	4	3	2	1	0
Reserved	MAX CONV2_2	MAX CONV2_1	MAX CONV2_0	MAX CONV1_3	MAX CONV1_2	MAX CONV1_1	MAX CONV1_0
R–x	RW–0	RW–0	RW–0	RW–0	RW–0	RW–0	RW–0

Note: *R = read access, W = write access, x = undefined, -0 = value after reset.*

Bits 15–7 Reserved

Bits 6–0 MAX CONVn. MAX CONVn bit field defines the maximum number of conversions executed in an auto-conversion session. The bit fields and their operation vary according to the sequencer modes (dual/cascaded).

- For SEQ1 operation, bits MAX CONV1_2 – 0 are used.
- For SEQ2 operation, bits MAX CONV2_2 – 0 are used.
- For SEQ operation, bits MAX CONV1_3 – 0 are used.

An auto-conversion session always starts with the initial state and continues sequentially until the end state if allowed. The result registers are filled in a sequential order. Any number of conversions between 1 and (MAX CONVn +1) can be programmed for a session.

Example: *MAXCONV Register Bit Programming*

If only five conversions are required, then MAX CONVn is set to four.

Case 1: Dual mode SEQ1 and cascaded mode
Sequencer goes from CONV00 to CONV04, and the five conversion results are stored in the registers Result 00 to Result 04 of the Conversion Result Buffer.

Case 2: Dual mode SEQ2
Sequencer goes from CONV08 to CONV12, and the five conversion results are stored in the registers Result 08 to Result 12 of the Conversion Result Buffer.

MAX CONV1 Value >7 for Dual-Sequencer Mode

If a value for MAX CONV1, which is greater than 7, is chosen for the dual-sequencer mode (i.e., two separate 8-state sequencers), then SEQ CNTR n will continue counting past seven, causing the sequencer to wrap around to CONV00 and continue counting.

5.2.6 Specifying ADC Input Channels and Conversion Order

The ADC input channels and conversion order are specified by the four Channel Select and Sequencing registers (CHSELSEQ1 through CHSELSEQ4). Each register selects four channels, which must be loaded in reverse order (from the least significant hex number to the most significant).

The ADC will perform conversions on the 16 channels in the order that is specified by the channel select sequence registers (CHSELSEQn). Channels must be written to the CHSELSEQ registers in reverse order (see Example 5.1).

CHSELSEQ1 controls and specifies conversions CONV00 through CONV03.
CHSELSEQ2 controls and specifies conversions CONV04 through CONV07.
CHSELSEQ3 controls and specifies conversions CONV08 through CONV11.
CHSELSEQ4 controls and specifies conversions CONV12 through CONV15.

Example 5.1: We want to perform conversions on channels 2, 4, 1, 5, 7, 1, and 4 in this order. We would load CHSELSEQ1 with "5142 h" and CHSELSEQ2 with "417 h". Since 7 conversions are needed, we would load MAXCONV with "6h".

ADC Input Channel Select Sequencing Control Registers (CHSELSEQn)

Each of the 4-bit fields, CONVnn, selects one of the 16 multiplexed analog input ADC channels for an auto-sequenced conversion.

	Bits 15-12	Bits 11-8	Bits 7-4	Bits 3-0	
70A3h	CONV03	CONV02	CONV01	CONV00	CHSELSEQ1
	RW–0	RW–0	RW–0	RW–0	

Note: *R = read access, W = write access, -0 = value after reset.*

	Bits 15-12	Bits 11-8	Bits 7-4	Bits 3-0	
70A4h	CONV07	CONV06	CONV05	CONV04	CHSELSEQ2
	RW–0	RW–0	RW–0	RW–0	

Note: *R = read access, W = write access, -0 = value after reset.*

	Bits 15-12	Bits 11-8	Bits 7-4	Bits 3-0	
70A5h	CONV11	CONV10	CONV09	CONV08	CHSELSEQ3
	RW–0	RW–0	RW–0	RW–0	

Note: *R = read access, W = write access, -0 = value after reset.*

	Bits 15-12	Bits 11-8	Bits 7-4	Bits 3-0	
70A6h	CONV15	CONV14	CONV13	CONV12	CHSELSEQ4
	RW–0	RW–0	RW–0	RW–0	

Note: *R = read access, W = write access, -0 = value after reset.*

5.2.7 Results of the ADC Conversion

After the ADC has finished performing the number of conversions specified by the MAXCONV register, the RESULTn registers can be read. Each result register contains a 10-bit conversion result in the 10 most significant bits (MSB) of the register. There are 16 total result registers, RESULT0 through RESULT15. These registers contain the conversion results in the sequential order that the conversions take place. For example, the result of the first conversion performed will be stored in RESULT0, the second in RESULT1 etc.

It is usually desired to right shift the contents of the result register by six places in order to truncate the extra zeros. This right shift can be performed easily by the SFR command. Once the ADC result has been shifted, it may be used in calculations or other purposes. The bit descriptions of the RESULT registers are given below.

ADC Conversion Result Buffer Registers (RESULTn)

Note: In the cascaded sequencer mode, registers RESULT8 through RESULT15 will hold the results of the ninth through sixteenth conversions.

15	14	13	12	11	10	9	8
D9	D8	D7	D6	D5	D4	D3	D2
7	6	5	4	3	2	1	0
D1	D0	0	0	0	0	0	0

Notes:
1) Buffer addresses = 70A8h to 70B7h (i.e., 16 registers)
2) The 10-bit conversion result (D9–D0) is left-justified

5.2.8 The Auto-Sequence Status Register

The Auto-Sequence Status Register contains information on the current state of the sequencer when running conversions. Its bits can be polled (read) to determine, for example, if the sequencer is near or closer to the end number of conversions.

Auto-sequence Status Register (AUTO_SEQ_SR) — Address 70A7h

15-12				11	10	9	8
Reserved				SEQ CNTR 3	SEQ CNTR 2	SEQ CNTR 1	SEQ CNTR 0
R–x				R–0	R–0	R–0	R–0
7	6	5	4	3	2	1	0
Reserved	SEQ2–State2	SEQ2–State1	SEQ2–State0	SEQ1–State3	SEQ1–State2	SEQ1–State1	SEQ1–State0
R–x	R–0	R–0	R–0	R–0	R–0	R–0	R–0

***Note:** R = read access, x = undefined, -0 = value after reset.*

Bits 15–12 Reserved

Bits 11–8 SEQ CNTR 3 – SEQ CNTR 0. Sequencing counter status bits

The SEQ CNTR n 4-bit status field is used by SEQ1, SEQ2, and the cascaded sequencer. SEQ2 is irrelevant in cascaded mode.

At the start of an auto-sequenced session, SEQ CNTR n is loaded with the value from MAX CONVn. The SEQ CNTR n bits can be read at any time during the countdown process to check the status of the sequencer. This value, together with the SEQ1 and SEQ2 busy bits, uniquely identifies the progress or state of the active sequencer at any point in time.

SEQ CNTR n (read only)	Number of conversions remaining
0000	1
0001	2
0010	3
0011	4
0100	5
0101	6
0110	7
0111	8
1000	9
1001	10
1010	11
1011	12
1100	13
1101	14
1110	15
1111	16

Bit 7 Reserved

Bits 6–4 SEQ2-State2 through SEQ2-State0

Reflects the state of SEQ2 sequencer at any point of time. If need be, user can poll these bits to read interim results before an EOS. SEQ2 is irrelevant in cascaded mode.

Bits 3–0 SEQ1-State3 through SEQ1-State0

Reflects the state of SEQ1 sequencer at any point of time. If need be, user can poll these bits to read interim results before an EOS.

ADC Register Addresses Summary (Mapped in Data Memory)

Address	Register	Name
70A0h	ADCTRL1	ADC control register 1
70A1h	ADCTRL2	ADC control register 2
70A2h	MAXCONV	Maximum conversion channels register
70A3h	CHSELSEQ1	Channel select sequencing control register 1
70A4h	CHSELSEQ2	Channel select sequencing control register 2
70A5h	CHSELSEQ3	Channel select sequencing control register 3
70A6h	CHSELSEQ4	Channel select sequencing control register 4
70A7h	AUTO_SEQ_SR	Autosequence status register
70A8h	RESULT0	Conversion result buffer register 0
70A9h	RESULT1	Conversion result buffer register 1
70AAh	RESULT2	Conversion result buffer register 2
70ABh	RESULT3	Conversion result buffer register 3
70ACh	RESULT4	Conversion result buffer register 4
70ADh	RESULT5	Conversion result buffer register 5
70AEh	RESULT6	Conversion result buffer register 6
70AFh	RESULT7	Conversion result buffer register 7
70B0h	RESULT8	Conversion result buffer register 8
70B1h	RESULT9	Conversion result buffer register 9
70B2h	RESULT10	Conversion result buffer register 10
70B3h	RESULT11	Conversion result buffer register 11
70B4h	RESULT12	Conversion result buffer register 12
70B5h	RESULT13	Conversion result buffer register 13
70B6h	RESULT14	Conversion result buffer register 14
70B7h	RESULT15	Conversion result buffer register 15
70B8h	CALIBRATION	Calibration result, used to correct subsequent conversions

5.3 Analog to Digital Converter Usage Exercise

The purpose of this exercise is to familiarize the reader with the practical usage of the ADC. As stated earlier, the ADC on the LF2407 produces a 10-bit binary number which represents the voltage of the sampled analog signal.

This 10-bit number is stored in a 16-bit register "RESULTn" n=0..15. When reading from the register, the least significant 6 bits (bits 0-5) need to be disregarded because the 10 ADC result bits are bits 15-6. This can be done with the repeat command and SFR command.

For example, after the initial value is loaded into the accumulator:

```
RPT  #5 SFR      ; the accumulator will be shifted right 6 (RPT+1) times
```

Create an assembly source file and project file called "lab5"which:

a. Turns on the ADC clock in SCSR1 during the general initialization.
b. Puts the ADC in Reset.
c. Configures the ADC for Cascaded Mode; Continuous Mode = OFF; brings ADC out of Reset mode.
d. Sets maximum conversions to one and selects channel 0 for conversion.
e. Triggers a SOC via BIT 13 in ADCTRL2.
f. Checks if the ADC is finished via BIT 12 in ADCCTRL2.
g. If the ADC is finished with the conversion, the accumulator is loaded with the value in RESULT0 (the ADC conversion).
h. Continuously loops in an endless loop, i.e., does nothing.

1. Run the code on the LF2407 EVM with a 1.5V battery connected to channel 0 of the ADC and analog ground.
2. Record the value from the accumulator (the accumulator has been loaded with the value from the first ADC RESULT register). The value should be approximately half of the full voltage value 0x3FFF or approximately "1FF" = (1.5V).
3. Modify the program to output the result data from address 60h to DAC channel 1 on the LF2407 EVM. The DAC onboard is a 12-bit DAC, so in order to get the correct voltage output, left shift the accumulator (ADC data) two places to account for the extra two least significant bits. Also, the EVM DAC has a voltage reference of 5V rather then 3.3V like the ADC. This will cause a voltage slightly higher than what the ADC sampled to be outputted on the DAC channel. For this academic exercise, this can be ignored because the voltage difference is somewhat small. When writing to the DAC Channel 1, the data must first be written to IO space address 0000h. Then, in order for the data to actually be "sent out" on the DAC, the IO space address 0004h must be written to. It does not matter what value is written to IO address 0004h, just as long as it is written to.
4. The data will then be sent from the buffers to the DAC outputs. (See the OUT command and the Spectrum Digital LF2407 EVM manual for more information on programming the on-board DAC.)
5. Rebuild the project, reset the DSP, and run the new code.
6. Measure the voltage output of the DAC. It should reflect the ADC input voltage.
7. Modify the program to have the ADC continuously sample and continuously output the sampled data on the DAC.
8. Rebuild the program, reset, and run the DSP.

This concludes the ADC usage exercise.

Chapter 6

THE EVENT MANAGERS (EVA, EVB)

This chapter explains the features and operation of the LF2407 Event Managers (EV1, EV2). There are two identical event managers on board the LF2407 DSP. All control orientated features of the LF2407 are centered in the EV. The event manager peripheral is made up of components such as timers and pulse width modulation (PWM) generators. We start with a brief overview of the EV without getting into too much detail. Since the EV consists of several sub-components, we discuss in detail the operation and functionality of each sub-component separately in subsequent sections.

6.1 Overview of the Event Manager (EV)

We start with the EV by reviewing the multiple functional modules of the peripheral. The two EVs (EVA/B) are identical to one another in terms of functionality and register/bit definition, but have different register names and addresses. Since both EV1 and EV2 are identical, only the functionality of EV1 will be explained.

Each EV module in the LF2407 contains the following sub-components:

- Interrupt logic
- Two general-purpose (GP) timers
- Three compare units
- PWM circuits that include space vector PWM circuits, dead-band generation units, and output logic
- Three Capture Units
- Quadrature encoder pulse (QEP) circuit

Figure 6.1 shows a block diagram of the EVA module. Similarly, Fig. 6.2 illustrates the block diagram of EVB.

Like all peripherals, the EV registers occupy a range of 16-bit memory addresses in data memory space. Most of these registers are programmable control and data registers, but read-only status registers are also present. EVA registers are located in the data memory range 7400h to 7431h. EVB registers are located in the range of 7500h to 7531h. Some of the EV memory allocation range is for use by the DSP only. These undefined registers and undefined bits of EV registers will just read zero when read by user software. Writes also have no effect on these registers. As a general rule, one should not write to reserved or illegal addresses in order to avoid an illegal address non-maskable interrupt (NMI) from occurring.

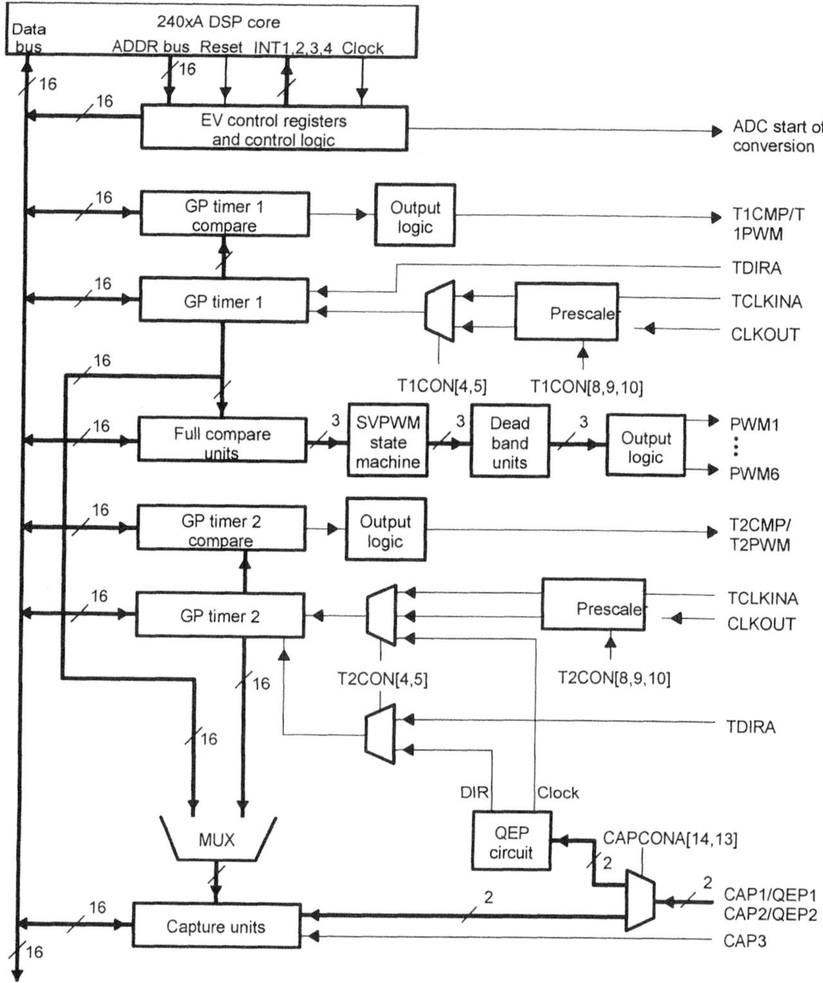

Figure 6.1 Event Manager A (EVA) block diagram. (Courtesy of Texas Instruments)

6.2 Event Manager Interrupts

The interrupt system in the EV will be discussed first because each of the sub-modules of the EVs have interrupt flags. The EV interrupt sub-system is slightly different from that of the main interrupt system. Each EV has its own "local" interrupt sub-system which includes its own interrupt mask and flag registers. After the EV interrupts pass through the sub-system, they flow into the PIE just like any other interrupt on the LF2407. The EV interrupts are arranged into three groups (A, B, C). Each group (A,B,C) has its own mask and flag register and is assigned to a particular CPU interrupt priority level at the PIE. EV interrupts happen to be only at the INT2, INT3, and INT4 CPU priority levels.

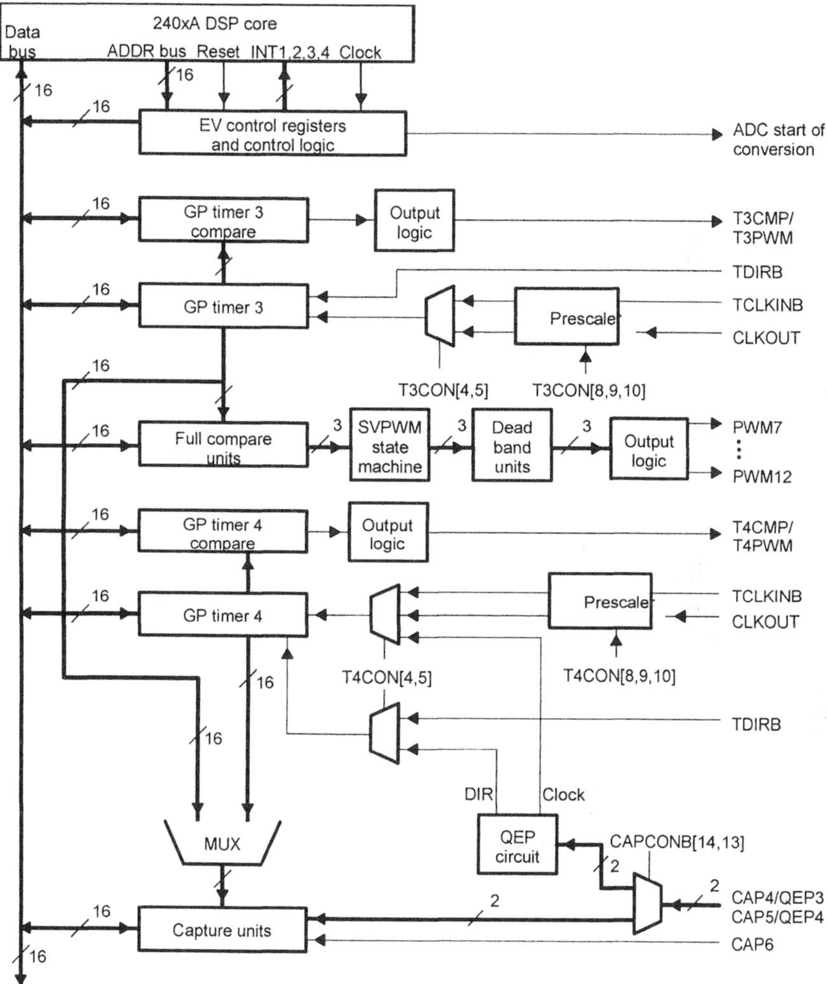

Figure 6.2 Event Manager B (EVB) block diagram. (Courtesy of Texas Instruments)

The following are the sequential steps for interrupt response within the EV:

1. ***Interrupt source***. When an EV interrupt condition occurs, the respective flag bits in registers EVxIFRA, EVxIFRB, or EVxIFRC (x = A or B) are set. As with other peripheral level flags, once set, these flags remain set until *explicitly* cleared by the software. In other words, you must clear

theses flags "manually" through your software in order for future interrupts to be recognized.

2. ***Interrupt enable.*** The EV interrupts can be individually enabled or disabled by the EV interrupt mask registers EVxIMRA, EVxIMRB, and EVxIMRC (x being either EV = A or B). To enable (unmask) an interrupt, the user must set the corresponding bit to "1". To disable (mask) the interrupt, clear the corresponding bit to "0". From now on, the interrupt is handled like other peripheral interrupts as discussed earlier in the text.

3. ***PIE request.*** If both interrupt flag bits and interrupt mask bits are set, then the interrupt request is passed to the PIE module. As with any other peripheral interrupts, the PIE module will send the CPU a request for a CPU level interrupt of the appropriate priority level based on the priority of the received interrupts.

4. ***CPU response.*** On receiving a CPU level interrupt request, the respective bit in the CPU interrupt flag register (IFR) will be set. If the corresponding interrupt mask register (IMR) bit is set and INTM bit is cleared, then the CPU recognizes the interrupt and issues an acknowledgement to the PIE module. Following this, the CPU finishes executing the current instruction and branches to the interrupt service routine via the interrupt vector. At this time, the respective IFR bit will be cleared and the INTM bit will be set disabling further interrupt recognition. The interrupt vector contains a branch instruction for the interrupt service routine. From here, the user software controls the interrupt servicing.

5. ***Interrupt software.*** The interrupt software can include two levels of response.

   a. **GISR:** The General Interrupt Service Routine (GISR) should do any context save and read the PIVR register to decide which specific interrupt occurred. Information on PIVR values and their corresponding interrupts can be found in Tables 6.1 and 6.2. Since the PIVR value for each interrupt is unique, it can be used to branch to the interrupt service routine specific to this interrupt condition.

   b. **SISR:** The Specific Interrupt Service Routine (SISR) level will normally reside as a sub-section of the GISR. After executing the interrupt specific service code, the routine should clear the interrupt flag in the EVxIFRA, EVxIFRB, or EVxIFRC that caused the serviced interrupt. Code will return the CPU to the pre-interrupt task after enabling the CPU's global interrupt bit INTM (clear INTM bit).

EVA Interrupts

Table 6.1 EVA Interrupts and Corresponding PIVR Values

Group	Interrupt	Priority within group	Vector (ID)	Description/Source	INT
	PDPINTA	1 (highest)	0020h	Power Drive Protection Interrupt A	1
A	CMP1INT	2	0021h	Compare Unit 1 compare interrupt	2
	CMP2INT	3	0022h	Compare Unit 2 compare interrupt	2
	CMP3INT	4	0023h	Compare Unit 3 compare interrupt	2
	T1PINT	5	0027h	GP Timer 1 period interrupt	2
	T1CINT	6	0028h	GP Timer 1 compare interrupt	2
	T1UFINT	7	0029h	GP Timer 1 underflow interrupt	2
	T1OFINT	8 (lowest)	002Ah	GP Timer 1 overflow interrupt	2
B	T2PINT	1 (highest)	002Bh	GP Timer 2 period interrupt	3
	T2CINT	2	002Ch	GP Timer 2 compare interrupt	3
	T2UFINT	3	002Dh	GP Timer 2 underflow interrupt	3
	T2OFINT	4	002Eh	GP Timer 2 overflow interrupt	3
C	CAP1INT	1 (highest)	0033h	Capture Unit 1 interrupt	4
	CAP2INT	2	0034h	Capture Unit 2 interrupt	4
	CAP3INT	3	0035h	Capture Unit 3 interrupt	4

EVB Interrupts

Table 6.2 EVB Interrupts and Corresponding PIVR Values

Group	Interrupt	Priority within group	Vector (ID)	Description/Source	INT
	PDPINTB	1 (highest)	0019h	Power Drive Protection Interrupt B	1
A	CMP4INT	2	0024h	Compare Unit 4 compare interrupt	2
	CMP5INT	3	0025h	Compare Unit 5 compare interrupt	2
	CMP6INT	4	0026h	Compare Unit 6 compare interrupt	2
	T3PINT	5	002Fh	GP Timer 3 period interrupt	2
	T3CINT	6	0030h	GP Timer 3 compare interrupt	2
	T3UFINT	7	0031h	GP Timer 3 underflow interrupt	2
	T3OFINT	8 (lowest)	0032h	GP Timer 3 overflow interrupt	2
B	T4PINT	1 (highest)	0039h	GP Timer 4 period interrupt	3
	T4CINT	2	003Ah	GP Timer 4 compare interrupt	3
	T4UFINT	3	003Bh	GP Timer 4 underflow interrupt	3
	T4OFINT	4	003Ch	GP Timer 4 overflow interrupt	3
C	CAP4INT	1 (highest)	0036h	Capture Unit 4 interrupt	4
	CAP5INT	2	0037h	Capture Unit 5 interrupt	4
	CAP6INT	3	0038h	Capture Unit 6 interrupt	4

EVA Interrupt Flag Register A (EVAIFRA) — Address 742Fh

15-11			10	9	8
Reserved			T1OFINT FLAG	T1UFINT FLAG	T1CINT FLAG
R–0			RW1C–0	RW1C–0	RW1C–0
7	6-4	3	2	1	0
T1PINT FLAG	Reserved	CMP3INT FLAG	CMP2INT FLAG	CMP1INT FLAG	PDPINTA FLAG
RW1C–0	R–0	RW1C–0	RW1C–0	RW1C–0	RW1C–0

Note: *R = read access, W1C = write 1 to clear, -0 = value after reset.*

Bits 15–11 **Reserved**. Reads return zero; writes have no effect.

Bit 10 **T1OFINT FLAG**. GP Timer 1 overflow interrupt.

Read: 0 Flag is reset
1 Flag is set
Write: 0 No effect
1 Resets flag

Bit 9 **T1UFINT FLAG**. GP Timer 1 underflow interrupt.

Read: 0 Flag is reset
1 Flag is set
Write: 0 No effect
1 Resets flag

Bit 8 **T1CINT FLAG**. GP Timer 1 compare interrupt.

Read: 0 Flag is reset
1 Flag is set
Write: 0 No effect
1 Resets flag

Bit 7 **T1PINT FLAG.** GP Timer 1 period interrupt.

Read: 0 Flag is reset
1 Flag is set
Write: 0 No effect
1 Resets flag

Bits 6–4 **Reserved**. Reads return zero; writes have no effect.

Bit 3 **CMP3INT FLAG**. Compare 3 interrupt.

Read: 0 Flag is reset
1 Flag is set
Write: 0 No effect
1 Resets flag

Bit 2 **CMP2INT FLAG.** Compare 2 interrupt.

Read:	0	Flag is reset
	1	Flag is set
Write:	0	No effect
	1	Resets flag

Bit 1 **CMP1INT FLAG.** Compare 1 interrupt.

Read:	0	Flag is reset
	1	Flag is set
Write:	0	No effect
	1	Resets flag

Bit 0 **PDPINTA FLAG.** Power drive protection interrupt.

Read:	0	Flag is reset
	1	Flag is set
Write:	0	No effect
	1	Resets flag

EVA Interrupt Flag Register B (EVAIFRB) — Address 7430h

15-4	3	2	1	0
Reserved	T2OFINT FLAG	T2UFINT FLAG	T2CINT FLAG	T2PINT FLAG
R–0	RW1C–0	RW1C–0	RW1C–0	RW1C–0

Note: *R = read access, W1C = write 1 to clear, -0 = value after reset.*

Bits 15–4 **Reserved.** Reads return zero; writes have no effect.

Bit 3 **T2OFINT FLAG.** GP Timer 2 overflow interrupt.

Read:	0	Flag is reset
	1	Flag is set
Write:	0	No effect
	1	Resets flag

Bit 2 **T2UFINT FLAG.** GP Timer 2 underflow interrupt.

Read:	0	Flag is reset
	1	Flag is set
Write:	0	No effect
	1	Resets flag

Bit 1 **T2CINT FLAG.** GP Timer 2 compare interrupt.

Read:	0	Flag is reset
	1	Flag is set
Write:	0	No effect
	1	Resets flag

Bit 0 **T2PINT FLAG**. GP Timer 2 period interrupt.

Read:	0	Flag is reset
	1	Flag is set
Write:	0	No effect
	1	Resets flag

EVA Interrupt Flag Register C (EVAIFRC) — Address 7431h

15-3	2	1	0
Reserved	CAP3INT FLAG	CAP2INT FLAG	CAP1INT FLAG
R–0	RW1C–0	RW1C–0	RW1C–0

Note: R = read access, W1C = write 1 to clear, -0 = value after reset.

Bits 15–3 **Reserved**. Reads return zero; writes have no effect.

Bit 2 **CAP3INT FLAG**. Capture 3 interrupt.

Read:	0	Flag is reset
	1	Flag is set
Write:	0	No effect
	1	Resets flag

Bit 1 **CAP2INT FLAG**. Capture 2 interrupt.

Read:	0	Flag is reset
	1	Flag is set
Write:	0	No effect
	1	Resets flag

Bit 0 **CAP1INT FLAG**. Capture 1 interrupt.

Read:	0	Flag is reset
	1	Flag is set
Write:	0	No effect
	1	Resets flag

EVA Interrupt Mask Register A (EVAIMRA) — Address 742Ch

15-11	10	9	8
Reserved	T1OFINT ENABLE	T1UFINT ENABLE	T1CINT ENABLE
R–0	RW–0	RW–0	RW–0

7	6-4	3	2	1	0
T1PINT ENABLE	Reserved	CMP3INT ENABLE	CMP2INT ENABLE	CMP1INT ENABLE	PDPINTA ENABLE
RW–0	R–0	RW–0	RW–0	RW–0	RW–1

Note: R = read access, W = write access, value following dash (–) = value after reset.

Bits 15–11 Reserved. Reads return zero; writes have no effect.

Bit 10 T1OFINT ENABLE
0 Disable
1 Enable

Bit 9 T1UFINT ENABLE
0 Disable
1 Enable

Bit 8 T1CINT ENABLE
0 Disable
1 Enable

Bit 7 T1PINT ENABLE
0 Disable
1 Enable

Bits 6–4 Reserved. Reads return zero; writes have no effect.

Bit 3 CMP3INT ENABLE
0 Disable
1 Enable

Bit 2 CMP2INT ENABLE
0 Disable
1 Enable

Bit 1 CMP1INT ENABLE
0 Disable
1 Enable

Bit 0 PDPINTA ENABLE. This is enabled (set to 1) following reset.
0 Disable
1 Enable

EVA Interrupt Mask Register B (EVAIMRB) — Address 742Dh

15-4	3	2	1	0
Reserved	T2OFINT ENABLE	T2UFINT ENABLE	T2CINT ENABLE	T2PINT ENABLE
R–0	RW–0	RW–0	RW–0	RW–0

Note: *R = read access, W = write access, -0 = value after reset.*

Bits 15–4 Reserved. Reads return zero; writes have no effect.

Bit 3 T2OFINT ENABLE
0 Disable
1 Enable

Bit 2 **T2UFINT ENABLE**
0 Disable
1 Enable

Bit 1 **T2CINT ENABLE**
0 Disable
1 Enable

Bit 0 **T2PINT ENABLE**
0 Disable
1 Enable

EVA Interrupt Mask Register C (EVAIMRC) — Address 742Eh

15-3	2	1	0
Reserved	CAP3INT ENABLE	CAP2INT ENABLE	CAP1INT ENABLE
R–0	RW–0	RW–0	RW–0

Note: R = read access, W = write access, -0 = value after reset.
Bits 15–3 **Reserved**. Reads return zero; writes have no effect.

Bit 2 **CAP3INT ENABLE**
0 Disable
1 Enable

Bit 1 **CAP2INT ENABLE**
0 Disable
1 Enable

Bit 0 **CAP1INT ENABLE**
0 Disable
1 Enable

EVB Interrupt Flag Register A (EVBIFRA) — Address 752Fh

15-11		10	9	8
Reserved		T3OFINT FLAG	T3UFINT FLAG	T3CINT FLAG
R–0		RW1C–0	RW1C–0	RW1C–0

7	6-4	3	2	1	0
T3PINT FLAG	Reserved	CMP6INT FLAG	CMP5INT FLAG	CMP4INT FLAG	PDPINTB FLAG
RW1C–0	R–0	RW1C–0	RW1C–0	RW1C–0	RW1C–0

Note: R = read access, W1C = write 1 to clear, -0 = value after reset.

Bits 15–11 Reserved. Reads return zero; writes have no effect.

Bit 10 T3OFINT FLAG. GP Timer 3 overflow interrupt.

Read:	0	Flag is reset
	1	Flag is set
Write:	0	No effect
	1	Resets flag

Bit 9 T3UFINT FLAG. GP Timer 3 underflow interrupt.

Read:	0	Flag is reset
	1	Flag is set
Write:	0	No effect
	1	Resets flag

Bit 8 T3CINT FLAG. GP Timer 3 compare interrupt.

Read:	0	Flag is reset
	1	Flag is set
Write:	0	No effect
	1	Resets flag

Bit 7 T3PINT FLAG. GP Timer 3 period interrupt.

Read:	0	Flag is reset
	1	Flag is set
Write:	0	No effect
	1	Resets flag

Bits 6–4 Reserved. Reads return zero; writes have no effect.

Bit 3 CMP6INT FLAG. Compare 6 interrupt.

Read:	0	Flag is reset
	1	Flag is set
Write:	0	No effect
	1	Resets flag

Bit 2 CMP5INT FLAG. Compare 5 interrupt.

Read:	0	Flag is reset
	1	Flag is set
Write:	0	No effect
	1	Resets flag

Bit 1 CMP4INT FLAG. Compare 4 interrupt.

Read:	0	Flag is reset
	1	Flag is set
Write:	0	No effect
	1	Resets flag

Bit 0 **PDPINTB FLAG**. Power drive protection interrupt.

Read:	0	Flag is reset
	1	Flag is set
Write:	0	No effect
	1	Resets flag

EVB Interrupt Flag Register B (EVBIFRB) — Address 7530h

15-4	3	2	1	0
Reserved	T4OFINT FLAG	T4UFINT FLAG	T4CINT FLAG	T4PINT FLAG
R–0	RW1C–0	RW1C–0	RW1C–0	RW1C–0

Note: *R = read access, W1C = write 1 to clear, -0 = value after reset.*

Bits 15–4 **Reserved**. Reads return zero; writes have no effect.

Bit 3 **T4OFINT FLAG**. GP Timer 4 overflow interrupt.

Read:	0	Flag is reset
	1	Flag is set
Write:	0	No effect
	1	Resets flag

Bit 2 **T4UFINT FLAG**. GP Timer 4 underflow interrupt.

Read:	0	Flag is reset
	1	Flag is set
Write:	0	No effect
	1	Resets flag

Bit 1 **T4CINT FLAG**. GP Timer 4 compare interrupt.

Read:	0	Flag is reset
	1	Flag is set
Write:	0	No effect
	1	Resets flag

Bit 0 **T4PINT FLAG**. GP Timer 4 period interrupt.

Read:	0	Flag is reset
	1	Flag is set
Write:	0	No effect
	1	Resets flag

EVB Interrupt Flag Register C (EVBIFRC) — Address 7531h

15-3	2	1	0
Reserved	CAP6INT FLAG	CAP5INT FLAG	CAP4INT FLAG
R–0	RW1C–0	RW1C–0	RW1C–0

Note: *R = read access, W1C = write 1 to clear, -0 = value after reset.*

Bits 15–3 **Reserved**. Reads return zero; writes have no effect.

Bit 2 **CAP6INT FLAG**. Capture 6 interrupt.

Read:	0	Flag is reset
	1	Flag is set
Write:	0	No effect
	1	Resets flag

Bit 1 **CAP5INT FLAG**. Capture 5 interrupt.

Read:	0	Flag is reset
	1	Flag is set
Write:	0	No effect
	1	Resets flag

Bit 0 **CAP4INT FLAG**. Capture 4 interrupt.

Read:	0	Flag is reset
	1	Flag is set
Write:	0	No effect
	1	Resets flag

EVB Interrupt Mask Register A (EVBIMRA) — Address 752Ch

15-11			10	9	8
Reserved			T3OFINT ENABLE	T3UFINT ENABLE	T3CINT ENABLE
R–0			RW–0	RW–0	RW–0
7	**6-4**	**3**	**2**	**1**	**0**
T3PINT ENABLE	Reserved	CMP6INT ENABLE	CMP5INT ENABLE	CMP4INT ENABLE	PDPINTB ENABLE
RW–0	R–0	RW–0	RW–0	RW–0	RW–1

Note: *R = read access, W = write access, -n = value after reset.*

Bits 15–11 **Reserved**. Reads return zero; writes have no effect.

Bit 10 **T3OFINT ENABLE**
- 0 Disable
- 1 Enable

Bit 9 **T3UFINT ENABLE**
- 0 Disable
- 1 Enable

Bit 8 **T3CINT ENABLE**
- 0 Disable
- 1 Enable

Bit 7 **T3PINT ENABLE**
- 0 Disable
- 1 Enable

Bits 6–4 **Reserved**. Reads return zero; writes have no effect.

Bit 3 **CMP6INT ENABLE**
- 0 Disable
- 1 Enable

Bit 2 **CMP5INT ENABLE**
- 0 Disable
- 1 Enable

Bit 1 **CMP4INT ENABLE**
- 0 Disable
- 1 Enable

Bit 0 **PDPINTB ENABLE.** This is enabled (set to 1) following reset.
- 0 Disable
- 1 Enable

EVB Interrupt Mask Register B (EVBIMRB) — Address 752Dh

15-4	3	2	1	0
Reserved	T4OFINT ENABLE	T4UFINT ENABLE	T4CINT ENABLE	T4PINT ENABLE
R–0	RW–0	RW–0	RW–0	RW–0

Note: *R = read access, W = write access, -0 = value after reset.*

Bits 15–4 **Reserved**. Reads return zero; writes have no effect.

Bit 3 **T4OFINT ENABLE**
0 Disable
1 Enable

Bit 2 **T4UFINT ENABLE**
0 Disable
1 Enable

Bit 1 **T4CINT ENABLE**
0 Disable
1 Enable

Bit 0 **T4PINT ENABLE**
0 Disable
1 Enable

EVB Interrupt Mask Register C (EVBIMRC) — Address 752Eh

15-3	2	1	0
Reserved	CAP6INT ENABLE	CAP5INT ENABLE	CAP4INT ENABLE
R–0	RW–0	RW–0	RW–0

Note: *R = read access, W = write access, -0 = value after reset.*

Bits 15–3 **Reserved**. Reads return zero; writes have no effect.

Bit 2 **CAP6INT ENABLE**
0 Disable
1 Enable

Bit 1 **CAP5INT ENABLE**
0 Disable
1 Enable

Bit 0 **CAP4INT ENABLE**
0 Disable
1 Enable

6.3 General Purpose (GP) Timers

A General Purpose (GP) timer is simply a 16-bit counter, which may be configured to count up, down, or continuously up and down. There are two GP Timers in each EV: Timer1 and Timer2 for EVA and Timer3 and Timer4 for EVB.

All timers use the CPU clock as a general timing reference, but each individual timer may use a “pre-scaled” or frequency reduced time base which is specified in each timer’s control register.

A GP Timer may also be configured to generate an interrupt or trigger another peripheral on certain events such as a timer overflow (timer reached period value), underflow (timer reached zero), or compare (timer value reached compare value). Some examples of uses for the GP Timers include: setting the sampling period for the ADC by triggering the start of conversion; or providing the switching period for the generation of a PWM signal.

Figure 6.3 shows a block diagram of a GP Timer. There are two cases that apply to Fig. 6.3:

1. When “x” = 2, “y” = 1 and “n” = 2
2. When “x” = 4, “y” = 3 and “n” = 4

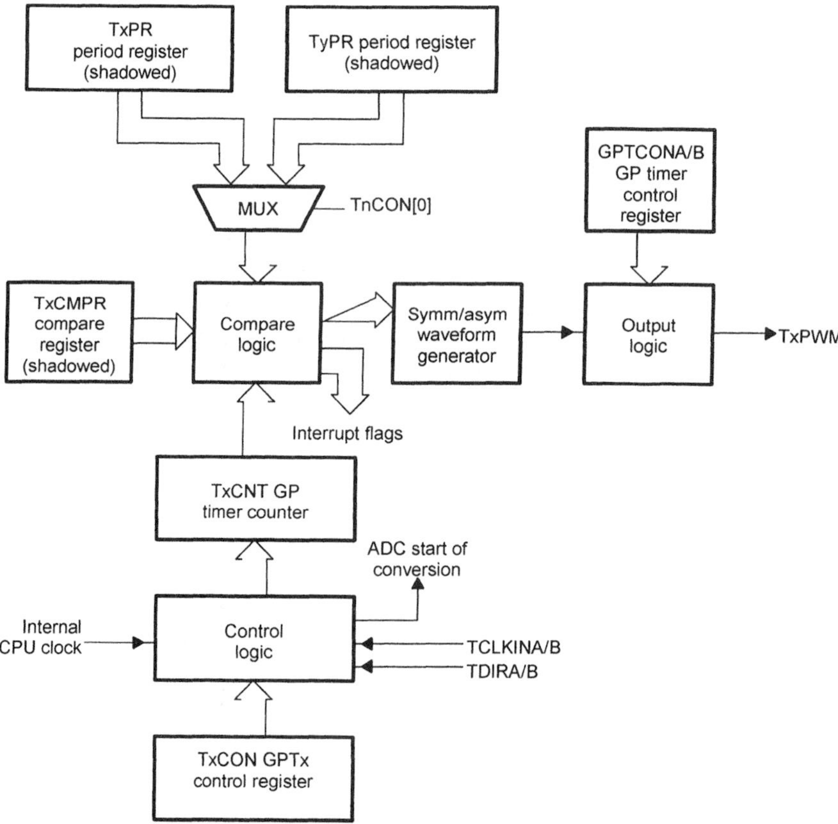

Figure 6.3 General purpose timer configuration diagram.

Each GP Timer consists of the following components:

- One readable and writeable (RW) 16-bit up and up/down counter register **TxCNT** (x = 1, 2, 3, 4). This register holds the current count value and increments or decrements depending on the direction of counting
- 16-bit timer compare register, **TxCMPR** (x = 1, 2, 3, 4)
- 16-bit timer period register, **TxPR** (x = 1, 2, 3, 4)
- 16-bit individual timer control register, **TxCON** (x = 1, 2, 3, 4)
- Programmable input clock divider (pre-scaler) applicable to both internal and external clock inputs
- One GP Timer compare output pin, **TxCMP** (x = 1, 2, 3, 4)
- Interrupt logic

6.3.1 GP Timer Inputs and Outputs

Each GP Timer has the following inputs:

- Clock Reference Inputs: (1) The internal device (CPU) clock and (2) external clock, TCLKINA/B, that has a maximum frequency of one-fourth that of the device clock
- Direction input, TDIRA/B, when a GP Timer is in directional up/down-counting mode
- Reset signal, RESET

The source of the GP Timer clock can be the internal CPU clock signal or the external clock input, TCLKINA/B. The frequency of the external clock must be less than or equal to one-fourth of that of the device clock. GP Timer 2 (EVA) and GP Timer 4 (EVB) can be used with the QEP circuits in directional up-/down-counting mode. In this case, the QEP circuits provide both the clock and direction inputs to the timer. A wide range of prescale factors are provided for the clock input to each GP Timer.

The QEP circuit, when selected, can generate the input clock and counting direction for GP Timer 2/4 in the directional up-/down-counting mode. A QEP signal may come from a rotary encoder which is attached to a motor shaft to provide speed/direction feedback. Via the QEP circuitry, it controls the clock input and direction of Timer 2/4. From this, the speed of the motor can be determined from the counting speed; the direction of count reflects the rotation direction. The QEP input clock cannot be scaled by GP Timer prescaler circuits (the prescaler of the selected GP Timer has no effect if the QEP circuit is selected as the clock source). Furthermore, the frequency of the clock generated by the QEP circuits is four times that of the frequency of each QEP input channel because the rising and falling edges of both QEP input channels are counted by the selected timer. In other words, the frequency of the incoming QEP signal must be less than or equal to one-fourth of that of the CPU clock.

Now that the inputs to the GP Timers have been discussed, we will next discuss the outputs associated with each GP Timer. Outputs are either connected to a data memory mapped register, another peripheral, or an external pin of the LF2407. Each GP Timer has the following outputs:

- GP Timer compare outputs TxCMP, x = 1, 2, 3, 4 (external pins on the LF2407)
- ADC start-of-conversion signal (connected to the ADC module)
- Underflow, overflow, compare match, and period match signals to its own compare logic and to the compare units (connected to the compare units of the EV)
- Counting direction indication bits (in the GPTCONA/B registers mapped to data memory)

The General Purpose Timer Control Register (GPTCONA/B), configures the action to be taken by the timers on different timer events, and indicates the counting directions of the GP Timers. GPTCONA/B is readable and writeable, although writing to the status bits in this register has no effect.

6.3.2 GP Counting Operation

GP Timers have four possible modes of counting operation:

1. Stop/Hold mode
2. Continuous Up-Counting mode
3. Directional Up/Down-Counting mode
4. Continuous Up/Down-Counting mode

Each timer is configured for desired counting mode in its corresponding Timer Control register (TxCON). Each GP Timer is enabled by setting the Timer Enable bit each timer's control register. When the timer is enabled, the timer counts according to the counting mode specified by the bits in the TxCON. The counting direction of the GP Timers are reflected by their respective bit in GPTCONA/B. When the timer is disabled (enable bit=0), counting is disabled and the prescaler of that timer is reset to the default value of "x/1".

Stop/Hold mode is like the "pause" button for the timer. In stop/hold mode the GP Timer stops and holds at its current state. The timer counter, the compare output, and the pre-scale select all remain unchanged.

Continuous Up-Counting Mode:

The continuous up-counting mode is useful in creating asymmetric PWM signals. In the continuous up-count mode the following events occur:

1. The GP Timer in this mode counts up in sync with the pre-scaled input clock until the value of the timer counter matches that of the period register.
2. On the next rising edge of the input clock after the match, the GP Timer resets to zero and starts counting up again.
3. The period interrupt flag of the timer is set one clock cycle after the match between the timer counter and period register. If the flag is not masked, a

peripheral interrupt request is generated. An ADC start is sent to the ADC module at the same time the flag is set if the period interrupt of this timer has been selected by the appropriate bits in GPTCONA/B to start the ADC.

4. One clock cycle after the GP Timer becomes 00001, the underflow interrupt flag of the timer is set. A peripheral interrupt request is generated by the flag if it is unmasked. An ADC start is sent to the ADC module at the same time if the underflow interrupt flag of this timer has been selected by the appropriate bits in the GPTCONA/B to start the ADC.

The duration of the timer period is (TxPR) + 1 cycles of the scaled clock input except for the first period. The duration of the first period is the same if the timer counter is zero when counting starts. The initial value of the GP Timer can be any value from 0h to FFFFh. When the initial value is greater than the value in the period register, the timer counts up to FFFFh, resets to zero, and continues the operation as if the initial value was zero. The overflow interrupt flag is set one clock cycle after the value in TxCNT matches FFFFh. A peripheral interrupt request is generated by the flag if it is unmasked.

When the initial value in the timer counter is the same as that of the period register, the timer sets the period interrupt flag, resets to zero, sets the underflow interrupt flag, and then continues the operation again as if the initial value was zero. If the initial value of the timer is between zero and the contents of the period register, the timer counts up to the period value and continues to finish the period as if the initial counter value was the same as that of the period register.

The counting direction indication bit in GPTCONA/B is "1" for the timer in this mode. Either the external or internal device clock can be selected as the input clock to the timer. The TDIRA/B input is ignored by the GP Timer in this mode since we are in an up-count only mode. The continuous up-count mode of the GP Timer is particularly useful for the generation of edge-triggered or asynchronous PWM waveforms and sampling periods in many motor and motion control systems. Figure 6.4 shows the continuous up-counting mode of the GP Timer.

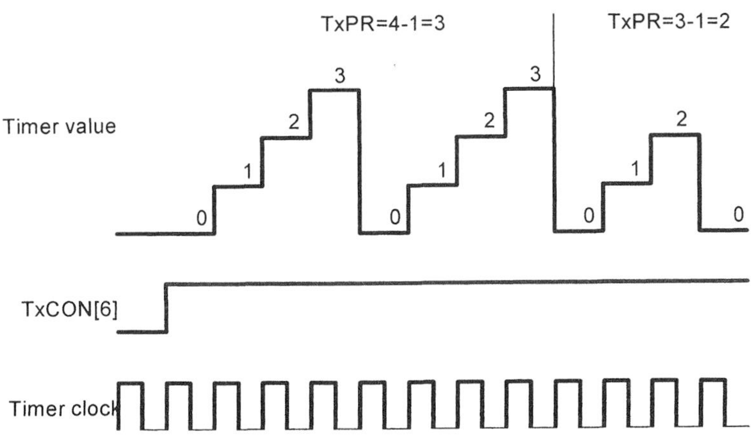

Figure 6.4 Operation of continuous up-counting mode (TxPR = 3 or 2).

Directional Up/Down-Counting Mode:

A GP Timer in directional up/down-counting mode counts either up or down according to the pre-scaled clock and TDIRA/B inputs. The input pin TDIRA/B determines the direction of counting when the GP Timer is in directional up/down-counting mode. When TDIRA/B is high, upward counting is specified; when TDIRA/B is low, downward counting is specified.

When the TDIRA/B pin is held high, the GP Timer will count up until it reaches the value of the period register. When the timer value equals that of its period register the timer will reset to zero and start counting up to the period again. The initial value of the timer can be any value between 0000h to FFFFh. In the case that the initial value of the timer counter is greater than that of the period register, the timer would count up to FFFFh before resetting itself to zero and continuing the counting operation. When TDIRA/B pin is held low, the GP Timer will count down from whatever initial value the counter was at until its count value becomes zero. When its count value becomes zero, the value of the period register is automatically loaded into the count value register and the timer begins counting down to zero.

In the directional up/down mode, the period, underflow, and overflow interrupt flags, interrupts, and associated actions are generated on respective events in the same manner as they are generated in the continuous up-counting mode. The direction of counting is indicated for the timer in this mode by the corresponding direction indication bit in GPTCONA/B: 1 means counting up; 0 means counting down. Either the external clock from the TCLKINA/B pin or the internal device clock can be used as the input clock for the timer in this mode. Figure 6.5 shows the directional up-/down-counting mode of the GP Timers.

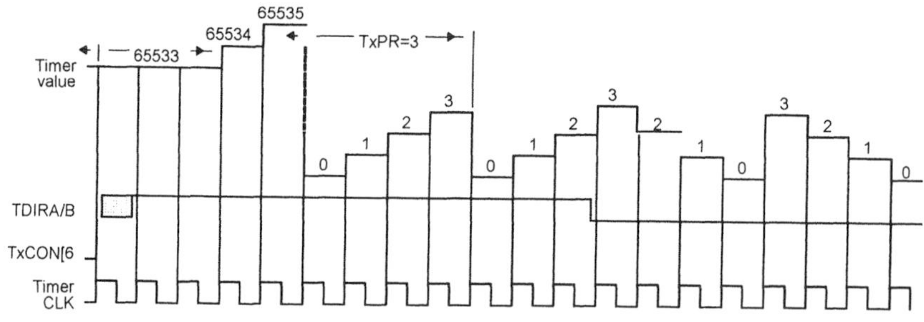

Figure 6.5 GP timer directional up/down-counting mode: prescale factor 1 and TxPR = 3A.

Additionally, the directional up-/down-counting mode of GP Timer 2 and 4 can also be used with the Quadrature Encoder Pulse (QEP) circuits in the EV module. While the QEP circuits are active, they provide both the counting clock and direction for GP Timers 2 or 4.

Continuous Up/Down-Counting Mode

The continuous up/down-counting mode is useful in generating symmetric PWM waveforms. This mode of operation is the same as the directional up-/down-counting mode, except for the fact that the TDIRA/B pin has no effect on the counting direction. The counting direction changes from up to down when the timer reaches the period value. The timer direction changes from down to up when the timer reaches zero. Continuous up/down-counting mode is particularly useful in generating centered or symmetric PWM waveforms.

The initial value of the GP Timer counter can be any value from 0h to FFFFh. When the initial value is greater than that of the period register (TxPR), the timer counts up to FFFFh, resets to zero, and continues the operation as if the initial value were zero. If the initial value of the timer counter is the same as that of the period register, the timer counts down to zero and continues again as if the initial value were zero. If the initial value of the timer is between zero and the contents of the period register, the timer will count up to the period value and continue to finish the period as if the initial counter value were the same as that of the period register.

The counting direction indication bit in the GPTCONA/B indicates "1" when the timer counts upward and "0" when the timer is counting downward. Either an external clock reference from the TCLKINA/B pin or the internal CPU clock can be selected as the input clock. Since the change of count direction is automatic in this mode, the TDIRA/B pin has no effect. The period, underflow, and overflow interrupt flags, interrupts, and associated actions are generated on the respective events in the same manner as they are generated in other counting modes. Figure 6.6 shows the continuous up-/down-counting mode of the GP Timer.

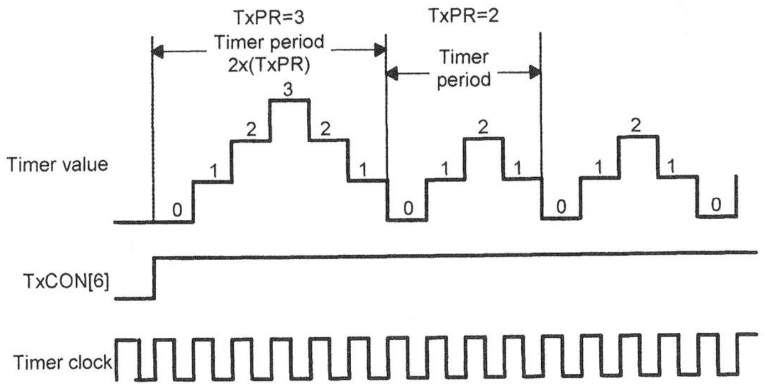

Figure 6.6 Continuous up/down counting mode (timer period register = 3 or 2).

Note: *The period of the timer in this mode is 2*(TxPR) cycles of the scaled clock input, except for the first period.*

6.3.3 Control Registers Associated with the General Purpose Timers

Individual Timer Control Registers (TxCON), where x=1,2,3,4

The operational mode of each GP Timer is controlled by the timer's corresponding control register (TxCON). The bits in the TxCON configure:

1. What counting mode the timer is set for
2. Whether the internal (CPU) or an external clock is to be used for the clock reference
3. Which of the eight input clock pre-scale factors (ranging from 1/1 to 1/128) is used
4. When (on which condition) the timer compare register is reloaded
5. Whether the timer is enabled or disabled
6. Whether the timer compare operation is enabled or disabled
7. Which period register is used by timer 2 (its own, or timer 1's period register (EVA))
8. Which period register is used by timer 4 (its own, or timer 3's period register (EVB))

In EVA, GP Timer 2 can be synchronized with GP Timer 1. Additionally, in EVB, GP Timer 4 can be synchronized with GP Timer 3 by configuring T2CON and T4CON, respectively, in the following ways:

EVA:

1. Set the T2SWT1 bit in T2CON to start GP Timer 2 counting with the TENABLE bit in T1CON (both timer counters start simultaneously)
2. Initialize the timer counter in GP Timers 1 and 2 with different values before starting synchronized operation
3. Specify that GP Timer 2 uses the period register of GP Timer 1 as its period register (ignoring its own period register) by setting SELT1PR in T2CON

EVB:

1. Set the T4SWT3 bit in T4CON to start GP Timer 4 counting with the TENABLE bit in T3CON (thus, both timer counters start simultaneously)
2. Initialize the timer counters in GP Timers 3 and 4 with different values before starting synchronized operation
3. Specify that GP Timer 4 uses the period register of GP Timer 3 as its period register (ignoring its own period register) by setting SELT3PR in T4CON

This allows the desired synchronization between GP Timer events. Since each GP Timer starts the counting operation from its current value in the counter register, one GP Timer can be programmed to start with a known delay after the other GP Timer.

Timer x Control Register Bit Descriptions (TxCON; x = 1, 2, 3, or 4) — Addresses: 7404h (T1CON), 7408h (T2CON), 7504h (T3CON), and 7508h (T4CON)

15	14	13	12	11	10	9	8
Free	Soft	Reserved	TMODE1	TMODE0	TPS2	TPS1	TPS0
RW–0	RW–0	RW–0	RW–0	RW–0	RW–0	RW–0	RW–0
7	6	5	4	3	2	1	0
T2SWT1/ T4SWT3†	TENABLE	TCLKS1	TCLKS0	TCLD1	TCLD0	TECMPR	SELT1PR/ SELT3PR†
RW–0	RW–0	RW–0	RW–0	RW–0	RW–0	RW–0	RW–0

† Reserved in T1CON andT3CON

***Note:** R = read access, W = write access, -0 = value after reset.*

Bits 15–14 Free, Soft. Emulation control bits.

- 00 Stop immediately on emulation suspend
- 01 Stop after current timer period is complete on emulation suspend
- 10 Operation is not affected by emulation suspend
- 11 Operation is not affected by emulation suspend

Bit 13 Reserved. Reads return zero, writes have no effect.

Bits 12–11 TMODE1–TMODE0. Count Mode Selection.

- 00 Stop/Hold
- 01 Continuous-Up/-Down Count Mode
- 10 Continuous-Up Count Mode
- 11 Directional-Up/-Down Count Mode

Bits 10–8 TPS2–TPS0.

Input Clock Prescaler.
000=x/1 , 001=x/2, 010=x/4, 011=x/8, 100=x/16, 101=x/32, 110=x/64
111=x/128 ; x = device (CPU) clock frequency

Bit 7 T2SWT1. In the case of EVA, this bit is T2SWT1. (GP Timer 2 start with GP Timer 1.) Start GP Timer 2 with GP Timer 1's timer enable bit. This bit is reserved in T1CON.

T4SWT3. In the case of EVB, this bit is T4SWT3. (GP Timer 4 start with GP Timer 3.) Start GP Timer 4 with GP Timer 3's timer enable bit. This bit is reserved in T3CON.

- 0 Use own TENABLE bit
- 1 Use TENABLE bit of T1CON (in case of EVA) or T3CON (in case of EVB) to enable and disable operation ignoring own TENABLE bit

Bit 6 **TENABLE**. Timer enable.

0	Disable timer operation (the timer is put in hold and the prescaler counter is reset)
1	Enable timer operations

Bits 5–4 TCLKS1, TCLKS0. Clock Source Select.

5	4	Source
0	0	Internal
0	1	External
1	0	Reserved
1	1	QEP Circuit† (in case of Timer 2/Timer 4) Reserved (in case of Timer 1/Timer 3)
†		This option is valid only if SELT1PR = 0

Bits 3–2 TCLD1, TCLD0. Timer Compare Register Reload Condition.

00	When counter is 0
01	When counter value is 0 or equals period register value
10	Immediately
11	Reserved

Bit 1 **TECMPR**. Timer compare enable.

0	Disable timer compare operation
1	Enable timer compare operation

Bit 0 **SELT1PR**. In the case of EVA, this bit is SELT1PR (Period register select).

When set to 1 in T2CON, the period register of Timer 1 is chosen for Timer 2 also, ignoring the period register of Timer 2. This bit is a reserved bit in T1CON. **SELT3PR**. In the case of EVB, this bit is SELT3PR (Period register select). When set to 1 in T4CON, the period register of Timer 3 is chosen for Timer 4 also, ignoring the period register of Timer 4. This bit is a reserved bit in T3CON.

0	Use own period register
1	Use T1PR (in case of EVA) or T3PR (in case of EVB) as period register ignoring own period register

Overall GP Timer Control Registers (GPTCONA/B)

The control register GPTCONA/B specifies the action to be taken by the timers on different timer events. This register also has timer direction status bits that display the current direction of the timers. Also, the polarity of the GP Timer compare outputs is configured here. Bits in GPTCONA/B can also configure specific timers to trigger an ADC start signal when an underflow, compare match, or period match occurs. This feature requires that the ADC also be configured to accept the start of conversion signal from the GP Timer. Having the GP Timer trigger provides for automatic synchronization between the GP Timer event and the ADC.

GP Timer Control Register A (GPTCONA) Bit Descriptions — Address 7400h

15	14	13	12-11	10-9	8-7
Reserved	T2STAT	T1STAT	Reserved	T2TOADC	T1TOADC
RW–0	R–1	R–1	RW–0	RW–0	RW–0

6	5-4	3-2	1-0
TCOMPOE	Reserved	T2PIN	T1PIN
RW–0	RW–0	RW–0	RW–0

Note: *R = read access, W = write access, -n = value after reset.*

Bit 15 **Reserved**. Reads return zero; writes have no effect.

Bit 14 **T2STAT**. GP Timer 2 Status. Read only.
- 0 Counting downward
- 1 Counting upward

Bit 13 **T1STAT**. GP Timer 1 Status. Read only.
- 0 Counting downward
- 1 Counting upward

Bits 12–11 **Reserved**. Reads return zero; writes have no effect.

Bits 10–9 **T2TOADC**. Start ADC with timer 2 event.
- 00 No event starts ADC
- 01 Setting of underflow interrupt flag starts ADC
- 10 Setting of period interrupt flag starts ADC
- 11 Setting of compare interrupt flag starts ADC

Bits 8–7 **T1TOADC**. Start ADC with timer 1 event.
- 00 No event starts ADC
- 01 Setting of underflow interrupt flag starts ADC
- 10 Setting of period interrupt flag starts ADC
- 11 Setting of compare interrupt flag starts ADC

Bit 6 **TCOMPOE**. Compare output enable. If PDPINTx is active this bit is set to zero.
- 0 Disable all GP Timer compare outputs (all compare outputs are put in the high-impedance state)
- 1 Enable all GP Timer compare outputs

Bits 5–4 Reserved. Reads return zero; writes have no effect.

Bits 3–2 T2PIN. Polarity of GP Timer 2 compare output.

00	Forced low
01	Active low
10	Active high
11	Forced high

Bits 1–0 T1PIN. Polarity of GP Timer 1 compare output.

00	Forced low
01	Active low
10	Active high
11	Forced high

GP Timer Control Register B (GPTCONB) Bit Descriptions — Address 7500h

15	14	13	12-11	10-9	8-7
Reserved	T4STAT	T3STAT	Reserved	T4TOADC	T3TOADC
RW–0	R–1	R–1	RW–0	RW–0	RW–0

6	5-4	3-2	1-0
TCOMPOE	Reserved	T4PIN	T3PIN
RW–0	RW–0	RW–0	RW–0

Note: R = read access, W = write access, -n = value after reset.

Bit 15 Reserved. Reads return zero; writes have no effect.

Bit 14 T4STAT. GP Timer 4 Status. Read only.

0	Counting downward
1	Counting upward

Bit 13 T3STAT. GP Timer 3 Status. Read only.

0	Counting downward
1	Counting upward

Bits 12–11 Reserved. Reads return zero; writes have no effect.

Bits 10–9 T4TOADC. Start ADC with timer 4 event.

00	No event starts ADC
01	Setting of underflow interrupt flag starts ADC
10	Setting of period interrupt flag starts ADC
11	Setting of compare interrupt flag starts ADC

Bits 8–7 T3TOADC. Start ADC with timer 3 event.

00	No event starts ADC
01	Setting of underflow interrupt flag starts ADC

10 Setting of period interrupt flag starts ADC
11 Setting of compare interrupt flag starts ADC

Bit 6 **TCOMPOE**. Compare output enable. If PDPINTx is active this bit is set to zero.

0 Disable all GP Timer compare outputs (all compare outputs are put in the high-impedance state)
1 Enable all GP Timer compare outputs

Bits 5–4 Reserved. Reads return zero; writes have no effect.

Bits 3–2 T4PIN. Polarity of GP Timer 4 compare output.

00 Forced low
01 Active low
10 Active high
11 Forced high

Bits 1–0 T3PIN. Polarity of GP Timer 3 compare output.

00 Forced low
01 Active low
10 Active high
11 Forced high

GP Timer Compare Registers (TxCMPR), x=1,2,3,4 – User Specified Value

Addresses 7402h (T1CMPR), 7406h (T2CMPR), 7502h (T3CMPR), 7506h (T4CMPR)

The compare register associated with each GP Timer stores the value that will be constantly compared with the current value of the GP Timer. When a compare match occurs, the following events also occur:

1. A transition occurs on the associated compare output according to the bit pattern in GPTCONA/B
2. The corresponding interrupt flag is set
3. A peripheral interrupt request is generated if the interrupt is unmasked
4. The compare operation of a GP Timer can be enabled or disabled by the appropriate bit in TxCON
5. The compare operation and outputs can be enabled in any of the timer counting modes, including the QEP circuit

GP Timer Period Registers (TxPR) – User Specified Value

Addresses 7403h (T1PR), 7407h (T2PR), 7503h (T3PR), 7507h (T4PR)

The period register determines the rate at which the timer resets itself or changes direction (the period of the timer). This register in combination with the input clock frequency (and clock pre-scale factor) determines the frequency of a

PWM signal created by the compare output pin. The corresponding timer either resets to "0", or starts counting downward (depending on the operating mode) when a match occurs between the period register and the timer counter (TxCNT).

CAUTION: *The period register of a GP Timer needs to be initialized before its counter is set to a non-zero value. Otherwise, the value of the period register will remain unchanged until the next underflow!*

Both the compare and period registers of the GP Timers are shadowed or double-buffered. This means that when either the period registers or compare registers are written to, the value is automatically stored first into a buffer register and then automatically written to the real register. The reason for this is to prevent the unacceptable situation such as a timer period register being written to and read from at the same time. Because of the register shadowing, a new value can be written to any of these registers at any time. The double buffering feature of the period and compare registers allows the user program to update the period and compare registers at any time in order to change the future timer period. Register shadowing is virtually transparent to the user. However, when configuring a compare unit, it is necessary to specify on what condition the actual compare register is reloaded from the buffer register. For the compare register, the content in the buffer register is loaded into the working (active) register **only** when the certain timer event specified by TxCON occurs. A compare register would be reloaded automatically either immediately after the shadow register is written, on underflow (GP Timer counter value equals "0"), on an underflow, or on period register match. In the case that the associated compare operation is disabled, any value written to the compare register is immediately loaded into the active register. The period register will be reloaded with the value in its buffer register only when the value of the counter register (TxCNT) becomes equal to "0". Except for the compare register reload condition, the user need not worry about register shadowing on the LF2407.

6.3.4 GP Timer Interrupts

There are 16 combined interrupt flags in the EVAIFRA, EVAIFRB, EVBIFRA, and EVBIFRB registers for the GP Timers. Each of the four GP Timers has the capability to generate up to four interrupts on the events listed in Table 6.3.

Table 6.3 General purpose timer interrupts

Interrupt Event	Interrupt Name (x=1,2,3,4)	Condition For Generation
Underflow	TxUFINT	When the counter reaches 0000h
Overflow	TxOFINT	When the counter reaches FFFFh
Compare Match	TxCINT	When the counter register contents match that of the compare register
Period Match	TxPINT	When the counter register contents match that of the period register

A timer compare "event" or match happens when the current count value of a GP Timer counter equals the value of the timer's compare register. The corresponding compare interrupt flag is set one clock cycle after the match if the compare operation is enabled. An overflow event occurs when the value of the timer counter reaches FFFFh. An underflow event occurs when the timer counter reaches 0000h. Similarly, a period event happens when the value of the timer counter is the same as that of the period register. The overflow, underflow, and period interrupt flags of the timer are set one clock cycle after the occurrence of each individual event. Note that these definitions of overflow and underflow are different from the conventional definitions the reader might be used to.

6.3.5 PWM Output and General Purpose Timer Compare Operation

A PWM waveform is a sequence of pulses with fixed frequency but varying pulse widths. The width of the pulse might vary from 0% to 100% of the fixed period. The pulse widths are modulated by another signal called the modulation signal. In order to generate a PWM signal digitally, a timer is set to continuously repeat a counting period. This period is known as the PWM carrier period. The inverse of the carrier period is called the carrier frequency.

The counting pattern of the timer will either be a "saw-tooth" (asymmetric) or "triangle" (symmetric) wave depending on what counting mode the timer has been configured for. As always, the compare value is constantly being compared with the value of the timer counter. When a match occurs, the output toggles High to Low, or Low to High. When the timer period value is reached or a second match occurs, the output toggles again. The on and off time of the pulse is directly dependent on the value loaded into the timer's compare register. By varying the number in the compare register by the modulation signal (usually a sinusoid), a PWM signal that represents the modulating signal can be produced.

The "output" discussed above refers to each GP Timer's associated PWM output pin (TxPWM). The logic level of the PWM output pin is determined automatically by hardware. This level is based on the value of the associated compare register and timer count value (see Fig. 6.7, note the compare match points and the output change at these points). If the compare operation is enabled in TxCON, the following events occur on a compare match:

1. The compare interrupt flag of the timer is set one clock cycle after the match.
2. A transition occurs on the associated PWM output pin one device clock cycle after the match according to the bit configuration in GPTCONA/B.
3. If the compare interrupt flag has been selected by the appropriate GPTCONA/B bits to start the ADC, an ADC start signal is generated at the same time the compare interrupt flag is set.
4. A peripheral interrupt request is generated by the compare interrupt flag if it is unmasked.

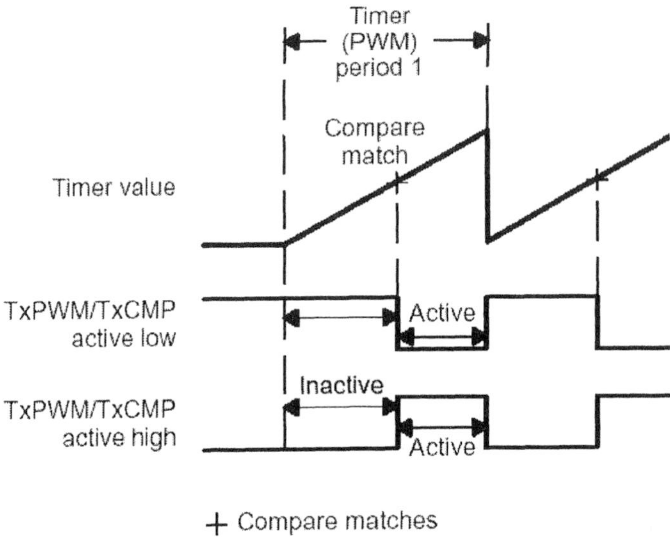

Figure 6.7 Timer compare match and associated change on TxPWM pin.

The polarity of the compare output (see diagram in Fig. 6.6) of a GP Timer can be specified active high, active low, forced high, or forced low. This polarity is determined by setting the bits in the GPTCONA/B register. If active low, the output changes from high to low on the first compare match. It then goes from low to high on the second compare match if the GP Timer is in an up/down-counting mode, or on period match if the GP Timer is in up-counting mode. If active high, the output changes from low to high on the first compare match. It then goes from high to low on the second compare match if the GP Timer is in an up-/down counting mode, or on period match if the GP Timer is in up-count mode. If forced low, the timer compare output becomes low immediately when it is specified. If forced high, the timer compare output becomes high immediately when it is specified.

By default (after a reset or power-on) all GP Timer PWM output pins are put in a high-impedance (HI-Z) state. The PWM output must be made active by configuring the GPTCONA/B registers. At anytime the PWM outputs will be made HI-Z whenever the power drive protection pin $\overline{PDPINTx}$ is active and is pulled low. Additionally, the corresponding PWM pin will be made HI-Z when bit 1 of the TxCON register is zeroed by software.

The transition on the PWM output pin is controlled by the asymmetric or symmetric timer waveform and the associated output logic. For an asymmetric wave form, the timer is set up in continuous up-count mode. To generate a symmetric waveform, the timer needs to be configured to continuous up/down counting.

Example 6.1 - Generation of an Asymmetric Waveform: The asymmetric waveform in Fig. 6.8 is generated when the GP Timer is in continuous up-counting mode. When in this mode the output changes in the following:

1. Output pin at "inactive level" before the counting operation starts
2. Output pin remains at "inactive level" until the compare match happens
3. Output toggles to "active level" on the compare match
4. Output remains unchanged at "active level" until the end of the period
5. At end of period, output resets to "inactive level"; that is if the new compare value is not zero

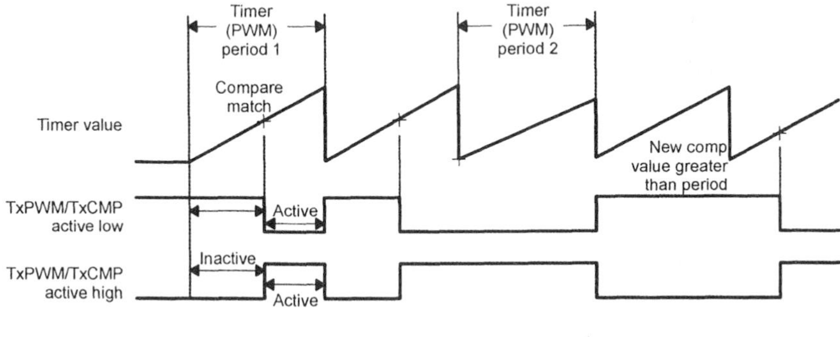

Figure 6.8 Asymmetric timer waveform generated by a GP timer in continuous up-count mode.

If the compare value is zero at the very beginning of the period, then a compare match is made at the very beginning and, consequently, the output is the active level for the period. If the output is "active" for the whole period and the new compare value for the next period is zero, then the output will stay at the active level so as not to cause a glitch. If the value in the compare register is greater than the value in the period register, then a compare match will never be made and consequently the output will be at the inactive level through the whole period.

The above allows the duty cycle of the PWM to range from 0 to 100% without glitches being present. If the compare value is the same as the period value, which causes a compare match, then the output pin will be at the active level for exactly one pre-scaled clock cycle.

Example 6.2 - Symmetric Waveform Generation: When the GP Timer is configured in continuous up/down-counting mode, a symmetric waveform is generated as in Fig. 6.9. The output changes in the following sequence:

1. "inactive level" before the counting operation starts
2. remains at "inactive level" until the compare match
3. toggles to "active level" on the first compare match

4. remains unchanged at "active level" until the second compare match
5. toggles to "inactive level" on the second compare match
6. remains unchanged at "inactive level" until the end of the period and remains unchanged until next compare match

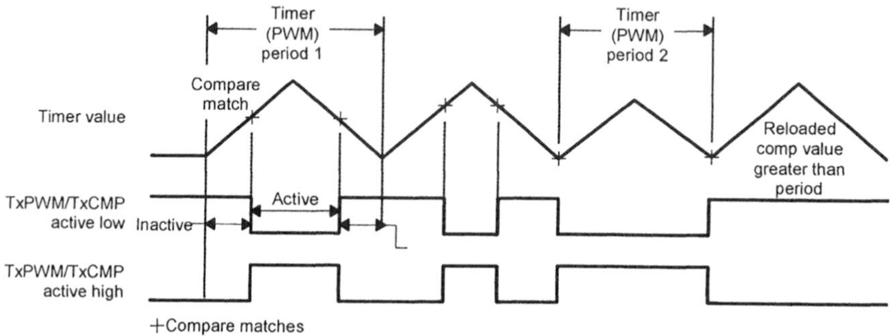

Figure 6.9 Symmetric timer waveform from continuous up/down count mode.

If the compare value is zero at the beginning of the period, the output is set to the active level at the beginning of a period and remains unchanged until the second compare match. After the first transition, the output remains at the active level until the end of the period if the compare value becomes zero for the second half of the period. When this happens, the output does not reset to zero if the new compare value for the following period is still zero.

This is done again to assure the generation of PWM pulses of 0% to 100% duty cycle without any glitches. The first transition does not happen if the compare value is greater than or equal to that of the period register for the first half of the period. However, the output still toggles when a compare match happens in the second half of the period. This error in output transition, often as a result of calculation error in the application routine, is corrected at the end of the period because the output resets to zero, unless the new compare value for the following period is zero. In this case, the output remains one, which again puts the output of the waveform generator in the correct state.

Calculations for Active and Inactive Time Periods

In order to utilize the GP Timer PWM outputs, it is sometimes necessary to calculate the active and inactive pulse times for the PWM output pins. We can find the active and inactive times for both the asymmetrical (Continuous Up-Count Mode) and symmetrical (Continuous Up/Down Count Mode). The calculation criteria for these times are as follows:

Continuous Up-Count Mode:

Active Output Pulse Time = [(TxPR) – (TxCMPR) + 1] cycles of the scaled input clock.

Inactive Output Pulse Time = (period of the scaled input clock) * (value of TxCMPR)

- When the value in TxCMPR is zero, the GP Timer compare output is active for the whole period.
- When TxCMPR is ≥ TxPR, the length of the active phase (the output pulse width) is zero.

Continuous Up/Down Counting Mode:

For the continuous up-/down-counting mode, the compare register can have different values while counting down and while counting up.

Active Output Pulse Time = [(TxPR) – (TxCMPR)**up** + (TxPR) – (TxCMPR)**dn**]** cycles of the scaled input clock

- If (TxCMPR) **up** is zero, the compare output is active at the beginning of the period. If (TxCMPR)dn is also zero, then output remains active until the end of the period.
- When (TxCMPR)**up** is ≥ (TxPR), the first transition is lost. Similarly, the second transition is lost when (TxCMPR)**dn** is ≥ (TxPR).
- If both (TxCMPR)up and TxCMPR)dn are greater than or equal to (TxPR), then the GP Timer compare output is inactive for the entire period.

**where (TxCMPR)up is the compare value on the timer's way up and (TxCMPR)dn is the compare value on the way down.

GP Timer PWM Generation -Practical Steps

To generate a PWM output signal on the GP Timer PWM outputs, make sure the following are configured to allow for PWM generation (also see Example 6.3):

1. Note what the PLL module is set to. The PLL provides the clock signal to the DSP and hence to the EV. In the timer control registers we have the option of pre-scaling (dividing) the clock signal to choose a time base for the GP Timers.
2. The corresponding EV pins need to be configured for their primary function in the appropriate MCRx register.
3. Initialize TxCNT (we usually set the count vale to zero)
4. Set TxPR according to the desired PWM (carrier) period. The TxPR value is calculated by the following formulas:

Asymmetric PWM:

$$\text{TxPR Value} = \left[\frac{\textit{desired PWM period}}{\textit{GP Timer prescaled clk period}} - 1\right] \tag{6.1}$$

Symmetric PWM:

$$\text{TxPR Value} = \left[\frac{desired\ PWM\ period}{2*(GP\ Timer\ prescaled\ clk\ period)}\right] \tag{6.2}$$

5. Initialize TxCMPR to first desired compare value
6. To create a PWM signal, the registers GPTCONA/B and TxCON need to be configured for TxCMP enabled, desired counting mode etc.
7. To create an asymmetric PWM signal, the timer is set to the Continuous-Up Count Mode. If a symmetric PWM signal is desired, then the Timer should be set to the Continuous-Up/Down Mode.
8. During run time, the GP Timer compare register (TxCMPR) will need to be periodically updated with new compare values corresponding to the modulation signal or new duty cycle. This can be done during an interrupt service routine.

Example 6.3 - Fixed Duty Cycle PWM

The following block of code is an example of generating a simple fixed-duty cycle PWM signal by using the GP Timer Compare function. The PLL needs to be set to CLKIN x 4, the watchdog needs to be disabled, and the wait state generator (WSGR) set for zero wait states.

```
LDP     #SCSR1>>7
        SPLK    #000Ch,SCSR1     ;EVA & EVB modules clock enable
        LDP     #0E1h            ;Set Mux pins for
        SPLK    #0FFFFh,MCRA     ;PWM function
        SPLK    #0FFFFh,MCRC     ;EVA PWM output initialization
        LDP     #GPTCONA >> 7h   ;Load EVA data-page
        SPLK    #00000h, T1CNT   ;this just zeros the counter T1 the
                                 ;counters are auto zeroed after a DSP
                                 ;reset
        SPLK    #0FFFFh, T1PR    ;the T1PR value sets the frequency in
                                 ;this case, it is 500 Hz cont up-cnt mod
        SPLK    #08000h, T1CMPR  ;50 % duty cycle PWM bits---
        SPLK    #0000000001000010b, GPTCONA
        SPLK    #1001000001000010b, T1CON
LOOP2   B       LOOP2            ;after the control registers are setup
                                 ;the program can loop endlessly while
                                 ;PWM is generated automatically
```

6.4 Compare Units

A PWM signal can also be generated using the compare unit (CMPRx). The compare units (CMPRx) in the LF2407 function identically to the GP Timer compare units (TxCMPR) discussed above. Unlike the GP Timer compare function, each compare unit has *two* associated PWM outputs which both toggle on the same compare match. The PWM outputs associated with the compare units allow for the generation of six PWM outputs per EV.

As shown in Fig. 6.10 the Compare Units Include:

- Three 16-bit compare registers (CMPR1, CMPR2, and CMPR3 for EVA; and CMPR4, CMPR5, and CMPR6 for EVB), all double-buffered
- One 16-bit compare control register (COMCONA for EVA, and COM-CONB for EVB)
- One 16-bit action control register (ACTRA for EVA, and ACTRB for EVB), with an associated buffer register
- Six PWM (3-state; Low, High, High Z) output (compare output) pins (PWMy, y = 1, 2, 3, 4, 5, 6 for EVA and PWMz, z = 7, 8, 9, 10, 11, 12 for EVB)

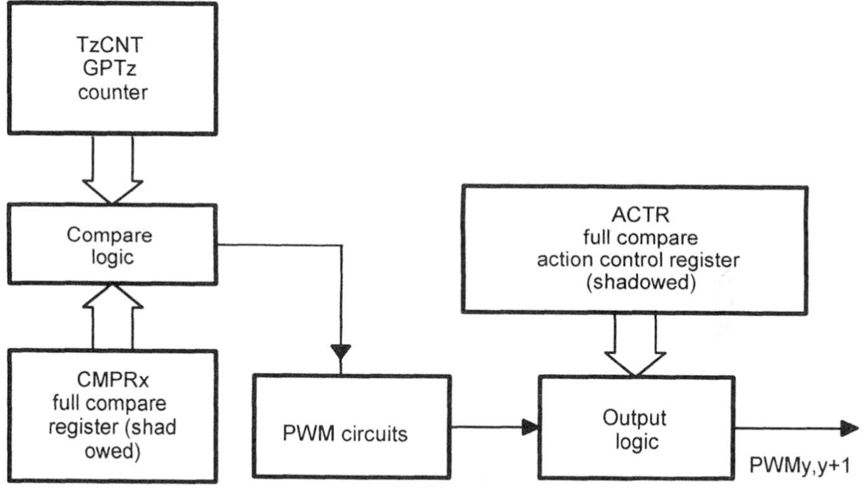

Figure 6.10 Compare unit block diagram.

For EVA: x = 1, 2, 3; y = 1, 3, 5; z = 1

For EVB: x = 4, 5, 6; y = 7, 9, 11; z = 3

6.4.1 *Inputs and Outputs of the Compare Units*

The inputs to a compare unit include:

- Control signals from compare control registers
- GP Timer 1/3 (T1CNT/T3CNT) count value, underflow, and period match signals
- System RESET
- The time base (counter value) for the compare units in EVA (CMPR1,2 ,3) is GP Timer 1, and for EVB (CMPR4, 5, 6) is GP Timer 3.

When any reset event occurs, all register bits associated with the compare units are reset to zero and all compare output pins are put in the high-impedance state.

The output of a compare unit is a compare match output, or in other words, a PWM output. If the compare operation is enabled, a compare match signal sets the corresponding interrupt flag and the two output pins associated with the compare unit to toggle. Either of the two outputs can be configured as either active high or active low, but will toggle upon the same event.

6.4.2 Operation of Compare Units

The sequence below is an example of the compare unit operation in EVA. For EVB operation, GP Timer 3 and ACTRB are used instead:

1. The value of the GP Timer 1 counter is continuously compared with that of the compare register.
2. When a compare match occurs, a transition appears on the two outputs of the compare unit according to the bits in the action control register (ACTRA). The bits in the ACTRA can individually specify each output to toggle active high or toggle active-low (if not forced high or low) on a compare match.
3. The compare interrupt flag associated with a compare unit is set when a compare match is made between GP Timer 1 and the compare register of a compare unit, if compare is enabled.
4. A peripheral interrupt request will then be generated if the interrupt is unmasked. The timing of output transitions, setting of interrupt flags, and the generation of interrupt requests are similar to the GP Timer compare operation.
5. The outputs of the compare units in compare mode are subject to modification by the output logic, dead band units, and the space vector PWM logic.

Having two outputs controlled by the same compare unit is useful in applications such as the control of a power inverter (see Fig. 6.11). With a power inverter, PWM signals can be used to gate the power transistors for creating currents through the legs of the inverter of any frequency or amplitude. This is useful in controlling electric motors their operation depends on the current flowing through the windings. By controlling the current flowing through motor windings, torque and speed control of the motor can be accomplished.

In inverter circuits such as those shown in Fig. 6.11, two power transistors are placed in series on each phase "leg" with the output being between them. This allows the output of the leg to be connected either to the DC supply voltage (Vdc) or ground. A potential hazard with these circuits is that if both transistors are turned on at the same time, a short circuit condition will exist through the leg and power transistors, causing the transistors to rapidly heat up and, in most cases, explode.

The solution to this problem is to make sure that only one transistor in each leg is on at a time. In theory, this is accomplished by feeding complementary PWM gating signals to each of the two transistors in a leg. So when one transistor is on, the other is off. In reality, all transistors turn on faster than they turn off. Therefore, it is necessary to add a time delay (dead-band) between the PWM signals to allow for the first transistor to fully turn off before the second one is turned on.

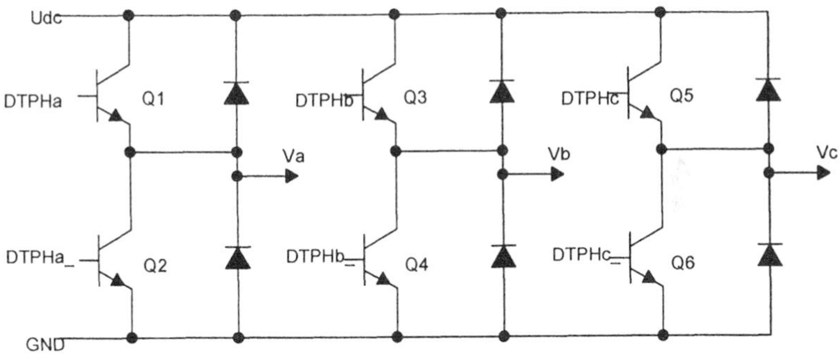

Figure 6.11 Basic three-phase inverter circuit.

6.4.3 *Dead Band Generation*

Unlike the GP Timer Compare PWM generation, the compare unit PWM outputs allow for a programmable dead band. Each EV on the LF2407 has its own programmable dead-band unit. The dead-band generators generate the dead-band delay between the toggling of the independent and dependent PWM outputs. Dead band solves the problem of inverter leg shoot through (short circuits). Figure 6.12 shows the interconnection between the dead band units and the compare units.

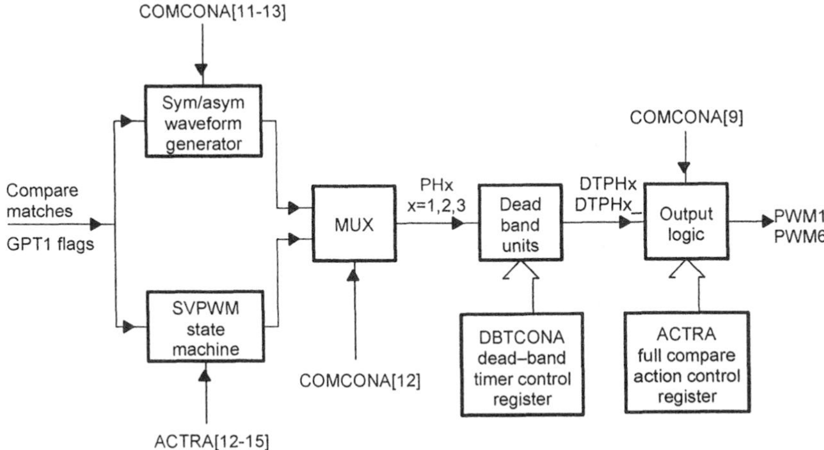

Figure 6.12 Block diagram of PWM outputs showing dead-band units.

Each programmable dead-band unit features:

- One 16-bit dead-band control register, DBTCONx (RW)
- One input clock prescaler: x/1, x/2, x/4, etc., to x/32
- Device (CPU) clock input
- Three 4-bit down-counting timers
- Control logic

Figures 6.13 and 6.14 illustrate the addition of a dead-band in both asymmetric and symmetric PWM outputs. The toggling sequence might go as follows: (1) toggle first output, (2) delay for a certain "dead-band" of clock cycles, (3) toggle the second output pin. This addition of a proper amount of dead-band prevents a short circuit across an inverter leg.

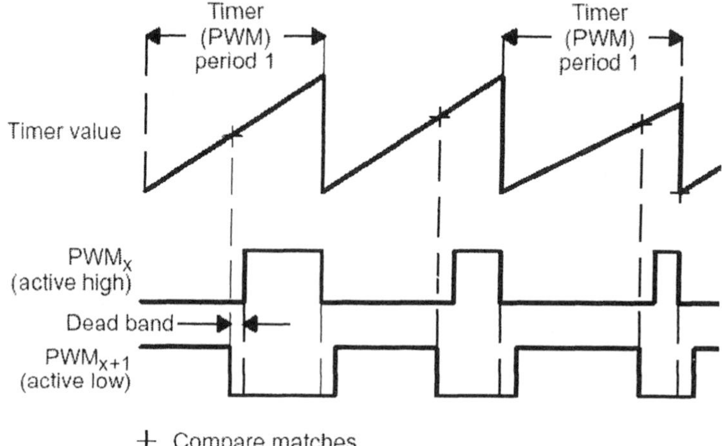

Figure 6.13 Dead-band with an asymmetric PWM output.

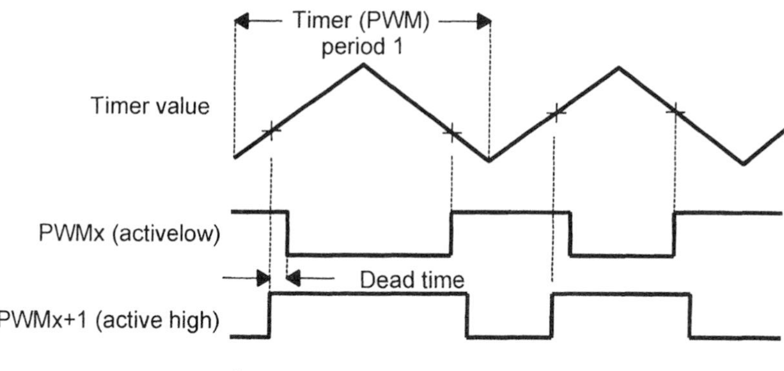

Figure 6.14 Dead-band with an asymmetric PWM output.

Depending on the switching device used, more or less dead-band might be needed. The use of dead-band should be experimented with when the inverter is supplied at a very low power level. This will ensure that if the current dead-band value is not sufficient, then the switching devices will not incur damage from the limited shoot through.

Table 6.4 lists the amounts of dead-band generated by the different bit combinations in DBTCONx. The values are based on a 25ns input clock signal. We can calculate the dead-band generated by the following simple formula:

$$\text{Dead Band (\# of CPU clock cycles)} = \frac{\textit{bits } [8 \rightarrow 11] \textit{ in DBTCONx}}{\textit{clock prescale value}}$$

Table 6.4 Dead-Band Values Generated by Bits [8 through 11] in DBTCONx Register

	(DBTCONx bits [4–2])					
DBTCONx bits [11–8]	110 and 1x1 (P=32)	100 (P=16)	011 (P=8)	010 (P=4)	001 (P=2)	000 (P=1)
0	0	0	0	0	0	0
1	0.8	0.4	0.2	0.1	0.05	0.025
2	1.6	0.8	0.4	0.2	0.1	0.05
3	2.4	1.2	0.6	0.3	0.15	0.075
4	3.2	1.6	0.8	0.4	0.2	0.1
5	4	2	1	0.5	0.25	0.125
6	4.8	2.4	1.2	0.6	0.3	0.15
7	5.6	2.8	1.4	0.7	0.35	0.175
8	6.4	3.2	1.6	0.8	0.4	0.2
9	7.2	3.6	1.8	0.9	0.45	0.225
A	8	4	2	1	0.5	0.25
B	8.8	4.4	2.2	1.1	0.55	0.275
C	9.6	4.8	2.4	1.2	0.6	0.3
D	10.4	5.2	2.6	1.3	0.65	0.325
E	11.2	5.6	2.8	1.4	0.7	0.35
F	12	6	3	1.5	0.75	0.375

Note: *Table values are given in μs.*

6.4.4 *Register Setup for Compare Unit Operation*

The following sequence should be used in setting up the Event Manager (EVA/B) for compare (PWM generation) operation:

EVA:

1. Set the T1PR for the desired period.
2. Configure ACTRA to select compare actions.
3. Configure DBTCONA, if dead band is to be used.
4. Initialize CMPRx to the first compare value.
5. Configure COMCONA for desired operation.
6. Configure T1CON to produce the desired operation for the time base and start the operation.
7. Load new compare values into CMPRx during program.

EVB:

1. Set the T3PR for the desired counting period.
2. Configure ACTRB to select compare actions.
3. Configure DBTCONA, if dead band is to be used.
4. Initialize CMPRx to the first compare value.
5. Configure COMCONB for desired operation.
6. Configure T3CON to produce the desired operation for the time base.
7. Load new compare values into CMPRx during program.

6.4.5 *Compare Unit Interrupts*

There is a maskable interrupt flag in EVIFRA and EVIFRC for each compare unit. If a compare operation is enabled, the interrupt flag of a compare unit is set one clock cycle after a compare match. A peripheral interrupt request will also be generated by the flag bit if the flag is unmasked.

6.4.6 *Data Memory Mapped Registers Associated with the Compare Units*

There are six main registers that control the functionality of the compare units on the LF2407: COMCONA/B, ACTRA/B, and DBTCONA/B. In addition to the control registers described in this section, the GP Timer registers should be thought of as being included because they provide the count value or time base in which the compare units operate.

Compare Control Registers (COMCONA and COMCONB)

These registers (COMCONA and COMCONB) control the operation of the compare units. They determine whether the compare operation is enabled, whether the compare outputs are enabled, the condition on which the compare registers are updated with the values in their buffer registers, and whether the Space Vector PWM (SVPWM) mode is enabled.

Compare Control Register A (COMCONA) — Address 7411h

15	14	13	12	11	10	9	8
CENABLE	CLD1	CLD0	SVENABLE	ACTRLD1	ACTRLD0	FCOMPOE	$\overline{PDPINTA}$ STATUS
RW–0	RW–0	RW–0	RW–0	RW–0	RW–0	RW–0	R–$\overline{PDPINTA}$ PIN

7-0
Reserved
R–0

Note: *R = read access, W = write access, -0 = value after reset.*

Bit 15 **CENABLE**. Compare enable.
- 0 Disables compare operation. All shadowed registers (CMPRx, ACTRA) become transparent
- 1 Enables compare operation

Bits14–13 **CLD1, CLD0**. Compare register CMPRx reload condition.
- 00 When T1CNT = 0 (that is, on underflow)
- 01 When T1CNT = 0 or T1CNT = T1PR (that is, on underflow or period match)
- 10 Immediately
- 11 Reserved; result is unpredictable

Bit 12 **SVENABLE**. Space vector PWM mode enable.
- 0 Disables space vector PWM mode
- 1 Enables space vector PWM mode

Bits 11–10 **ACTRLD1, ACTRLD0**. Action control register reload condition.
- 00 When T1CNT = 0 (on underflow)
- 01 When T1CNT = 0 or T1CNT = T1PR (on underflow or period match)
- 10 Immediately
- 11 Reserved

Bit 9 **FCOMPOE**. Compare output enable. Active PDPINTA clears this bit to zero.
- 0 PWM output pins are in high-impedance state; that is, they are disabled
- 1 PWM output pins are not in high-impedance state; that is, they are enabled

Bit 8 **PDPINTA STATUS**. This bit reflects the current status of the PDPINTA pin. (This bit is applicable to 240xA devices only — it is reserved on 240x devices and returns a zero when read.)

Bits 7–0 **Reserved**. Read returns zero; writes have no effect.

Compare Control Register B (COMCONB) — Address 7511h

15	14	13	12	11	10	9	8
CENABLE	CLD1	CLD0	SVENABLE	ACTRLD1	ACTRLD0	FCOMPOE	$\overline{PDPINTB}$ STATUS
RW–0	RW–0	RW–0	RW–0	RW–0	RW–0	RW–0	R–$\overline{PDPINTB}$ PIN
7-0							
Reserved							
R–0							

Note: R = read access, W = write access, -0 = value after reset.

Bit 15 **CENABLE**. Compare enable.

0 Disable compare operation. All shadowed registers (CMPRx, ACTRB) become transparent

1 Enable compare operation

Bits14–13 **CLD1, CLD0**. Compare register CMPRx reload condition.

00 When T3CNT = 0 (that is, on underflow)

01 When T3CNT = 0 or T3CNT = T3PR (that is, on underflow or period match)

10 Immediately

11 Reserved; result is unpredictable

Bit 12 **SVENABLE**. Space vector PWM mode enable.

0 Disables space vector PWM mode

1 Enables space vector PWM mode

Bits 11–10 **ACTRLD1, ACTRLD0**. Action control register reload condition.

00 When T3CNT = 0 (on underflow)

01 When T3CNT = 0 or T3CNT = T3PR (on underflow or period match)

10 Immediately

11 Reserved

Bit 9 **FCOMPOE**. Compare output enable. Active PDPINTB clears this bit to zero.

0 PWM output pins are in high-impedance state; that is, they are disabled

1 PWM output pins are not in high-impedance state; that is, they are enabled

Bit 8 **PDPINTB STATUS**. This bit reflects the current status of the PDPINTB pin. (This bit is applicable to 240xA devices only — it is reserved on 240x devices and returns a zero when read.)

Bits 7–0 **Reserved**. Read returns zero; writes have no effect.

Compare Action Control Registers (ACTRA and ACTRB)

The double buffered, compare action control registers (ACTRA and ACTRB) determine what action occurs on each of the six compare output pins when a compare event occurs (if the compare operation is enabled by COMCONx[15]). The compare output pins are PWMx, where x = 1–6 for ACTRA, and x = 7–12 for ACTRB. The condition on which ACTRA and ACTRB are reloaded is defined by the bits in COMCONx.

Compare Action Control Register A (ACTRA) — Address 7413h

15	14	13	12	11	10	9	8
SVRDIR	D2	D1	D0	CMP6ACT1	CMP6ACT0	CMP5ACT1	CMP5ACT0
RW–0	RW–0	RW–0	RW–0	RW–0	RW–0	RW–0	RW–0
7	6	5	4	3	2	1	0
CMP4ACT1	CMP4ACT0	CMP3ACT1	CMP3ACT0	CMP2ACT1	CMP2ACT0	CMP1ACT1	CMP1ACT0
RW–0	RW–0	RW–0	RW–0	RW–0	RW–0	RW–0	RW–0

Note: *R = read access, W = write access, -0 = value after reset.*

Bit 15 **SVRDIR**. Space vector PWM rotation direction. Used only in space vector PWM output generation.

0 Positive (CCW)
1 Negative (CW)

Bits 14–12 **D2–D0**. Basic space vector bits. Used only in space vector PWM output generation.

Bits 11–10 **CMP6ACT1–0**. Action on compare output pin 6, CMP6.

00 Forced low
01 Active low
10 Active high
11 Forced high

Bits 9–8 **CMP5ACT1–0**. Action on compare output pin 5, CMP5.

00 Forced low
01 Active low
10 Active high
11 Forced high

Bits 7–6 CMP4ACT1–0. Action on compare output pin 4, CMP4.

00	Forced low
01	Active low
10	Active high
11	Forced high

Bits 5–4 CMP3ACT1–0. Action on compare output pin 3, CMP3.

00	Forced low
01	Active low
10	Active high

Bits 3–2 CMP2ACT1–0. Action on compare output pin 2, CMP2.

00	Forced low
01	Active low
10	Active high
11	Forced high

Bits 1–0 CMP1ACT1–0. Action on compare output pin 1, CMP1.

00	Forced low
01	Active low
10	Active high
11	Forced high

Compare Action Control Register B (ACTRB) — Address 7513h

15	14	13	12	11	10	9	8
SVRDIR	D2	D1	D0	CMP12ACT1	CMP12ACT0	CMP11ACT1	CMP11ACT0
RW–0	RW–0	RW–0	RW–0	RW–0	RW–0	RW–0	RW–0
7	6	5	4	3	2	1	0
CMP10ACT1	CMP10ACT0	CMP9ACT1	CMP9ACT0	CMP8ACT1	CMP8ACT0	CMP7ACT1	CMP7ACT0
RW–0	RW–0	RW–0	RW–0	RW–0	RW–0	RW–0	RW–0

Note: *R = read access, W = write access, -0 = value after reset.*

Bit 15 SVRDIR. Space vector PWM rotation direction. Used only in space vector PWM output generation.

0	Positive (CCW)
1	Negative (CW)

Bits 14–12 D2–D0. Basic space vector bits. Used only in space vector PWM output generation.

Bits 11–10 CMP12ACT1–0. Action on compare output pin 12, CMP12.

00	Forced low
01	Active low

10 Active high
11 Forced high

Bits 9–8 CMP11ACT1–0. Action on compare output pin 11, CMP11.
00 Forced low
01 Active low
10 Active high
11 Forced high

Bits 7–6 CMP10ACT1–0. Action on compare output pin 10, CMP10.
00 Forced low
01 Active low
10 Active high
11 Forced high

Bits 5–4 CMP9ACT1–0. Action on compare output pin 9, CMP9.
00 Forced low
01 Active low
10 Active high
11 Forced high

Bits 3–2 CMP8ACT1–0. Action on compare output pin 8, CMP8.
00 Forced low
01 Active low
10 Active high
11 Forced high

Bits 1–0 CMP7ACT1–0. Action on compare output pin 7, CMP7.
00 Forced low
01 Active low
10 Active high
11 Forced high

Dead-Band Timer Control Register A (DBTCONA) — Address 7415h

15-12				11	10	9	8
Reserved				DBT3	DBT2	DBT1	DBT0
R–0				RW–0	RW–0	RW–0	RW–0
7	6	5	4	3	2	1-0	
EDBT3	EDBT2	EDBT1	DBTPS2	DBTPS1	DBTPS0	Reserved	
RW–0	RW–0	RW–0	RW–0	RW–0	RW–0	R–0	

Note: *R = read access, W = write access, -0 = value after reset.*

Bits 15–12 Reserved. Reads return zero; writes have no effect.

Bits 11–8 DBT3 (MSB)–DBT0 (LSB). Dead-band timer period. These bits define the period value of the three 4-bit dead-band timers.

Bit 7 **EDBT3**. Dead-band timer 3 enable (for pins PWM5 and PWM6 of Compare Unit 3).
0 Disable
1 Enable

Bit 6 **EDBT2**. Dead-band timer 2 enable (for pins PWM3 and PWM4 of Compare Unit 2).
0 Disable
1 Enable

Bit 5 **EDBT1**. Dead-band timer 1 enable (for pins PWM1 and PWM2 of Compare Unit 1).
0 Disable
1 Enable

Bits 4–2 DBTPS2 to DBTPS0. Dead-band timer prescaler.
000 x/1
001 x/2
010 x/4
011 x/8
100 x/16
101 x/32
110 x/32
111 x/32
x = Device (CPU) clock frequency

Bits 1–0 Reserved. Reads return zero; writes have no effect.

Dead-Band Timer Control Register B (DBTCONB) — Address 7515h

15-12				11	10	9	8
Reserved				DBT3	DBT2	DBT1	DBT0
R–0				RW–0	RW–0	RW–0	RW–0
7	6	5	4	3	2	1-0	
EDBT3	EDBT2	EDBT1	DBTPS2	DBTPS1	DBTPS0	Reserved	
RW–0	RW–0	RW–0	RW–0	RW–0	RW–0	R–0	

Note: *R = read access, W = write access, -0 = value after reset.*

Bits 15–12 Reserved. Reads return zero; writes have no effect.

Bits 11–8 **DBT3 (MSB)–DBT0 (LSB)**. Dead-band timer period. These bits define the period value of the three 4-bit dead-band timers.

Bit 7 **EDBT3**. Dead-band timer 3 enable (for pins PWM11 and PWM12 of Compare 6).
0 Disable
1 Enable

Bit 6 **EDBT2**. Dead-band timer 2 enable (for pins PWM9 and PWM10 of Compare 5).
0 Disable
1 Enable

Bit 5 **EDBT1**. Dead-band timer 1 enable (for pins PWM7 and PWM8 of Compare 4).
0 Disable
1 Enable

Bits 4–2 DBTPS2 to DBTPS0. Dead-band timer prescaler.
000 x/1
001 x/2
010 x/4
011 x/8
100 x/16
101 x/32
110 x/32
111 x/32
x = Device (CPU) clock frequency

Bits 1–0 Reserved. Reads return zero; writes have no effect.

6.5 Capture Units and Quadrature Encoded Pulse (QEP) Circuitry

The Capture Units on the LF2407 allow an event (rising/falling edge) on the capture pin to be time stamped by a selected GP Timer. There are three Capture Units in each EV, each with its own capture input pin (CAPx). Capture Units 1, 2, and 3 are associated with EVA while Capture Units 4, 5, and 6 are associated with EVB. Each EV module contains the following (shown in Figs. 6.14 and 6.15):

- One 16 bit capture control register per EV (CAPCOMA for EVA, CAPCOMB for EVB) is used for configuring the Capture Unit functionality.
- Three 16-bit, 2-level-deep First-In-First-Out (FIFO) stacks per EV (CAPxFIFO); one FIFO stack for each Capture Unit; the “captured” timer count value is stored here.
- One 16-bit capture status register (CAPFIFOA for EVA, CAPFIFOB for EVB); provides information on the number of timer captures in each capture FIFO.

- Inputs of either GP Timer 1 or 2 (for EVA) and GP Timer 3 or 4 (for EVB) as the time base.
- One capture pin per Capture Unit with user-specified transition detection (rising edge, falling edge, or both edges). CAP1 through CAP3 for EVA, and CAP4 through CAP6 for EVB.
- Six maskable interrupt flags, one for each Capture Unit.

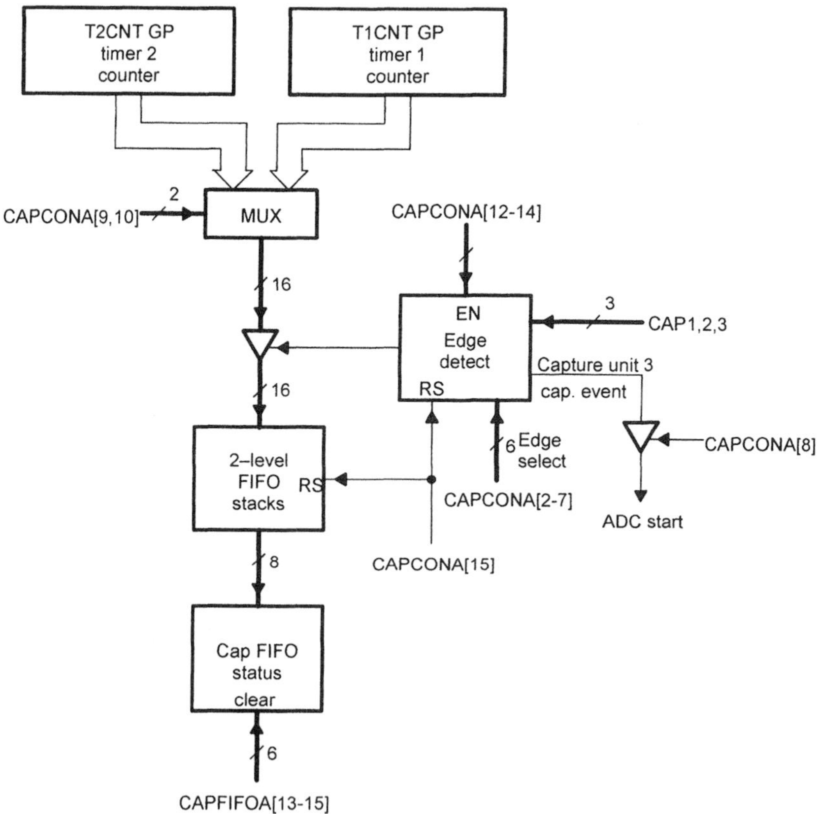

Figure 6.15 EVA capture unit diagram. (Courtesy of Texas Instruments)

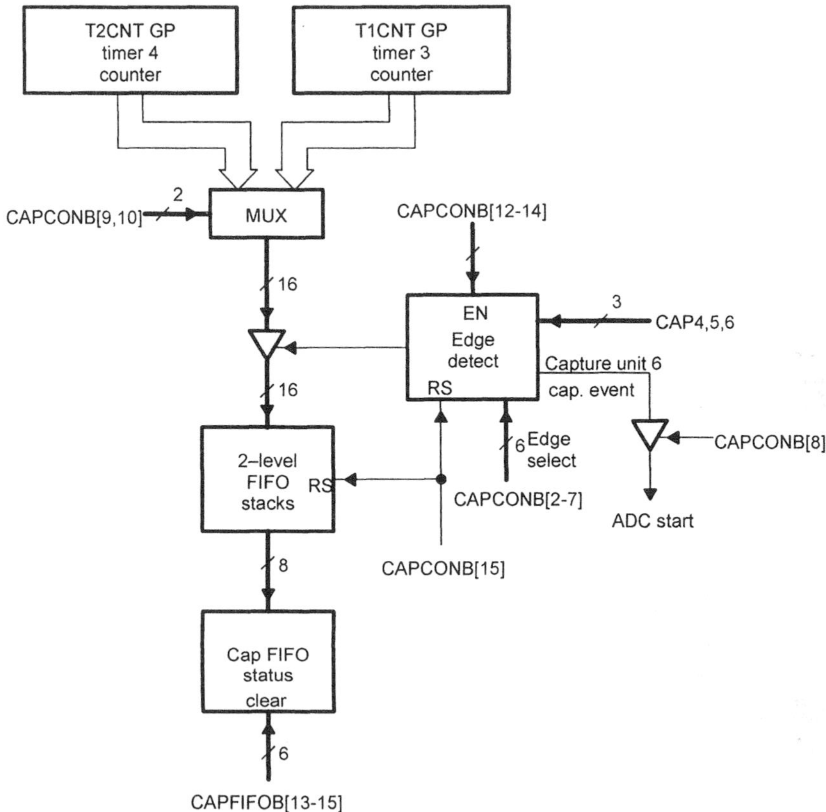

Figure 6.16 EVB Capture unit diagram. (Courtesy of Texas Instruments)

The Capture Units are useful in applications where the time of an external trigger needs to be "captured". For example, if we want to measure the time between the rising edges of two pulses, we would configure the appropriate registers for capture operation on a specific capture pin. At each rising edge, the Capture Unit would then store the corresponding timer values. The user program could then subtract the second capture value from the first value and determine the time between the pulses.

The Capture Units are accompanied by the Quadrature Encoded Pulse (QEP) circuitry which uses the GP Timers to "decode" a QEP signal. When the QEP mode is selected, pins CAP1 and CAP2 (CAP4 and CAP5 in case of EVB) are used as QEP inputs. More on the QEP circuitry will be discussed shortly.

6.5.1 Operation of the Capture Unit

When a Capture Unit is enabled, when either a rising or falling edge is detected on the capture input pin (CAPx), the current value of the selected GP Timer counter is copied and stored in the corresponding capture FIFO. In order for a transition to be captured, the input must hold at its current level for the duration of at least two CPU clock cycles. After the GP Timer value is recorded in the capture FIFO, an interrupt could also be generated, and software may then read the FIFO value. The value from the capture FIFO can then be used in an algorithm.

While we can think of the capture FIFO as being a two-level deep single register, each capture FIFO stack actually consists of two registers, CAPxFIFO and CAPxBOT (for EVA x=1,2,3; EVB x=4,5,6). When a new value is stored in the FIFO during a capture, in reality it first goes to the bottom register. When the top register of the FIFO stack is empty (either because this is the first capture, or the FIFO was just read), the value in the bottom register is automatically shifted into the top register (CAPxFIFO). Because of the above operation, when reading from the FIFO, the FIFO will always return the oldest stored value first. When the FIFO contains two values and is read, the oldest value will be read and removed from the FIFO. On the next read, the next oldest value will be read and removed from the FIFO. Usually, only read from the CAPxFIFO register, but the bottom register of the stack (CAPxBOT) can also be read.

When no FIFO reads have been performed, after two captures the corresponding capture FIFO will have two timer values stored in it and will be full. In the case that the FIFO has still not been read from and a third capture is recorded, the first capture value will be pushed out of the FIFO and lost.

The bits in the FIFO status registers indicate how many values are currently stored in each FIFO. When a value is read from the CAPxFIFO (or bottom register, CAPxBOT), the status bits will indicate one less value in the FIFO. The two status bits corresponding to a particular FIFO should normally indicate “00”, “01”, and “10”. If a third capture occurs and the previous two values have not been read from the FIFO, the status bits will indicate “11”, indicating that the oldest value was lost. In this case, after the next FIFO read, the status bits return to their usual values of “00”, “01” or “10”.

The following steps should be taken to configure the Capture Units for operation:

1. Initialize the CAPFIFOx and clear the appropriate status bits.
2. Set the selected GP Timer in the desired counting mode.
3. Set the associated GP Timer compare register or GP Timer period register, if necessary.
4. Set up CAPCONA or CAPCONB as appropriate for desired operation.

6.5.2 Capture Stack Interrupt Flag Operation:

Because the FIFO stack is two levels deep, the corresponding interrupt flag is set as soon as there are two values in the stack (the FIFO is full). This means that if there is one (or two) previous values in the FIFO (indicated by CAPxFIFO bits not equal to zero) and another capture takes place, the interrupt flag will be set. Like all interrupts, if the flag is unmasked, then an interrupt is generated. If interrupt operation is not desired, either the interrupt flag or the status bits can be polled continuously to determine if capture events have occurred.

6.5.3 Quadrature Encoded Pulse (QEP) Circuitry

QEPs are two sequences of pulses which have a variable frequency and are 90° out of phase with one another (see Fig. 6.17). QEP signals are usually generated by a position/speed sensing device such as a rotary optical encoder. When the encoder is rotated, the direction of rotation can be determined by which sequence of pulses leads the other. Rotational speed and position can be determined from the count and frequency of the pulses.

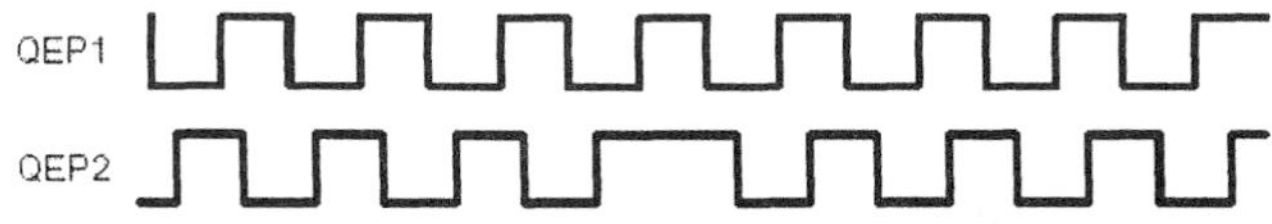

Figure 6.17 A pair of quadrature encoded pulses.

Each EV module has a QEP circuit associated with the Capture Units (see Figs. 6.18 and 6.19). The QEP circuit, when enabled, decodes and counts the quadrature encoded input pulses on the QEP input pins. The input pins consist of CAP1/QEP1 and CAP2/QEP2 for EVA or CAP4/QEP3 and CAP5/QEP4 for EVB. When the QEP function is enabled, the compare function of the pins is disabled and the pins are configured for QEP input.

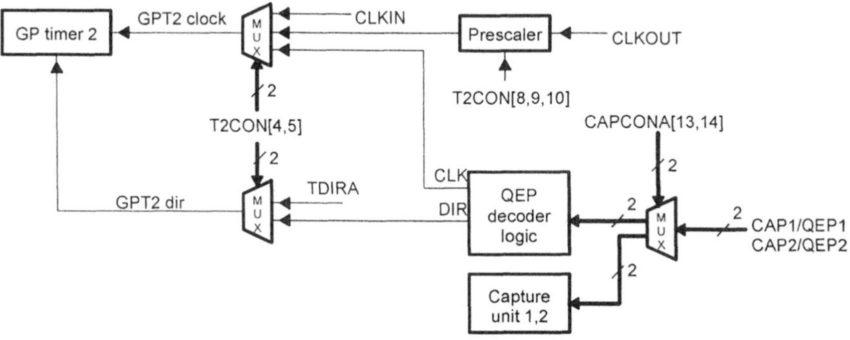

Figure 6.18 QEP circuit block for EVA. (Courtesy of Texas Instruments)

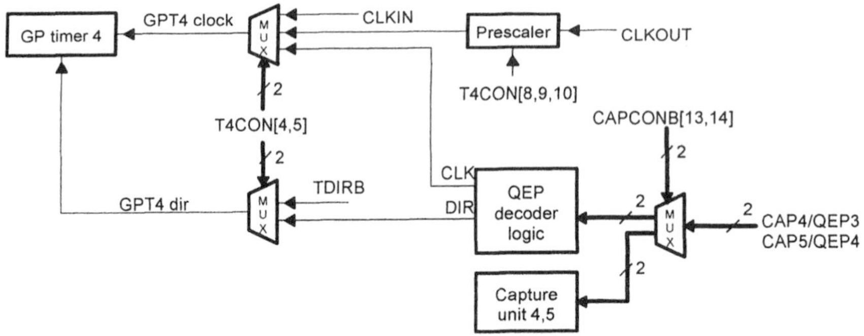

Figure 6.19 QEP circuit for EVB. (Courtesy of Texas Instruments)

QEP Circuit Operation

The counter for the QEP circuit is provided by GP Timer 2 for EVA and GP Timer 4 for EVB. The GP Timer must be configured for directional-up/down count mode. When the QEP circuit is selected as the clock source, the timer ignores the direction and clock (TDIRA/B and TCLKINA/B) input pins. The QEP circuit will act as the clock reference and the direction input for the timer. The QEP circuit determines which one of the sequences is the leading sequence. It then generates a direction signal as the direction input to the GP Timer. The timer counts up if CAP1/QEP1 (CAP4/QEP3 for EVB) input is the leading sequence, and counts down if CAP2/QEP2 (CAP5/QEP4 for EVB) is the leading sequence. Both edges of the pulses of the two quadrature encoded inputs are counted by the QEP circuit. Therefore, the frequency of the clock generated by the QEP logic to GP Timer 2 (or 4) is four times that of each input sequence. This quadrature clock is connected to the clock input of GP Timer 2 (or 4).

Note: *Upon a DSP RESET, the QEP logic will miss the first QEP edge.*

Configuring for QEP Operation:

EVA:

1. Load GP Timer 2's counter, period, and compare registers if desired; for simple QEP decoding, this is not required.
2. Configure T2CON to set GP Timer 2 in directional-up/down mode with the QEP circuits as clock source, and enable the selected timer.
3. Configure CAPCONA to enable the QEP circuit.

EVB:

1. Load GP Timer 4's counter, period, and compare registers with desired values; for simple QEP decoding, this is not required.
2. Configure T4CON to set GP Timer 4 in directional-up/down mode with the QEP circuits as clock source, and enable the selected timer.
3. Configure CAPCONB to enable the QEP circuit.

Interrupt flags normally associated with the timer operation are still operational with the QEP. Period, underflow, overflow, and compare interrupt flags for a GP Timer with a QEP circuit clock are generated on respective matches. If the respective interrupt flags are unmasked, timer interrupt requests will be generated.

6.5.4 Capture Unit / QEP Control Registers

Upon a RESET, all capture registers are cleared to zero. There are four 16-bit registers that control the functionality of the Capture Units. These registers are CAPCONA, CAPCONB, CAPFIFOA, and CAPFIFOB. In addition to these four registers the individual timer control registers (TxCON, x = 1, 2, 3, or 4) control the selected timer which acts as the time base for the Capture Unit. CAPCONA and CAPCONB also control the QEP functionality.

Capture Control Register A (CAPCONA) — Address 7420h

15	14-13	12	11	10	9	8
CAPRES	CAPQEPN	CAP3EN	Reserved	CAP3TSEL	CAP12TSEL	CAP3TOADC
RW–0	RW–0	RW–0	R–0	RW–0	RW–0	RW–0

7-6	5-4	3-2	1-0
CAP1EDGE	CAP2EDGE	CAP3EDGE	Reserved
RW–0	RW–0	RW–0	R–0

Note: *R = read access, W = write access, -0 = value after reset.*

Bit 15 **CAPRES**. Capture reset. Always reads zero.

Note: This bit is not implemented as a register bit. Writing a 0 simply clears the capture registers.

0 Clear all registers of Capture Units and QEP circuit to 0

1 No action

Bits 14–13 **CAPQEPN**. Capture Units 1 and 2 control.

00 Disables Capture Units 1 and 2; FIFO stacks retain their contents

01 Enables Capture Units 1 and 2

10 Reserved

11 Reserved

Bit 12 **CAP3EN**. Capture Unit 3 control.
- 0 Disables Capture Unit 3; FIFO stack of Capture Unit 3 retains its contents
- 1 Enable Capture Unit 3

Bit 11 **Reserved**. Reads return zero; writes have no effect.

Bit 10 **CAP3TSEL**. GP Timer selection for Capture Unit 3.
- 0 Selects GP Timer 2
- 1 Selects GP Timer 1

Bit 9 **CAP12TSEL**. GP Timer selection for Capture Units 1 and 2.
- 0 Selects GP Timer 2
- 1 Selects GP Timer 1

Bit 8 **CAP3TOADC**. Capture Unit 3 event starts ADC.
- 0 No action
- 1 Starts ADC when the CAP3INT flag is set

Bits 7–6 **CAP1EDGE**. Edge detection control for Capture Unit 1.
- 00 No detection
- 01 Detects rising edge
- 10 Detects falling edge
- 11 Detects both edges

Bits 5–4 **CAP2EDGE**. Edge detection control for Capture Unit 2.
- 00 No detection
- 01 Detects rising edge
- 10 Detects falling edge
- 11 Detects both edges

Bits 3–2 **CAP3EDGE**. Edge detection control for Capture Unit 3.
- 00 No detection
- 01 Detects rising edge
- 10 Detects falling edge
- 11 Detects both edges

Bits 1–0 **Reserved**. Reads return zero; writes have no effect.

Capture Control Register B (CAPCONB) — Address 7520h

15	14-13	12	11	10	9	8
CAPRES	CAPQEPN	CAP6EN	Reserved	CAP6TSEL	CAP45TSEL	CAP6TOADC
RW–0	RW–0	RW–0	R–0	RW–0	RW–0	RW–0

7-6	5-4	3-2	1-0
CAP4EDGE	CAP5EDGE	CAP6EDGE	Reserved
RW–0	RW–0	RW–0	RW–0

***Note:** R = read access, W = write access, -0 = value after reset.*

Bit 15 **CAPRES**. Capture reset. Always reads zero.
Note: This bit is not implemented as a register bit. Writing a 0 simply clears the capture registers.
- 0 Clears all registers of Capture Units and QEP circuit to 0
- 1 No action

Bits 14–13 **CAPQEPN**. Capture Units 4 and 5 and QEP circuit control.
- 00 Disables Capture Units 4 and 5 and QEP circuit. FIFO stacks retain their contents
- 01 Enables Capture Units 4 and 5, disable QEP circuit
- 10 Reserved
- 11 Enables QEP circuit. Disable Capture Units 4 and 5; bits 4–7 and 9 are ignored

Bit 12 **CAP6EN**. Capture Unit 6 control.
- 0 Disables Capture Unit 6; FIFO stack of Capture Unit 6 retains its contents
- 1 Enables Capture Unit 6

Bit 11 **Reserved**. Reads return zero; writes have no effect.

Bit 10 **CAP6TSEL**. GP Timer selection for Capture Unit 6.
- 0 Selects GP Timer 4
- 1 Selects GP Timer 3

Bit 9 **CAP45TSEL**. GP Timer selection for Capture Units 4 and 5.
- 0 Selects GP Timer 4
- 1 Selects GP Timer 3

Bit 8 **CAP6TOADC**. Capture Unit 6 event starts ADC.
- 0 No action
- 1 Starts ADC when the CAP6INT flag is set

Bits 7–6 CAP4EDGE. Edge detection control for Capture Unit 4.

00	No detection
01	Detects rising edge
10	Detects falling edge
11	Detects both edges

Bits 5–4 CAP5EDGE. Edge detection control for Capture Unit 5.

00	No detection
01	Detects rising edge
10	Detects falling edge
11	Detects both edges

Bits 3–2 CAP6EDGE. Edge detection control for Capture Unit 6.

00	No detection
01	Detects rising edge
10	Detects falling edge
11	Detects both edges

Bits 1–0 Reserved. Reads return zero; writes have no effect.

Capture Status Registers

The ability to write to the CAPFIFOx registers can be used as a programming advantage. For example, if a "01" is written to the CAPnFIFO bits by user code, the EV module is led to believe that there is already an entry in the FIFO. Subsequently, every time the FIFO gets a new value, a capture interrupt will be generated. If a write occurs to the CAPnFIFOA status bits at the same time as they are being updated by hardware (because of a capture event), the user written data takes precedence.

Capture FIFO Status Register A (CAPFIFOA) — Address 7422h

15-14	13-12	11-10	9-8
Reserved	CAP3FIFO	CAP2FIFO	CAP1FIFO
R-0	RW-0	RW-0	RW-0

7-0
Reserved
R-0

Note: *R = read access, W = write access, -0 = value after reset.*

Bits 15–14 Reserved. Reads return zero; writes have no effect.

Bits 13–12 CAP3FIFO. CAP3FIFO Status

00	Empty
01	Has one entry
10	Has two entries

11	Had two entries and captured another one; first entry has been lost

Bits 11–10 **CAP2FIFO**. CAP2FIFO Status

00	Empty
01	Has one entry
10	Has two entries
11	Had two entries and captured another one; first entry has been lost

Bits 9–8 **CAP1FIFO**. CAP1FIFO Status

00	Empty
01	Has one entry
10	Has two entries
11	Had two entries and captured another one; first entry has been lost

Bits 7–0 **Reserved**. Reads return zero; writes have no effect.

Capture FIFO Status Register B (CAPFIFOB) — Address 7522h

15-14	13-12	11-10	9-8
Reserved	CAP6FIFO	CAP5FIFO	CAP4FIFO
R–0	RW–0	RW–0	RW–0
7-0			
Reserved			
R–0			

Note: *R = read access, W = write access, -0 = value after reset.*

Bits 15–14 **Reserved**. Reads return zero; writes have no effect.

Bits 13–12 **CAP6FIFO**. CAP6FIFO Status

00	Empty
01	Has one entry
10	Has two entries
11	Had two entries and captured another one; first entry has been lost

Bits 11–10 **CAP5FIFO**. CAP5FIFO Status

00	Empty
01	Has one entry
10	Has two entries
11	Had two entries and captured another one; first entry has been lost

Bits 9–8 **CAP4FIFO**. CAP4FIFO Status

00	Empty
01	Has one entry
10	Has two entries
11	Had two entries and captured another one; first entry has been lost

Bits 7–0 **Reserved**. Reads return zero; writes have no effect.

6.6 General Event Manager Information

Table 6.5 Event Manager A (EVA) Pins

Pin Name	Description
CAP1/QEP1	Capture Unit 1 input, QEP circuit input 1
CAP2/QEP2	Capture Unit 2 input, QEP circuit input 2
CAP3	Capture Unit 3 input
PWM1	Compare Unit 1 output 1
PWM2	Compare Unit 1 output 2
PWM3	Compare Unit 2 output 1
PWM4	Compare Unit 2 output 2
PWM5	Compare Unit 3 output 1
PWM6	Compare Unit 3 output 2
T1CMP/T1PWM	Timer 1 compare/PWM output
T2CMP/T2PWM	Timer 2 compare/PWM output
TCLKINA	External clock-in for timers in EVA (*when configured to operate from external clock*)
TDIRA	External timer direction input in EVA (*when timer is in directional up/down mode*)

Table 6.6 Event Manager B (EVB) Pins

Pin Name	Description
CAP4/QEP3	Capture Unit 4 input, QEP circuit input 3
CAP5/QEP4	Capture Unit 5 input, QEP circuit input 4
CAP6	Capture Unit 6 input
PWM7	Compare Unit 4 output 1
PWM8	Compare Unit 4 output 2
PWM9	Compare Unit 5 output 1
PWM10	Compare Unit 5 output 2
PWM11	Compare Unit 6 output 1
PWM12	Compare Unit 6 output 2
T3CMP/T3PWM	Timer 3 compare/PWM output
T4CMP/T4PWM	Timer 4 compare/PWM output
TCLKINB	External clock-in for timers in EVB (*when configured to operate from external clock*)
TDIRB	External timer direction input in EVB (*when timer is in directional up/down mode*)

NOTE: *Most of the EV pins are mapped with a second function. In order to use the EV, you must configure the appropriate pins to their EV function. For more information on how pin sharing works and how to configure pins refer to Chapter 4.*

Event Manager (EV) Register Addresses

Table 6.7 Addresses of EVA Timer Registers

Address	Register	Name	
7400h	GPTCONA	GP Timer control register A	Timer 1
7401h	T1CNT	Timer 1 counter register	Timer 1
7402h	T1CMPR	Timer 1 compare register	Timer 1
7403h	T1PR	Timer 1 period register	Timer 2
7404h	T1CON	Timer 1 control register	Timer 2
7405h	T2CNT	Timer 2 counter register	Timer 2
7406h	T2CMPR	Timer 2 compare register	Timer 2
7407h	T2PR	Timer 2 period register	Timer 2
7408h	T2CON	Timer 2 control register	Timer 2

Table 6.8 Addresses of EVB Timer Registers

Address	Register	Name	
7500h	GPTCONB	GP Timer control register B	Timer 3
7501h	T3CNT	Timer 3 counter register	Timer 3
7502h	T3CMPR	Timer 3 compare register	Timer 3
7503h	T3PR	Timer 3 period register	Timer 3
7504h	T3CON	Timer 3 control register	Timer 3
7505h	T4CNT	Timer 4 counter register	Timer 4
7506h	T4CMPR	Timer 4 compare register	Timer 4
7507h	T4PR	Timer 4 period register	Timer 4
7508h	T4CON	Timer 4 control register	Timer 4

Table 6.9 Addresses of EVA Compare Control Registers

Address	Register	Name
7411h	COMCONA	Compare control register
7413h	ACTRA	Compare action control register
7415h	DBTCONA	Dead-band timer control register
7417h	CMPR1	Compare register 1
7418h	CMPR2	Compare register 2
7419h	CMPR3	Compare register 3

Table 6.10 Addresses of EVB Compare Control Registers

Address	Register	Name
7511h	COMCONB	Compare control register
7513h	ACTRB	Compare action control register
7515h	DBTCONB	Dead-band timer control register
7517h	CMPR4	Compare register 4
7518h	CMPR5	Compare register 5
7519h	CMPR6	Compare register 6

Table 6.11 Addresses of EVA Capture Registers

Address	Register	Name
7420h	CAPCONA	Capture control register
7422h	CAPFIFOA	Capture FIFO status register
7423h	CAP1FIFO	Two-level-deep capture FIFO stack 1
7424h	CAP2FIFO	Two-level-deep capture FIFO stack 2
7425h	CAP3FIFO	Two-level-deep capture FIFO stack 3
7427h	CAP1FBOT	Bottom registers of FIFO stacks; allows most recent CAPTURE value to be read
7428h	CAP2FBOT	
7429h	CAP3FBOT	

Table 6.12 Addresses of EVB Capture Registers

Address	Register	Name
7520h	CAPCONB	Capture control register
7522h	CAPFIFOB	Capture FIFO status register
7523h	CAP4FIFO	Two-level-deep capture FIFO stack 4
7524h	CAP5FIFO	Two-level-deep capture FIFO stack 5
7525h	CAP6FIFO	Two-level-deep capture FIFO stack 6
7527h	CAP4FBOT	Bottom registers of FIFO stacks, allows most recent CAPTURE value to be read
7528h	CAP5FBOT	
7529h	CAP6FBOT	

Table 6.13 Addresses of EVA Interrupt Registers

Address	Register	Name
742Ch	EVAIMRA	Interrupt mask register A
742Dh	EVAIMRB	Interrupt mask register B
742Eh	EVAIMRC	Interrupt mask register C
742Fh	EVAIFRA	Interrupt flag register A
7430h	EVAIFRB	Interrupt flag register B
7431h	EVAIFRC	Interrupt flag register C

Table 6.14 Addresses of EVB Interrupt Registers

Address	Register	Name
752Ch	EVBIMRA	Interrupt mask register A
752Dh	EVBIMRB	Interrupt mask register B
752Eh	EVBIMRC	Interrupt mask register C
752Fh	EVBIFRA	Interrupt flag register A
7530h	EVBIFRB	Interrupt flag register B
7531h	EVBIFRC	Interrupt flag register C

6.7 Exercise: PWM Signal Generation

As discussed in the previous sections, there are two ways to generate a PWM signal on the LF2407: through the GP Timer compare operation, or the Compare Units. This exercise will allow you to use your knowledge of the LF2407 DSP to write code that will generate PWM signals on both the GP Timer and Compare Unit outputs.

Procedure:

1. Write a program that outputs a fixed duty cycle "PWM" on a GP Timer 2 compare pin. Create the program so that the period of the PWM signal is 1 kHz and the duty cycle (on time/period) is fixed at 75%. The information on the GP Timer compare operation in the previous section will be very useful in writing this code.
2. View the output (1 kHz fixed duty cycle signal) on the T1PWM/T1CMP/IOPB4 pin. The Spectrum Digital LF2407 EVM schematic will be helpful in determining the location of this pin connection on the EVM.
3. If available, connect this fixed duty cycle signal to a dc voltage converter and use it to control the speed of a dc motor by varying the duty cycle of the waveform.
4. Modify the above program to now create a sinusoidally modulated PWM signal on the GP Timer Compare pin. To do this, a sinusoidal look-up

table can be created separately and then included with the source code. To modulate the signal, the timer compare register needs to be repeatedly updated with the modulation signal at a desired rate for a particular sinusoidal output frequency.

5. Write another program that creates the sinusoidal PWM, but instead uses the Compare Units.
6. If available, connect the two PWM outputs of the compare unit to a power inverter and run a single-phase induction motor. Vary the speed of the motor by manually varying the magnitude and rate at which the compare registers are updated with the modulation signal. Maintain a constant voltage/frequency (V/Hz) ratio to the induction motor.

Chapter 7

DSP-BASED IMPLEMENTATION OF DC-DC BUCK-BOOST CONVERTERS

7.1 Introduction

In a large number of industrial applications, it is required to convert a dc voltage to a different dc voltage level, often with a regulated output. To perform this task, a dc-dc converter is needed. A dc-dc converter directly converts a dc voltage of one level to another. It can be used to step-down (buck), or step-up (boost) a dc voltage source. In this chapter, the DSP-based control of a buck-boost, a specific type of dc-dc converter, is explained.

7.2 Converter Structure

The buck-boost converter has the structure shown in Fig. 7.1. The principle of operation is that when the transistor T is turned on, the input voltage V_{in} is applied across the inductor L and the current i_L in the inductor rises. Then the transistor is turned off. The current in the inductor must continue to flow somehow, and consequently finds its path through the load resistor R, and back to the inductor through the diode D. This discharges the inductor, and the current through it decreases. The capacitor C filters the output voltage ripple. The description given in the above is with the continuous conduction mode, meaning the inductor current never goes discontinuous. The continuous mode will be discussed further in the next section.

This converter has two dominant characteristics: the output voltage is always negative with respect to the input voltage and the output voltage may be higher or lower than the input voltage. This is why this converter may also be referred to as a step-up/step-down converter.

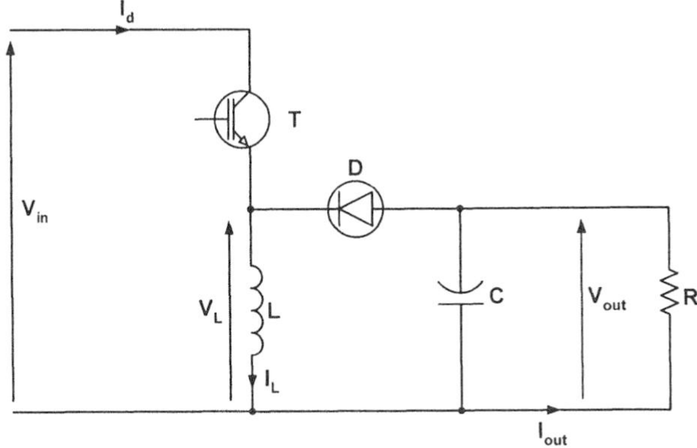

Figure 7.1 Buck-boost converter structure.

7.3 Continuous Conduction Mode

The input and output voltages are related by the following equation:

$$V_{out} = -\frac{d}{1-d} \times V_{in} \tag{7.1}$$

In this equation, d is the transistor or switch duty cycle. Figure 7.2 shows the switching pattern command to turn on or off, which must be fed to the transistor for proper operation of the buck-boost converter.

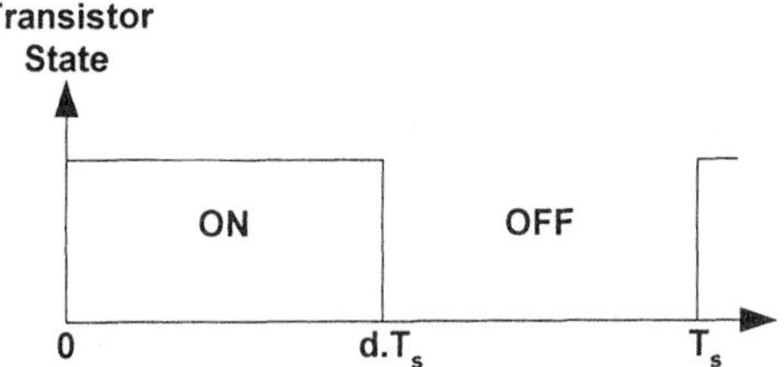

Figure 7.2 Transistor switching pattern.

Obviously, the duty cycle may vary only from 0 to 1. The resulting values for the converter voltage gain are:

$$d = 0 \Rightarrow G = \frac{V_{out}}{V_{in}} = 0 \tag{7.2}$$

$$d = 1 \Rightarrow G = \frac{V_{out}}{V_{in}} = -\infty \tag{7.3}$$

The theoretical gain range achievable is potentially very large. Practically, it is limited by the parasitic characteristics of the converter. In addition, it is often desirable to keep the duty cycle between 0.1 (10%) and 0.9 (90%) for practical engineering considerations. The relationship between the converter duty cycle and its gain, shown in Fig. 7.3, is non-linear.

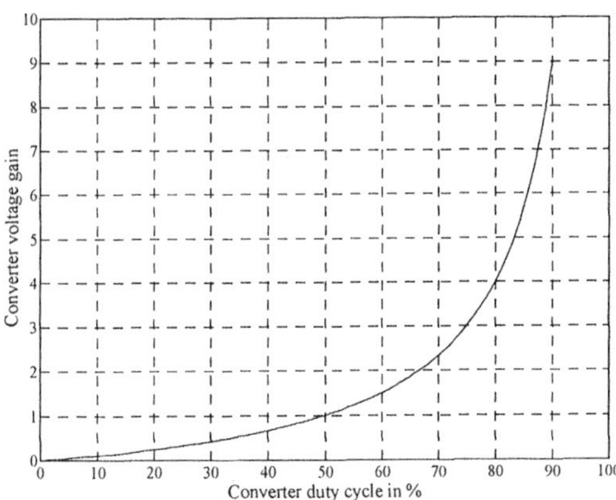

Figure 7.3 Converter voltage gain versus converter duty cycle.

7.4 Discontinuous Conduction Mode

The switching results in a cyclic current increase and decrease in the inductor. This current ripple has a non-negligible influence on the operation of the converter. If during the switching period T_S (shown in Fig. 7.2) the current never goes to zero, then the converter is said to operate in continuous conduction mode.

However, if the current does go to zero at any time, then the conduction is said to be discontinuous. In discontinuous conduction mode, the voltage gain of the converter is not solely a function of the duty cycle, but also of the output current. An example of a discontinuous conduction current waveform is shown in Fig. 7.4.

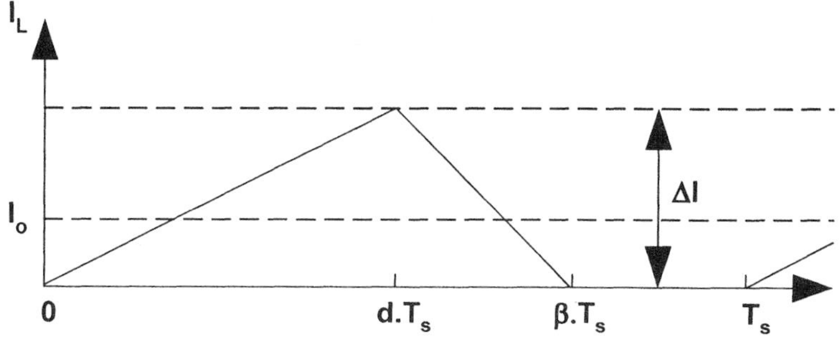

Figure 7.4 Discontinuous conduction mode current waveform.

7.5 Connecting the DSP to the Buck-Boost Converter

To fully control the buck-boost converter voltage and current with a DSP, one digital output and two analog inputs are required from the DSP. Figure 7.5 shows a conceptual connection diagram between the DSP and converter. The DSP will output a PWM switching waveform to the converter. The DSP will also receive information of the instantaneous current and voltage from the load via the analog to digital converter inputs. The following subsections describe the circuits necessary for interfacing the DSP to the dc-dc converter.

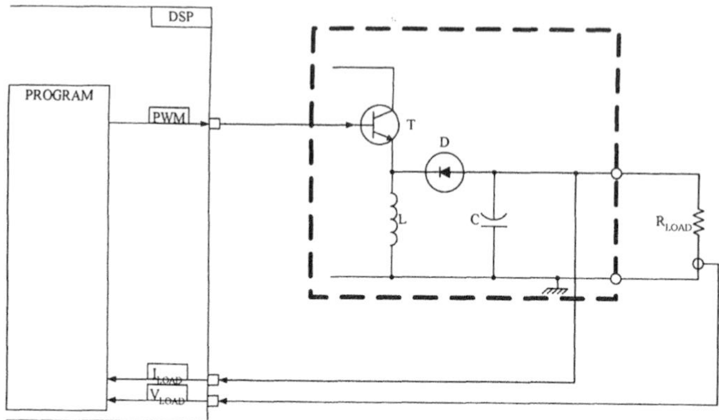

Figure 7.5 Physical implementation.

7.5.1 *Gate Driver*

The gate driver for this example is shown in Fig. 7.6 and is an integrated driver; it includes an opto-isolator, NPN transistor, PNP transistor, and the necessary logic to control them, all within an integrated package. Only the addition of two resistors is required to complete the gate driver circuit. The transistor that is being driven here is a MOSFET.

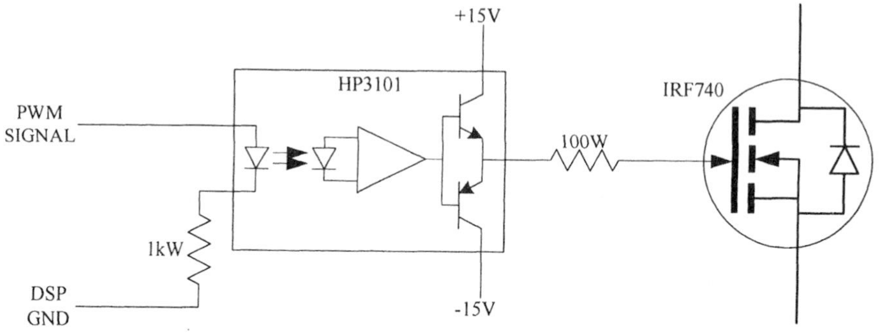

Figure 7.6 Gate driver circuit schematic.

7.5.2 *Current Sensor*

Current measurement can be performed using a shunt resistor in series with the output. This solution is more adapted to sensing small currents than a Hall-effect sensor and is also less expensive. The voltage across the 1Ω shunt resistor shown in Fig. 7.7 is buffered by a non-inverting amplifier, which provides infinite input impedance for the ADC input of the DSP. Due to the topology of the dc-dc converter, the output voltage is negative and must therefore be inverted. Because the shunt resistor is of a low value, the voltage across it will be small and must be amplified. A variable gain inverting amplifier provides for both these needs. The variable gain of the amplifier is used to adjust the gain of the sensed shunt resistor voltage signal. The output of the amplifier is connected to an opto-isolator, which changes the signal path from electrical to optical, then back to an electrical signal. The optical transmission provides the necessary galvanic insulation between the power side of the converter and the DSP. This isolation is necessary because the DSP is a very sensitive device, while the converter is a major source of voltage surges and interferences. The operational amplifier feeds the input, or luminescent diode, of the opto-isolator. The output of the opto-isolator feeds into the DSP analog input. A variable collector resistor is used to set the offset voltage of the ADC input.

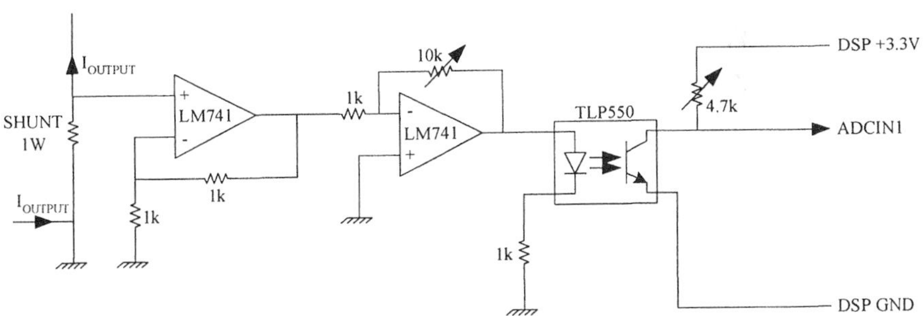

Figure 7.7 Current sensor circuit schematic.

7.5.3 *Voltage Sensor*

The voltage sensor uses the same circuit as the current sensor, but with a few differences. The output voltage of the converter is directly measurable and is directly fed into the infinite impedance buffer. The inverting amplifier is also slightly different in that it uses a 1kΩ resistor instead of a 10kΩ resistor. The difference is that while the current signal has to be amplified, the voltage level of the converter output must be attenuated to match the acceptable voltage range of the ADC input.

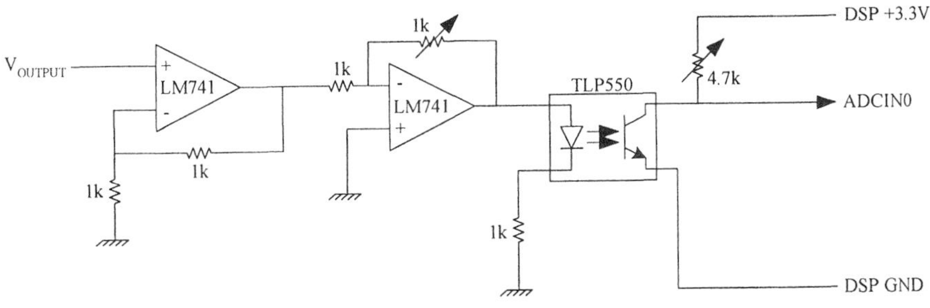

Figure 7.8 Voltage sensor circuit schematic.

7.6 Controlling the Buck-Boost Converter

The controller of a buck-boost converter usually has two objectives:

- Controlling the output voltage to a predetermined value
- Protecting the converter by limiting the output current to a predetermined value

Obviously, simply regulating the output voltage is the normal mode for the controller. If the load is such that the converter output voltage causes the current to go beyond the limit, then the controller must also control the converter to prevent the current from exceeding the maximum limit. Maintaining the current below the maximum is necessary in order to keep the converter and load from overheating. Because current regulation is necessary for safety, it must have the highest priority over all the other tasks in the control system, including the voltage regulation. This means that under any variation of the load, the current will be kept below its maximum.

The proposed control scheme has the following properties:

- The voltage will be regulated using a closed-loop PI regulator.

- The current will be checked every switching cycle, and if its value is above the limit, then the voltage regulation will be suspended. This will be achieved by setting the error signal to zero, which will disable the proportional action and disable the integration of the voltage error. The integrator is decremented by 1 every cycle in order to smoothly bring the voltage to a value, which will keep the current at its limit.

- The current regulation is effective if the current is only slightly above the limit. It is not effective against sudden surges such as a short circuit. If the current is above two times the maximum, then the controller will reset the PWM generation (thus shutting down the transfer of power from the source to the load) and reset the integrator. This will cause the converter restart from zero voltage in voltage regulation mode. Since the voltage is very low, the current will be slightly beyond the current limit, thus causing the controller to enter the current regulation mode. This is effective because output filter capacitor will be quickly discharged by a short circuit, and thus the output voltage will be zero. The voltage necessary for keeping the current below the limit in a short circuit condition is very low (in the order of a few mV), so the recovery from a short circuit should happen quickly.

Regulating the output voltage can be achieved easily by means of an integral regulator in a negative feedback loop. The integral regulator outputs a command of gain for the converter. For optimum performance, this command should be linearized; that is, converted into a useful duty cycle using the following equation:

$$d = \frac{Gain}{Gain + 1} \tag{7.4}$$

However, this is not mandatory because the negative feedback and the integral regulator ultimately ensure the convergence of the output voltage toward the reference. In addition, the DSP is very fast in reacting to variations in the system. Equation (7.4) is theoretical and does not take into account the parasitic elements of the system, which make the voltage gain a different function of the duty cycle. The theoretical voltage gain as a function of the duty cycle and the actual gain are plotted in Fig. 9. One sees that the two curves differ slightly. A look-up table would be the most accurate and simplest way to linearize the function. The voltage regulation is described using the block diagram shown in Fig. 7.10.

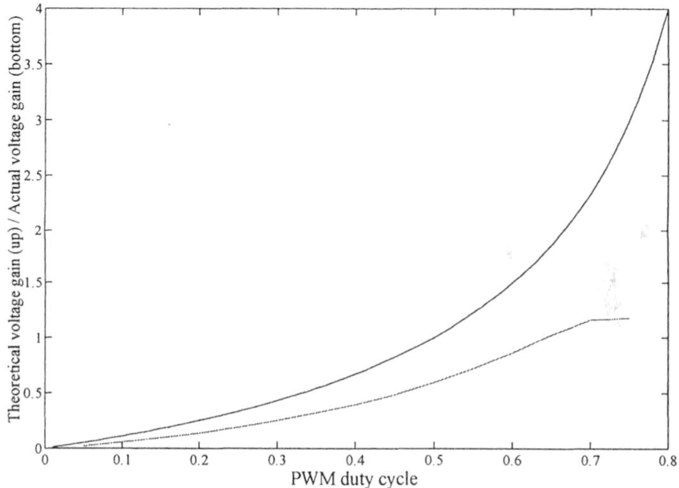

Figure 7.9 Divergence between theoretical and actual voltage gains.

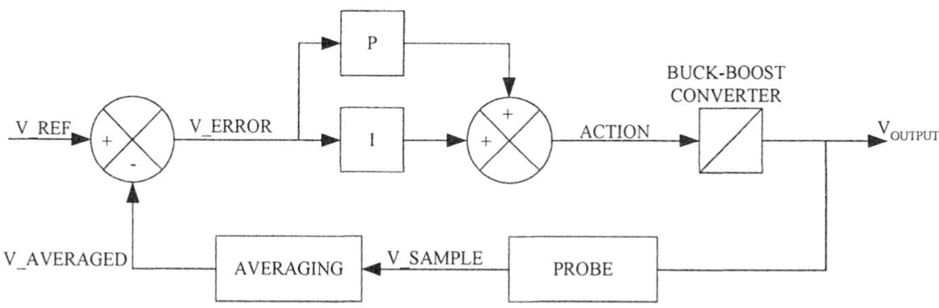

Figure 7.10 Voltage regulation block diagram.

This block diagram will be calculated once every switching cycle, which is the maximum speed at which parameters may be updated. Calculating the regulation more often would be useless because the actuator, the PWM generator, would not react until the next cycle.

The block diagram shown in Fig. 7.11 is implemented within the voltage regulation code in two sequences: current regulation and regulation reset.

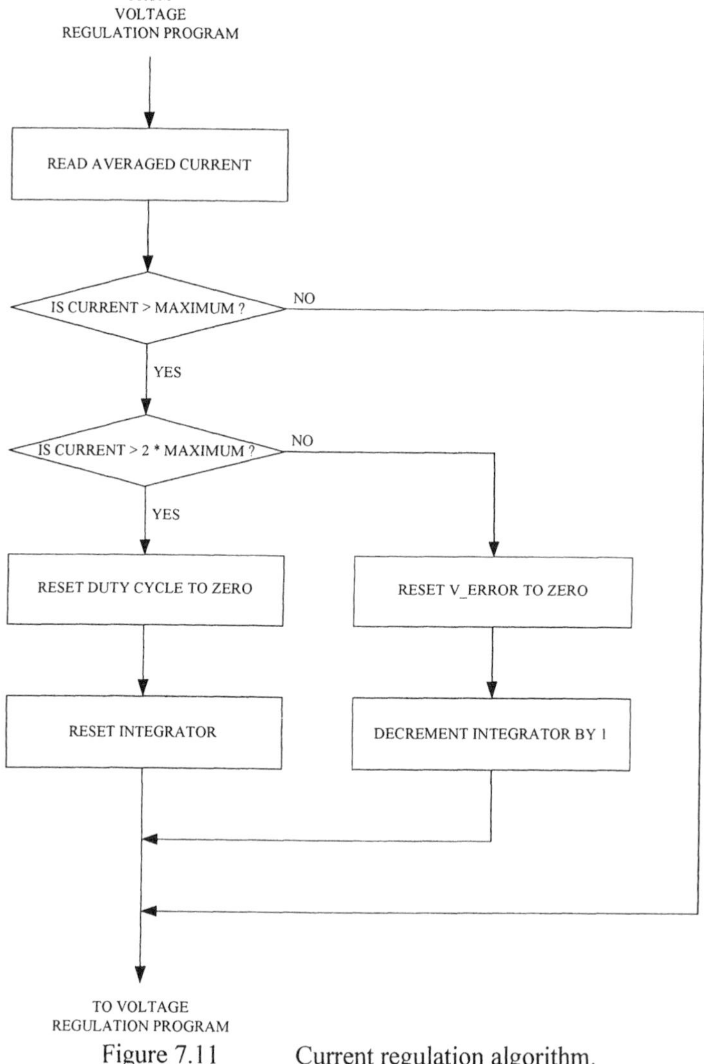

Figure 7.11 Current regulation algorithm.

Figure 7.12 illustrates the flow chart of the program developed in this chapter. Notice that several loops are used in the program.

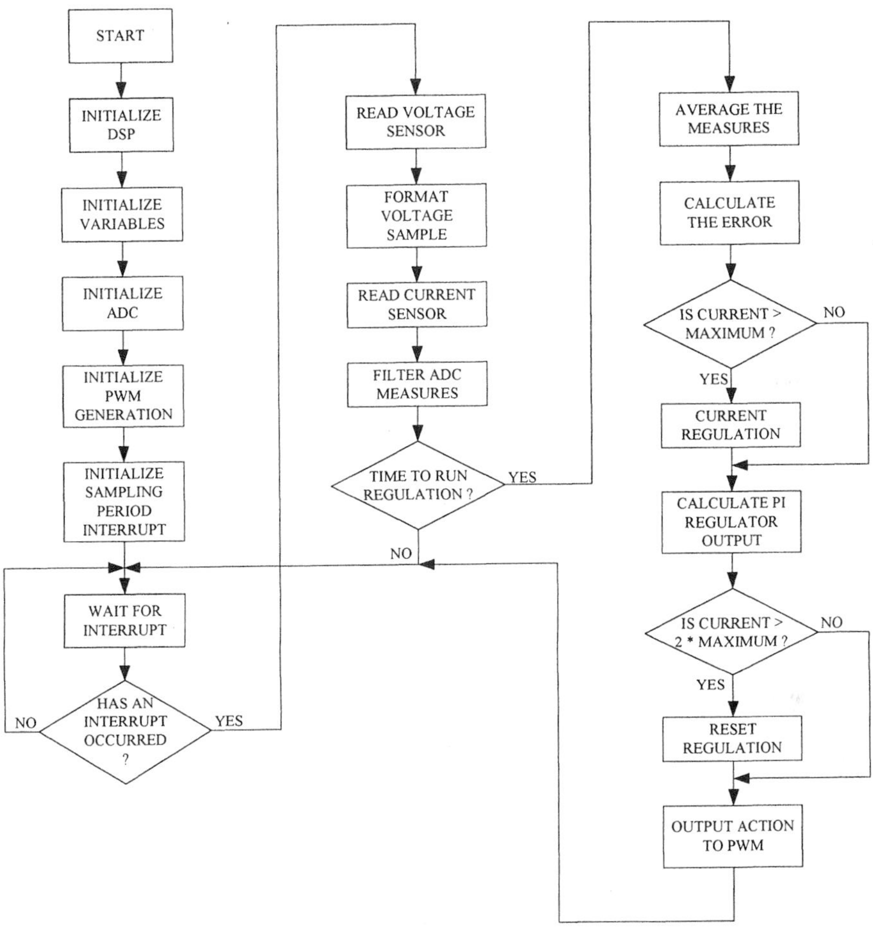

Figure 7.12 General program flow-chart.

7.7 Main Assembly Section Code Description

7.7.1 Variables Initialization

The block of code below initializes the defined variables with constants.

```
LDP    #06h
SPLK   #0900h,   V_OFFSET      ;Voltage probe offset * 2^6
SPLK   #0880h,   I_OFFSET      ;Current probe offset * 2^6
SPLK   #0000h,   V_SAMPLE      ;Voltage sample
SPLK   #0000h,   I_SAMPLE      ;Current sample
SPLK   #0000h,   V_SUM         ;Voltage sum
SPLK   #0000h,   I_SUM         ;Current sum
SPLK   #0000h,   V_AVERAGED    ;Voltage averaged
```

```
SPLK    #0000h,   I_AVERAGED   ;Current averaged
SPLK    #0138h,   V_REF        ;Voltage reference=5V=5000d/16d
SPLK    #0000h,   ACTION       ;Action output by the regulator
SPLK    #00FAh,   I_LIMIT      ;Current limit FAh = 250d= 25mA
SPLK    #0000h,   V_ERROR      ;Error of voltage regulation
SPLK    #0000h,   INTEGRAL     ;Regulation integrator
SPLK    #0000h,   INT_CNT      ;Interrupt counter
```

7.7.2 Initialization of the ADC

This code initializes the Analog-to-Digital Converter hardware of the LF2407 that performs the analog current and voltage sensing.

```
LDP     #ADCTRL1>>7h             ;Set data page corresponding to
                                 ;ADCcontrol registers
SPLK    #0100000000000000b, ADCTRL1
                                 ;Reset ADC
SPLK    #0011000000000000b, ADCTRL1
                                 ;Set ADC for Bit 6=0 start-stop
                                 ;mode
SPLK    #0001h,   MAXCONV        ;Set for 2 conversions (2
                                 ;channels)
SPLK    #0010h,   CHSELSEQ1      ;Set for conversion on channel 1
                                 ;and 0
SPLK    #0100000000000000b, ADCTRL2
                                 ;Reset sequencer
SPLK    #0000000100000000b, ADCTRL2
                                 ;Enables ADC to be started by an
                                 ;event(timer 2 period here)
```

7.7.3 Initialization of GP Timer 1 for PWM Generation

This code sets the parameters of the PWM waveform generation. The period is set for a 1kHz carrier frequency. The duty cycle is set to a near zero value.

```
LDP     #0E1h                    ;Set data page corresponding to
                                 ;GPIO pin registers
SPLK    #0FFFFh,  MCRA           ;Set GPIO pins for primary
                                 ;function (IO)
LDP     #GPTCONA>>7h             ;Set data page corresponding to
                                 ;general purpose timer control
                                 ;register
SPLK    #00000h,  T1CNT          ;Reset timer 1 counter
SPLK    #3FFFh,   T1PR           ;Set timer 1 period to ~= 1ms
SPLK    #3FF0h,   T1CMPR         ;Set duty cycle
```

7.7.4 Sampling Period Interrupt Initialization

This code initializes GP Timer 2 and the interrupt operation for the Timer 2 period interrupt.

```
        SPLK    #00000h,  T2CNT         ;Reset timer 2 counter
        LACL    T1PR                    ;Load timer 1 (PWM) period
                                        ;register
        SACL    T2PR,     4             ;Divide by 16 (right shift by 4
                                        ;bits) and store in timer 2
                                        ;period register
        SPLK    #0000010001000010b, GPTCONA
                                        ;Set general purpose timer
                                        ;control register for: Bit 10,9
                                        ;= 10 start ADC upon timer 2
                                        ;period occurrence, Bit 6=1
                                        ;enables timer 1 compare output
                                        ;for PWM generation, Bit 2,1=10
                                        ;sets output pin polarity high
        SPLK    #1000100001000010b, T1CON
                                        ;Sets timer 1 control register
                                        ;for: Bit 12,11 = 01 continuous
                                        ;up down count mode, Bit 6 = 1
                                        ;enables timer, Bit 1 = 1
                                        ;enables timer compare operation
        SPLK    #1000100001000000b, T2CON
                                        ;Sets timer 2 control register
                                        ;for: Bit 12,11=01 continuous up
                                        ;down count mode, Bit 6 = 1
                                        ;enables timer
        SPLK    #0000000000000001b, EVAIMRB
                                        ;Enables interrupt upon timer 2
                                        ;period occurrence
        SPLK    #0000000000000000b, EVAIFRB
                                        ;Reset corresponding interrupt
                                        ;flags
        LDP     #0h                     ;Set data page corresponding to
                                        ;registers
        SPLK    #0000000000000100b, IMR
                                        ;Enable level 3 (INT3)
                                        ;interrupts
        CLRC    INTM                    ;Enable interrupts
LOOP:   B       LOOP                    ;Program infinitely loops here
                                        ;while waiting for interrupts
```

7.8 Interrupt Service Routine

Once an interrupt has occurred, the algorithm will perform several tasks as shown in Fig. 7.12. After the sensor voltage and current values are obtained, the algorithm either returns to the main wait loop or branches to the regulation code.

The code sequence below is in charge of identifying what event caused the interrupt. Reading the PIVR register obtains contains the identification number of the occurring interrupt. If the PIVR number corresponds to the Timer 2 period match interrupt (#002Bh), then the DSP branches to the regulation code. Otherwise, it branches back to the wait loop (LOOP:) in the main code.

```
PERIOD:LDP    #PIVR>>7h              ;Set data page corresponding to
                                     ;PIVR register
       LACL   PIVR                   ;Load content of PIVR register
                                     ;to accumulator
       SUB    #002Bh                 ;Subtract number of timer 2
                                     ;period match interrupt
       BCND   REGULATION, EQ         ;If content matches, then branch
                                     ;to regulation code
       CLRC   INTM                   ;Otherwise clear interrupt mask
                                     ;to re-enable interrupts
       RET                           ;And return to wait loop in main
                                     ;code
```

7.8.1 Reading Voltage Sensors

This section of code contains protection against negative values that may occur because of physical sensor drift. A negative value must be eliminated. The probe offset is determined manually when physically connecting the DSP to the converter. The block of code below reads in the voltage from the ADC result register.

```
       LDP    #RESULT0>>7h           ;Set data page corresponding to
                                     ;ADC registers
       LACL   RESULT0                ;Load result register 1 content
                                     ;(i.e. current sample) to
                                     ;accumulator
       LDP    #06h                   ;Set data page corresponding to
                                     ;variables
       SUB    V_OFFSET               ;Subtract voltage probe offset
       BCND   S1, GEQ                ;If the result is positive or
                                     ;zero, then branch to proceed
                                     ;normally
       LACL   #0000h                 ;Otherwise, set result to zero
S1:    SACL   V_SAMPLE, 10           ;Right shift result by 10 bits
                                     ;and store it. The 6-bit shift
```

```
                                          ;is because the 6 LSBs of the
                                          ;result register are
                                          ;insignificant. The 4-bit shift
                                          ;is for formatting purposes.
```

7.8.2 Formatting the Voltage Sample

The voltage sample read from the A/D converter needs to be multiplied by the value 14d (=Eh) in order to change the value into the 16mV per digit format.

```
        LDP     #06h                      ;Set data page corresponding to
                                          ;variables
        LT      V_SAMPLE                  ;Load voltage sample to
                                          ;multiplier
        MPY     #00Eh                     ;Multiply by 14d
        SPL     V_SAMPLE                  ;Store 16 least significant bits
                                          ;of the result to V_SAMPLE
```

7.8.3 Reading the Current Sensors

This code is similar to the code that reads the voltage sensors, with the exception of the channel read, which is channel 1 here instead of channel 0. As with the voltage reading code, the code below reads the result register of the ADC that contains the result from the current measurement.

```
        LDP     #RESULT1>>7h              ;Set data page corresponding to
                                          ;ADC registers
        LACL    RESULT1                   ;Load result register 1 content
                                          ;(i.e. current sample) to
                                          ;accumulator
        LDP     #06h                      ;Set data page corresponding to
                                          ;variables
        SUB     I_OFFSET                  ;Subtract current probe offset
        BCND    S2, GEQ                   ;If result is positive or zero
                                          ;then branch to proceed normally
        LACL    #0000h                    ;Otherwise, set result to zero
S2:     SACL    I_SAMPLE, 6               ;Right shift result by 6 bits
                                          ;and store it. Right shift is
                                          ;because the 6 LSBs are
                                          ;insignificant
```

7.8.4 Filtering the ADC Readings

This code accumulates the voltage and current samples from every interrupt in order to calculate their averages once every PWM cycle.

```
        LDP     #06h                      ;Set data page corresponding to
```

```
                                ;variables
LACL    V_SUM                   ;Load voltage sample sum to
                                ;accumulator
ADD     V_SAMPLE                ;Add voltage sample to
                                ;accumulator
SACL    V_SUM                   ;Store result as voltage sum
LACL    I_SUM                   ;Load current sample sum to
                                ;accumulator
ADD     I_SAMPLE                ;Add current sample to
                                ;accumulator
SACL    I_SUM                   ;Store result as current sum
```

7.9 The Regulation Code Sequences

The following sequences described in this section execute once every Timer 2 period interrupt. The Timer 2 period interrupt is set to occur at a frequency of 16 times that of the PWM switching frequency. This ensures that the regulation is calculated only once every PWM cycle.

The code checks the counter of interrupt occurrence (INT_CNT). Every time the four least significant bits are equal to 15 (every 16 interruptions), the DSP branches to the regulation code. Otherwise, it returns directly from the interrupt service routine.

```
LDP     #06h                    ;Set data page corresponding to
                                ;variables
LACL    INT_CNT                 ;Load interrupt occurrence
                                ;counter into accumulator
ADD     #1h                     ;Increment by 1
SACL    INT_CNT                 ;Store back as interrupt
                                ;occurrence counter
AND     #000Fh                  ;Discard all bits but 4 LSBs
XOR     #000Fh                  ;Check for equality with 16d=Fh
BCND    S_RET, NEQ              ;If not, then branch to return
                                ;from interrupt
```

7.9.1 Calculating the Voltage and Current Average Values

This sequence takes the sum of 16 voltage and current samples and divides it by 16. The division is performed with a 4-bit right shift. The results are the averaged values of the load voltage and current.

```
LDP     #06h                    ;Set data page corresponding to
                                ;variables
LACL    V_SUM                   ;Load sum of voltage sample to
                                ;accumulator
SACL    V_AVERAGED, 4           ;Shift right by 4 bits and store
```

```
                                    ;as the average load voltage
      SPLK     #0000h, V_SUM        ;Reset sum of voltage samples
      LACL     I_SUM                ;Load sum of current samples to
                                    ;accumulator
      SACL     I_AVERAGED, 4        ;Shift right by 4 bits and store
                                    ;as the average load current
      SPLK     #0000h, I_SUM        ;Reset sum of current samples
```

7.9.2 *Voltage Comparator*

This code simply outputs the difference between the voltage reference (V_REF) and the averaged load voltage (V_AVERAGED) as the error (V_ERROR).

```
      LDP      #06h                 ;Set data page corresponding to
                                    ;variables
      ACL      V_REF                ;Load voltage reference in the
                                    ;accumulator
      SUB      V_AVERAGED           ;Subtract the averaged load
                                    ;voltage. The accumulator now
                                    ;contains the difference
      SACL     V_ERROR              ;Store the result as V_ERROR
```

7.9.3 *Current Regulation*

This sequence checks the averaged load current versus the predefined limit and, if necessary, stops the integration of the voltage error and decrements the integrator.

```
      LDP      #06h                 ;Set data page corresponding to
                                    ;variables
      LACL     I_LIMIT              ;Load maximum current value to
                                    ;accumulator
      SUB      I_AVERAGED           ;Subtract actual averaged
                                    ;current value
      BCND     S3, GEQ              ;If actual current below maximum
                                    ;then branch to proceed normally
      SPLK     #0000h, V_ERROR      ;Otherwise, stop integrating
                                    ;voltage error
      LACL     INTEGRAL             :And load integral value
      SUB      #1h                  ;Decrement it by 1
      SACL     INTEGRAL             ;Store it back as integral value
```

7.9.4 *PI Regulator*

The code below is actually only an integral regulator, which proved to be sufficient for the application.

```
S3:     LDP     #06h            ;Set data page corresponding to
                                ;variables
        LACL    INTEGRAL        ;Load integral value to
                                ;accumulator
        ADD     V_ERROR         ;Integrate the error
        SACL    INTEGRAL        ;Store result as integral value
        SACL    ACTION, 3       ;Right shift by 3 bits for
                                ;formatting purposes and output
                                ;result as action
```

7.9.5 Short Circuit Protection

The protection algorithm is activated if the average current rises to twice the limit. This sequence executes along with the rest of the regulation code. The protection algorithm must immediately reset the PWM output and the integrator to zero in order to protect the buck-boost converter and load against sudden surges of current. It should be noted that in case of a short-circuit, this protection will be activated only a few times. Once the filter capacitor is discharged, the load current will drop to acceptable levels and the LF2407 will work in current regulation mode.

```
        LDP     #06h               ;Set data page corresponding to
                                   ;variables
        LACL    I_LIMIT            ;Load current limit to
                                   ;accumulator
        SFL                        ;Multiply by 2
SUB     I_AVERAGED                 ;Subtract the average load
                                   ;current value
        BCND    S_PWM, GEQ         ;If the average load current is
                                   ;below the critical limit, then
                                   ;proceed normally to outputting
                                   ;the action to PWM
        SPLK    #0000h, ACTION     ;Otherwise, reset the PWM
                                   ;output through the ACTION
                                   ;signal
        SPLK    #0000h, INTEGRAL   ;And reset the PI regulator
                                   ;integrator. The proceed to
                                   ;outputting the action to the
                                   ;PWM hardware
```

7.9.6 Output Action to PWM

The action signal from the PI regulator should be linearized (i.e., transformed into a corresponding duty cycle) for optimum dynamic performance. However, dynamic performance is ensured by the processing speed of the DSP. In addition, the actual relationship between the duty cycle and the voltage gain is dependent upon the load and converter characteristics and would therefore have to be identified for each specific application. The duty cycle effectively output by the PWM hardware as it is programmed here is the following function:

$$d = \frac{T1PR - T1CMPR}{T1PR} \tag{7.5}$$

Therefore, the code must adapt the action signal from the PI regulator (ACTION) into a value to be stored in the T1CMPR register. The equation used is

$$\frac{ACTION}{T1PR} = \frac{T1PR - T1CMPR}{T1PR} \tag{7.6}$$

which yields the simple transformation implemented in the code sequence below.

$$T1CMPR = T1PR - ACTION \tag{7.7}$$

```
LDP     #GPTCONA>>7h        ;Set data page corresponding to
                            ;PWM timer registers
LACL    T1PR                ;Load period timer value
LDP     #06h                ;Set data page corresponding to
                            ;variables
SUB     ACTION              ;Subtract the action signal from
                            ;the PI regulator. The result is
                            ;the PWM timer compare value
LDP     #GPTCONA>>7h        ;Set data page corresponding to
                            ;the PWM timer registers
SACL    T1CMPR              ;Store calculated compare value.
                            ;It is now effectively the value
                            ;that will be used by the PWM
                            ;hardware for the next PWM cycle
```

7.9.7 Return to Main Code

This code clears the flag corresponding to the Timer 2 period match interrupt in EVAIFRB and re-enables the interrupts. It branches back to where the program was before the interrupt, i.e., the wait loop in the main code.

```
S_RET:  LDP     #EVAIFRB>>7h    ;Set data page corresponding to
                                ;event manager A registers
        LACL    EVAIFRB         ;Load value of event manager A
                                ;interrupt flags register
        SACL    EVAIFRB         ;Store it back to the register.
                                ;This effectively clears the
                                ;interrupt flags
        CLRC    INTM            ;Re-enable the interrupts
        RET                     ;Return from interrupt
```

7.10 Results

The following waveforms were captured from a physical buck-boost converter under the control of the DSP algorithm described in this chapter. The waveform of the diode current is shown first in Fig. 7.13.

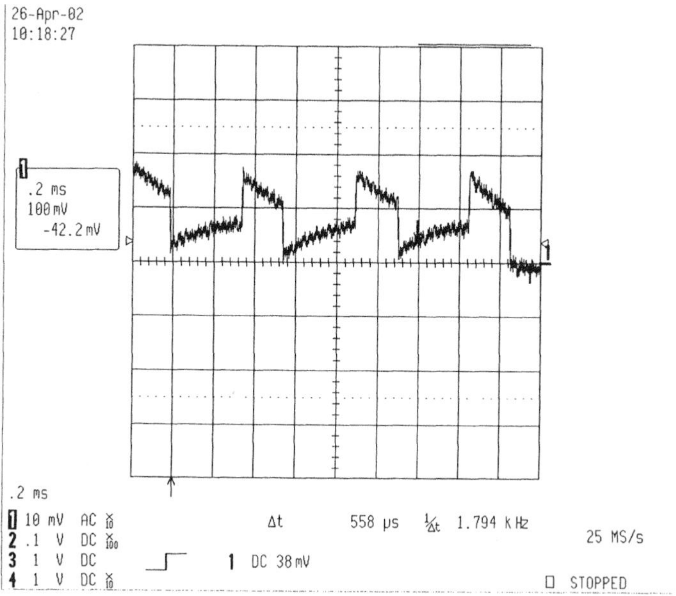

Figure 7.13 Diode current ripple.

The current waveform displays a ripple that emphasizes the need for an averaging filter. The ripple observed has an amplitude of about 15mA, with an average value of 17mA. The waveform also displays high frequency noise, which further reduces the precision of the measurements, requiring the use of a filter.

The voltage waveform in Fig. 7.14 shows the ripple due to the charge of the capacitor during the switch-off period and the discharge of the capacitor by the load during the switch-on period. The amplitude of the ripple is only 10mV, but its high frequency makes an output filter mandatory. Furthermore, the load used in these waveforms is a low current load. A larger load would make the ripple much more significant, thus also requiring an averaging filter.

The controlled buck-boost converter was tested in normal load conditions, with an increased load, and in sudden short circuit. It held the voltage at the predetermined reference value (5V) under normal load conditions (300Ω). A current limit was set at 25mA. When the load was decreased to 150Ω, the control entered in current regulation mode. The current was held precisely at 25mA, resulting in a voltage of 3.76V. When the output was suddenly short-circuited, the converter reacted immediately and held the current to 28mA instead of 25mA. The reason for this is that the output voltage was only a few mV, which is very close to the

3.224mV of the ADC. It is impossible for the controller to see the difference between 25 and 28mA, and hence impossible to regulate the current to exactly 25mA.

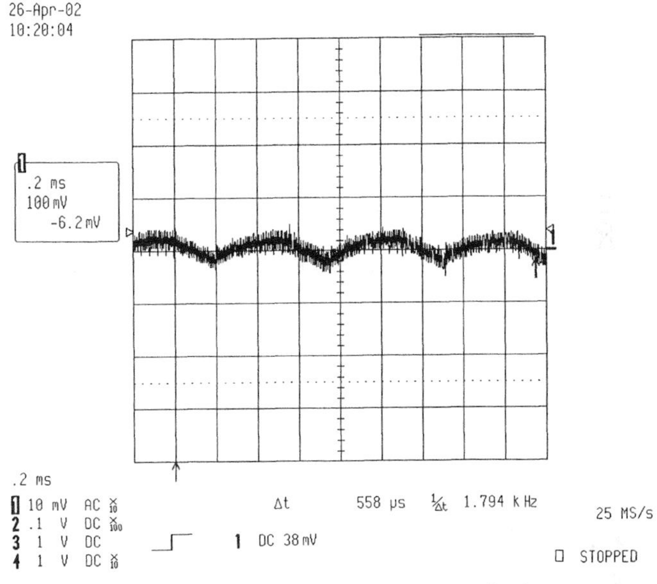

Figure 7.14 Load voltage ripple.

Chapter 8

DSP-BASED CONTROL OF STEPPER MOTORS

8.1 Introduction

A stepper motor is an electric machine that rotates in discrete angular increments or steps. Stepper motors are operated by applying current pulses of a specific frequency to the inputs of the motor. Each pulse applied to the motor causes its shaft to move a certain angle of rotation, called a stepping angle. Since the input signal is converted directly into a requested shaft position without any rotor position sensors or feedback, the stepper motor has the following advantages:

- Rotational speed proportional to the frequency of input pulses
- Digital control of speed and position
- No need of feedback sensor for open loop control
- Excellent acceleration and deceleration responses to step commands

The stepper motor also possesses drawbacks such as the possibility of losing synchronism, harmonic resonance, and small oscillations at the end of each step. With the above parameters in mind, the stepper motor is used in applications such as printers, plotters, X-Y tables, facsimile machines, barcode scanners, image scanners, copiers, medical apparatus, and other devices.

The stepper motor has salient poles on both the stator and the rotor, and normally only the stator poles hold the poly-phase windings called the control windings. Usually stepper motors are classified as

- Active rotor (permanent magnet rotor)
- Reactive rotor (reluctance type)
- Hybrid motors (combining the operating principles of the permanent magnet (PM) and reluctance stepper motor)

While each of these types of stepper motors has merit, hybrid stepper motors are becoming more popular in industrial applications. In this chapter, we focus on the principles and implementation of a hybrid stepper motor control system using the LF2407 DSP controller.

8.2 The Principle of Hybrid Stepper Motor

8.2.1 The Structure of Hybrid Stepper Motor

Figure 8.1 shows a simplified construction of a unipolar hybrid stepper motor. The rotor of this machine consists of two star-shaped milled steel pieces with three teeth on each. A cylindrical, axially magnetized PM is placed between the milled pieces making the end of each rotor either a north or a south pole. The teeth are offset at the north and south ends as shown in Fig. 8.1. The stator has four poles, each of which has a center-tapped winding. Since all the windings have the common connection V+, only five wires, A, B, C, D, and V+, leave the motor. A winding is excited by sending current into the V+ wire and out one of the other wires. The windings are wound in the stator teeth in such a way so that the motor behaves in the following way:

- If winding A or C is excited, pole 1 or pole 3 is energized as south.
- If winding B or D is excited, pole 2 or pole 4 is energized as north.

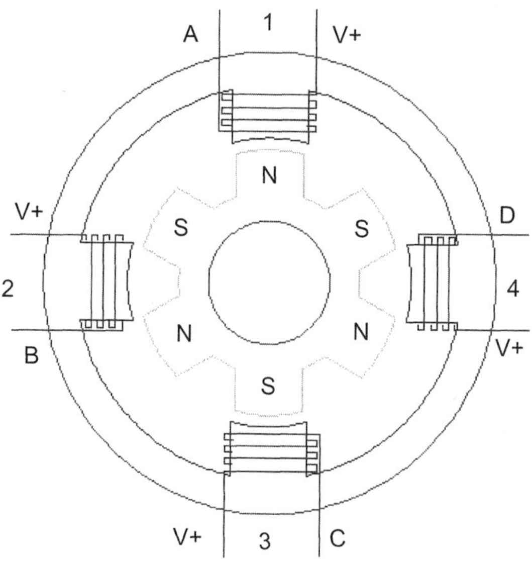

Figure 8.1 The four-phase, six-pole stepper motor.

Stepper motors are also classified with respect to the stator windings as being either bipolar or unipolar. Bipolar stepper motors have two windings with an opposing magnetizing effect in each pole, while unipolar stepper motors use only one winding per pole.

8.3 The Basic Operation

The operation of the stepper motor relies on the simple principle of magnetic attraction. This principle states that opposite magnetic poles attract while like poles repel each other. If the windings are excited in the correct sequence, the rotor will rotate following a certain direction. The basic operation of a stepper motor can be classified generally as either full step mode or half step mode. These modes are discussed in detail in the following section using the simplified stepper motor construction shown in Fig. 8.1.

8.3.1 Full-step Mode

If none of the stator windings are excited, an attraction between the stator poles and rotor teeth still exists because the PM rotor is trying to minimize the reluctance of the magnetic flux path from one end to the other. As a result, the rotor will tend to rest at one of the rest equilibrium positions. From Fig. 8.1, a rest position exists when a pair of rotor teeth are aligned with two of the stator poles. In the case of Fig. 8.1, the rotor is aligned with pole 1 and pole 3 on the stator. There are a total of 12 possible equilibrium positions for a 4-phase, 6-pole stepper motor. The force or torque that holds the rotor in one of these positions is called the detent torque. The value of the detent torque is usually small because no current flows through the stator windings.

Consider the case of the stator windings being excited according to Table 8.1. Assume at the beginning we are in mode 1 and the rotor aligns with poles 1 and 3 as shown in Fig. 8.2(a). When the excited sequences switch from mode 1 to mode 2, the north and south stator poles become pole 2 and pole 4. When this happens, the teeth of opposite polarity on the rotor will experience an attractive force, creating a torque on the rotor. Since this torque is much greater than the detent torque, the rotor will turn 30^0 counterclockwise, corresponding to one full step. Following the sequence of modes 1, 2, 3, and 4, the stator field rotates 90^0, attracting the corresponding rotor poles when the mode switches from one to the next. After switching four times, the rotor has moved four steps (120^0) and the rotor and stator fields return back to the initial condition or mode 1. A complete revolution requires 12 steps. The clockwise direction will be obtained if the reverse excited sequence of the stator winding is applied.

Table 8.1 Full step, single-phase excited sequence

	Winding A	**Winding B**	**Winding C**	**Winding D**	**Rotor Position**
Mode 1	On	Off	Off	Off	0
Mode 2	Off	On	Off	Off	θ
Mode 3	Off	Off	On	Off	2θ
Mode 4	Off	Off	Off	On	3θ

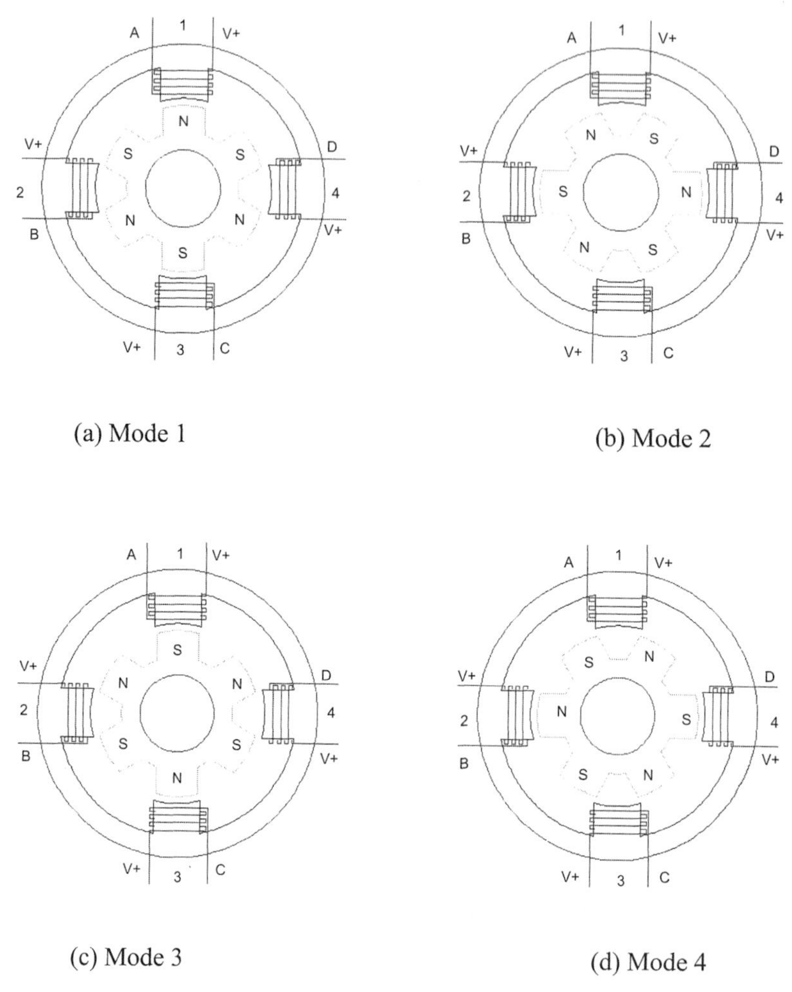

Figure 8.2 The principle of single-phase full-step mode.

For the full-step operation, greater torque can be produced if the two windings are excited simultaneously. The excited sequence of the stator winding is given in Table 8.2. During this operation, the rotor takes up an intermediate position because it experiences an equal attraction to the two stator poles as shown in Fig. 8.3. As in the single-phase full-step operation, a switch between two adjacent modes will cause a 90^0 shifting of the stator field. This results in a 30^0 rotation of the rotor. Twelve steps are required for a complete revolution in this mode as well. The sequence in Table 8.2 will rotate the motor counterclockwise, while reversing the sequence will run the motor clockwise.

Table 8.2 Full step, two-phase excited sequence

	Winding A	Winding B	Winding C	Winding D	Rotor Position
Mode 1	On	On	Off	Off	θ/2
Mode 2	Off	On	On	Off	3θ/2
Mode 3	Off	Off	On	On	5θ/2
Mode 4	On	Off	Off	On	7θ/2

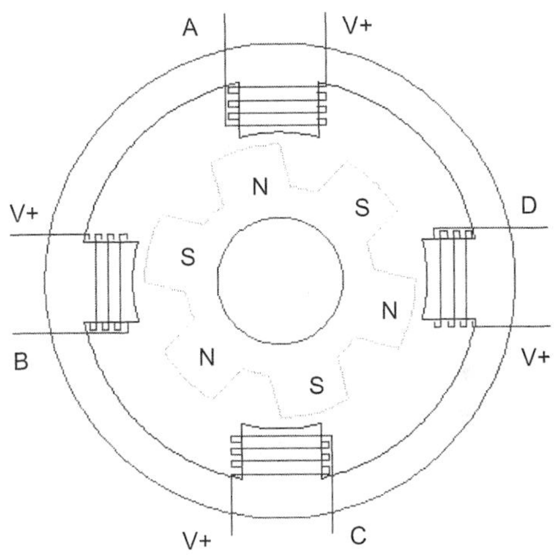

Figure 8.3 The rest equilibrium position of two-phase full-step mode.

8.3.2 Half-Step Mode

The stepper motor operation discussed rotates 30^0 per step. In the half step mode, alternately exciting one winding, then exciting two windings, will cause the rotor to move through only 15 degree per step. Though there is a slight loss of the torque while the single winding is being excited, half-step operation allows for smoother operation at lower speeds and less overshoot at the end of each step. The excitation sequence of the stator windings in half-step mode is given in Table 8.3.

During this operation, each switch between the two nearest modes will cause a 45^0 shift of stator field which results in a 15^0 rotation of the rotor. A total of 24 steps are required for a complete revolution, double of what is required for full step modes.

Table 8.3 Half-step, two-phase excited sequence.

	Winding A	Winding B	Winding C	Winding D	Rotor Position
Mode 1	On	Off	Off	Off	0
Mode 2	On	On	Off	Off	θ/2
Mode 3	Off	On	Off	Off	θ
Mode 4	Off	On	On	Off	3θ/2
Mode 5	Off	Off	On	Off	2θ
Mode 6	Off	Off	On	On	5θ/2
Mode7	Off	Off	Off	On	3θ
Mode8	On	Off	Off	On	7θ/2

8.3.3 Micro-Step Mode

For the operating modes discussed previously, the same amount of current flows through the energized stator windings. However, if the currents are not equal, the rotor will be shifted toward the stator pole with the higher current. The amount of deviation is proportionate to the values of the currents in each winding. This principle is utilized in the micorstep mode. During this mode, each basic full mode step can be divided into as many as 500 microsteps, providing the proper current profile is applied.

8.4 The Stepper Motor Drive System

An open loop stepper motor control system is shown in Fig. 8.4. The total control system consists of the power electronic drive circuit and controller. These components will be discussed in detail in following sections.

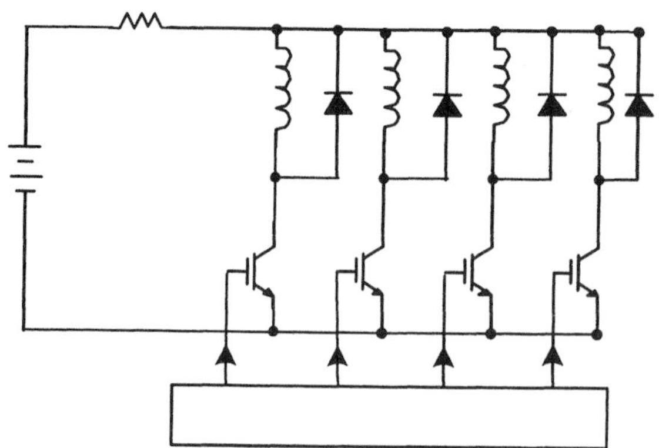

Figure 8.4 The stepper motor speed control system.

8.4.1 Power Electronic Drive Circuit

The drive circuit of a stepper motor is displayed in Fig. 8.4. Wires A, B, C, and D are connected to the power switch device T_1, T_2, T_3, and T_4. The V+ wire is connected to a +12V power supply through a series resistor. When one of the switches turns on, the corresponding winding is excited.

The windings in a stepper motor also have inductance. When the switch turns on, the winding inductance will increase the amount of time it takes for the current to reach its full value. Since the speed of the stepper motor is proportionate to the switching frequency, this effect limits the maximum motor speed. A series resistance (R_s), as shown in Fig. 8.4, is added to reduce this problem. Assuming the winding's inductance and resistance are L and R, when the switch turns on, the winding current can be calculated by:

$$i(t) = \frac{V_{dc}}{R + R_s}(1 - e^{-\frac{R+R_s}{L}t}) \tag{8.1}$$

From (8.1), it can be seen that the series resistance reduces the time constant so that the current can increase faster. However, the resistance causes a voltage drop, which requires a larger power supply to compensate for the resistor losses so that the same current can be applied to the motor windings.

The winding inductance also leads to another problem when the switch turns off. If no additional current path is provided to dissipate the energy stored in the inductance, a voltage spike will be generated across the switching devices and may damage them. To solve this problem, a freewheeling diode (D_1-D_2) parallel to the winding is employed. In addition, a series resistor may also be added to the circuit to limit the voltage spike.

8.4.2 Controller

The LF2407 DSP controller is used to implement the speed control of a stepper motor drive system. The interface of the LF2407 is illustrated in Fig. 8.5. Since this control scheme is an open-loop control system, no feedback information is required. The four I/O ports on the DSP provide the gating signals to the transistors, which provide current to the windings in the specified sequence. The speed rate at which the switching sequence is applied determines the speed of the motor.

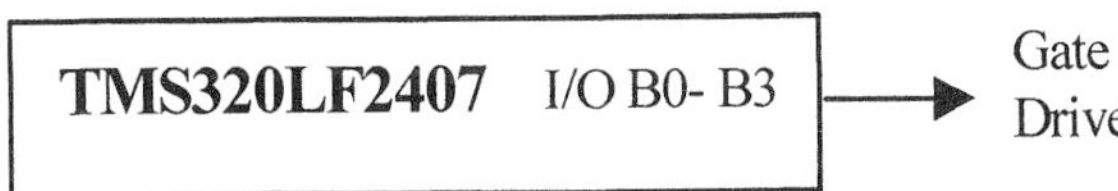

Figure 8.5 DSP interface.

8.5 The Implementation of Stepper Motor Control System Using the LF2407 DSP

The assembly code associated with the LF2407 was developed to implement the open loop speed control system discussed previously. The flowchart for the DSP software is shown in Fig. 8.6.

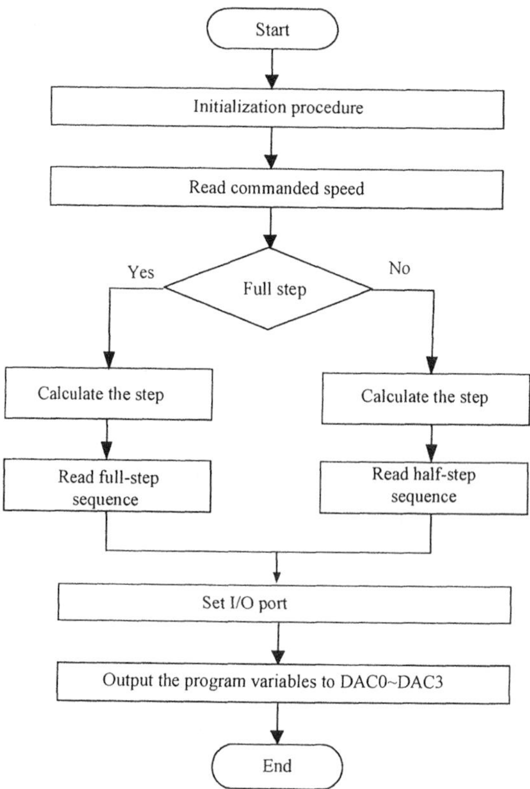

Figure 8.6 Flowchart of the stepper motor control algorithm.

It can be seen from Fig. 8.6 that the control algorithm of the stepper motor drive system consists of one main routine and includes four subroutine modules:

- Initialization procedure
- Speed control module
- Output signals via I/O port
- DAC module

Only the speed control module is specific to the stepper motor control system and will be dicussed in detail.

8.6 The Subroutine of Speed Control Module

The Timer 1 period interrupt is used for the speed control subroutine. This subroutine performs the task of reading the commanded speed and then converting it to a pulse output on the I/O ports. Hence, the motor speed is determined by the time interval of this interrupt. The block of assembly code below shows the Timer 1 Interrupt Service Routine (ISR), which executes all subroutines upon every interrupt.

```
T1_ISR:        NOP
;------------------*
; Context Saving    *
;------------------*
;Context save regs
        MAR     *,      AR1             ;AR1 is stack pointer
        MAR     *+                      ;skip one position
        SST     #1,     *+              ;save ST1
        SST     #0,     *+              ;save ST0
        SACH    *+                      ;save acc high
        SACL    *                       ;save acc low
        POINT_EV
SPLK    #0FFFFh ,       EVIFRA

POINT_B0

RUN_MODE POINT_B0
        CALL    SPEED_PROFILE
        CALL    VTIMER_SEC
        LACC    STEP_FLG
        BCND    HALF_MODE,GT
        CALL     FULL_STEP
        B       END_MODE

HALF_MODE
        CALL    HALF_STEP

END_MODE
        CALL   DAC_VIEW_Q15I
;-------------------------------*
;* Context restore and Return   *
;-------------------------------*
END_ISR:
        MAR     *,      AR1             ;make stack pointer active
        LACL    *-                      ;Restore Acc low
        ADDH    *-                      ;Restore Acc high
        LST     #0,     *-              ;load ST0
        LST     #1,     *-              ;load ST1
        POINT_PG0
        CLRC    INTM
        RET
```

8.6.1 Full-Step Mode

Two-phase full-step mode described in Section 8.3.1 is implemented in the full-step subroutine as shown in the code on the next page. The commanded speed is converted first to a pulse with a certain frequency in this routine. According to Table 8.2, the different sequence is read and then the corresponding I/O ports (IOPB0, 1, 2, 3 – A, B, C D) are set high/low to control the turn on/off of the switches.

```
FULL_STEP:          POINT_B0
        ...
        ...
        POINT_B0
        LACC     #MODE_FUL
        sub              #3
        BCND     SET_MODE_FUL,NEQ
        SPLK     #0,MODE_CNTL_FUL
        B        FUL_EXIT
        ...
        ...
        RET
```

8.6.2 Half-Step Mode

Following the same procedure as described above, two-phase half-step mode strategy described in Section 8.3.2 is implemented in the code block shown below.

```
HALF_STEP:     POINT_B0
        POINT_B0
        ...
        ...
        POINT_B0
        LACC     MODE_CNTL_HLF
        sub      #7
        BCND     SET_MODE_HLF,NEQ
        SPLK     #0,MODE_CNTL_HLF
        B        HLF_EXIT
        ...
        ...
        RET
```

Reference

1. Digital Signal Processing Control of Electric Machines and Drives Laboratory Manual, Department of Electrical Engineering, The Ohio State University, March 2002.

Chapter 9

DSP-BASED CONTROL OF PERMANENT MAGNET BRUSHLESS DC MACHINES

9.1 Introduction

Permanent magnet alternating current (PMAC) motors are synchronous motors that have permanent magnets mounted on the rotor and poly-phase, usually three-phase, armature windings located on the stator. Since the field is provided by the permanent magnets, the PMAC motor has higher efficiency than induction or switched reluctance motors. It also draws better power factor and has higher power density. The advantages of PMAC motors, combined with a rapidly decreasing cost of permanent magnets, have led to their widespread used in many variable-speed drives such as robotic actuators, computer disk drives, domestic appliances, automotive applications, and heating-ventilating-air conditioning (HVAC) equipment.

In general, PMAC motors are categorized into two types. The first type of motor is referred to as PM synchronous motor (PMSM). These motors produce a sinusoidal back-EMF shown in Fig. 9.1(a), and should be supplied with sinusoidal current/voltage. The PMSM's electronic control and drive system uses continuous rotor position feedback and pulse-width-modulation (PWM) to supply the motor with the sinusoidal voltage or current. With this, constant torque is produced with very little ripple. A detailed discussion of the PMSM drive system is given in Chapter 12.

The second type of PMAC motor has a trapezoidal back-EMF and is referred to as the brushless DC (BLDC) motor. The back-EMF of the BLDC motor is shown in Fig. 9.1(b). The BLDC drive system is based on the feedback of rotor position, which is not continuous as with the PMSM, but rather obtained at fixed points typically every 60 electrical degrees for commutation of the phase currents. The BLDC motor requires that quasi-rectangular shaped currents are fed into the machine. Alternatively, the voltage may be applied to the motor every 120^0, with a current limit to hold the currents within the motor's capabilities. Because the phase currents are excited in synchronism with the constant part of the back-EMF, constant torque is generated.

The objective of this chapter is to introduce the principles of the BLDC motor control system, and then discuss the procedure of its implementation using the LF2407 DSP.

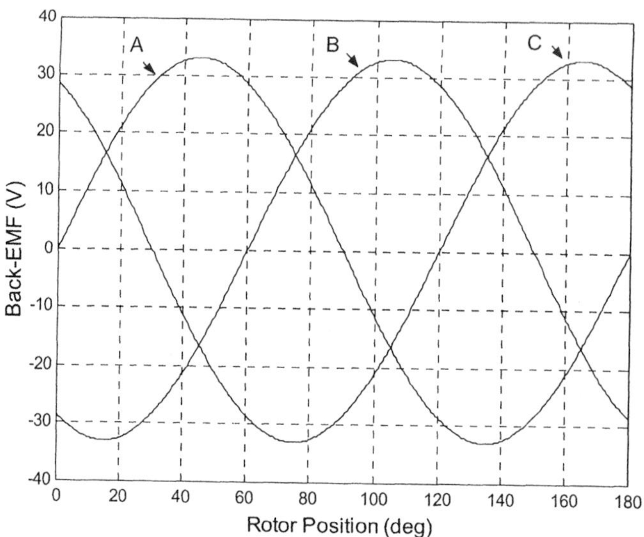

(a) Three-phase back-EMF of PMSM.

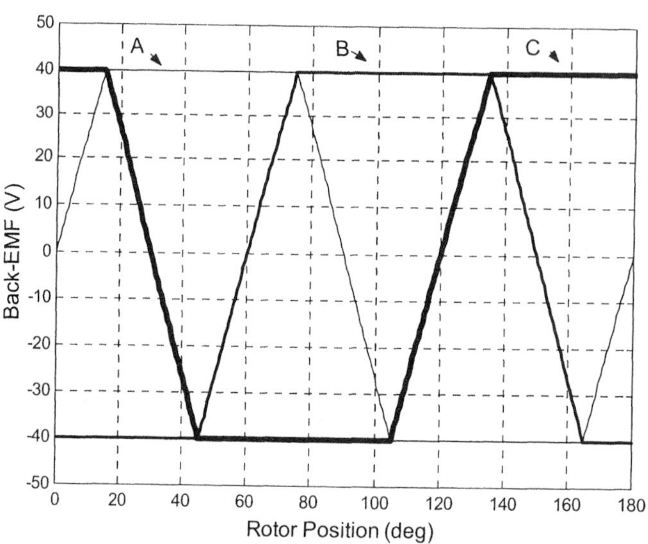

(b) Three-phase back-EMF of BLDC motors.

Figure 9.1 The back-EMF of PMAC motors.

9.2 Principles of the BLDC Motor

9.2.1 Mathematical Model

The phase variables are used to model the BLDC motor due to its non-sinusoidal back-EMF and phase current. The terminal voltage equation of the BLDC motor can be written as

$$\begin{bmatrix} v_a \\ v_b \\ v_c \end{bmatrix} = \begin{bmatrix} R+pL_s & 0 & 0 \\ 0 & R+pL_s & 0 \\ 0 & 0 & R+pL_s \end{bmatrix} . \begin{bmatrix} i_a \\ i_b \\ i_c \end{bmatrix} + \begin{bmatrix} e_a \\ e_b \\ e_c \end{bmatrix} \tag{9.1}$$

where v_a, v_b, v_c are the phase voltages, i_a, i_b, i_c are the phase currents, e_a, e_b, e_c are the phase back-EMF voltages, R is the phase resistance, L_s is the synchronous inductance per phase and includes both leakage and armature reaction inductances, and p represents d/dt. The electromagnetic torque is given by

$$T_e = (e_a i_a + e_b i_b + e_c i_c)/\omega_m \tag{9.2}$$

where ω_m is the mechanical speed of the rotor. The equation of motion is

$$\frac{d}{dt}\omega_m = (T_e - T_L - B\omega_m)/J \tag{9.3}$$

where T_L is the load torque, B is the damping constant, and J is the moment of inertia of the rotor shaft and the load.

9.3 Torque Generation

From (9.2), the electromagnetic torque of the BLDC motor is related to the product of the phase back-EMF and current. The back-EMFs in each phase are trapezoidal in shape and are displaced by 120 electrical degrees with respect to each other in a three-phase machine. A rectangular current pulse is injected into each phase so that current coincides with the crest of the back-EMF waveform, hence the motor develops an almost constant torque. This strategy, commonly called six-step current control, is illustrated by Fig. 9.2 and explained by (9.4). The amplitude of each phase's back-EMF is proportional to the rotor speed, and is given by

$$E = k\phi\omega_m \tag{9.4}$$

where k is a constant and depends on the number of turns in each phase, ϕ is the permanent magnet flux, and ω_m is the mechanical speed. In Fig. 9.2, during any 120^0 interval, the instantaneous power converted from electrical to mechanical is the sum of the contributions from two phases in series, and is given by

$$P_o = \omega_m T_e = 2EI \tag{9.5}$$

where T_e is the output torque and I is the amplitude of the phase current. From (9.4) and (9.5), the expression for output torque can be written as

$$T_e = 2k\phi I = k_t I \tag{9.6}$$

where k_t is the torque constant. Since the electromagnetic torque is only proportional to the amplitude of the phase current in (9.6), torque control of the BLDC motor is essentially accomplished by phase current control.

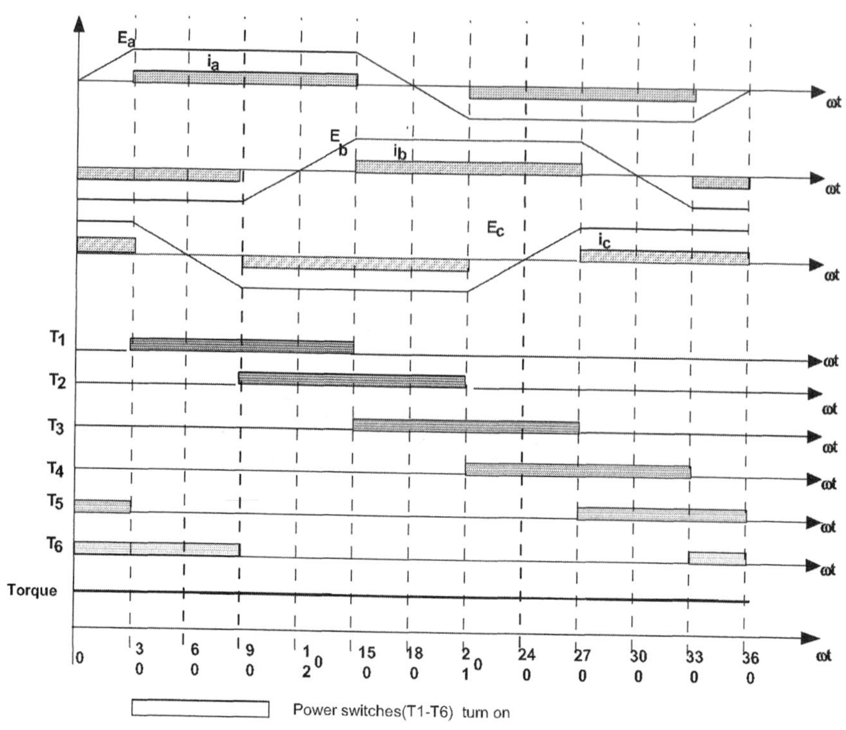

Figure 9.2 The principle of the six-step current control algorithm. T1-T6 are the gate signals, E_a, E_b, and E_c are the motor phase back-EMF, I_a, I_b, and I_c are the motor phase currents.

9.4 BLDC Motor Control System

Based on the previously discussed concept, a BLDC motor drive system is shown in Fig. 9.3. It can be seen that the total drive system consists of the BLDC motor, power electronics converter, sensor, and controller. These components are discussed in detail in the following sections.

9.4.1 BLDC Machine

BLDC motors are predominantly surface-magnet machines with wide magnet pole-arcs. The stator windings are usually concentrated windings, which produce a square waveform distribution of flux density around the air-gap. The design of the BLDC motor is based on the crest of each half-cycle of the back-EMF waveform. In order to obtain smooth output torque, the back-EMF waveform should be wider than 120 electrical degrees. A typical BLDC motor with 12 stator slots and 4 poles on the rotor is shown in Fig. 9.4.

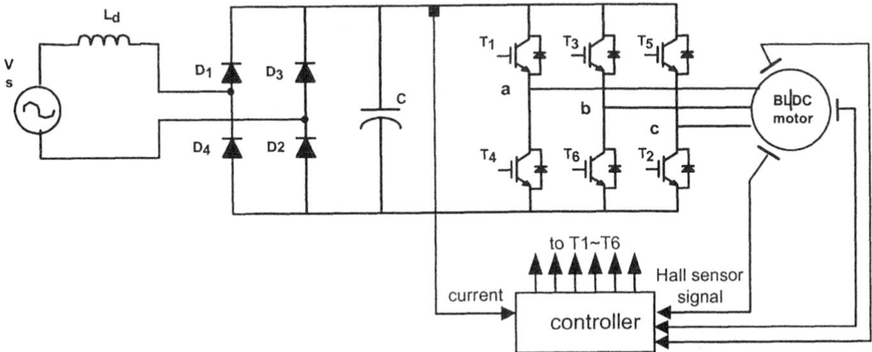

Figure 9.3 BLDC motor control system.

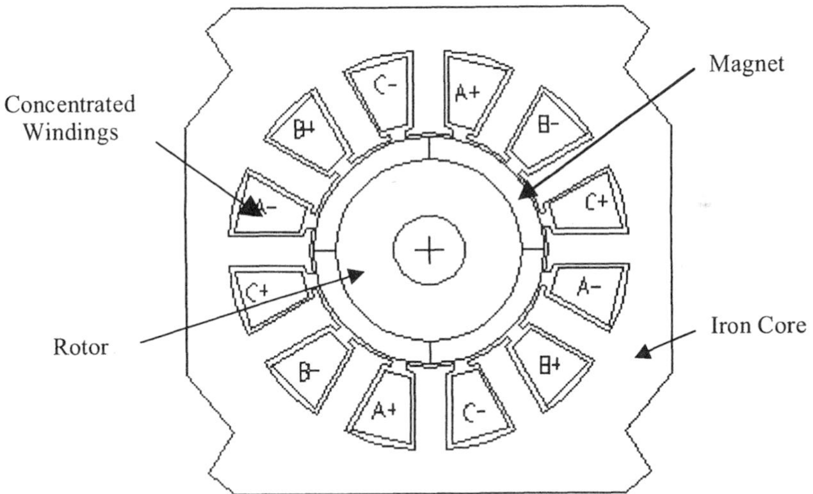

Figure 9.4 The 4-pole 12-slot BLDC motor.

9.4.2 Power Electronic Converter

As shown in Fig. 9.3, the power electronic converter in the BLDC motor drive system consists of two parts: a front-end rectifier and a three-phase full-bridge inverter. The front-end rectifier is usually a full-bridge diode rectifier unless a switching rectifier is used to provide regeneration capability.

The inverter is usually responsible for the electronic commutation and current regulation. For the six-step current control, if the motor windings are Y connected without the neutral connection, only two of the three phase currents flow through the inverter in series. This results in the amplitude of the DC link current always being equal to that of the phase currents. As far as the inverter goes, there are only two switches per leg, one upper and one lower switch which conduct at any moment. PWM current controllers are typically used to regulate the actual machine currents in order to match the rectangular current reference waveforms shown in Fig. 9.2. For example, during one 60° interval when switches T_1 and T_6 are active, phases A and B conduct. The lower switch T_6 is always turned on and the upper switch T_1 is chopped on/off using either a hysteresis current controller with variable switch frequency or a PI controller with fixed switch frequency. When T_1 and T_6 are conducting, current builds up in the path shown in the dashed line of Fig. 9.5(a). When switch T_1 is turned off, the current decays through diode D_4 and switch T_6 as shown in the dashed line of Fig. 9.5(b). In the next interval, switch T_2 is on, and T_1 is chopped so that phase A and phase C conduct. During the commutation interval, the phase B current rapidly decreases through the freewheeling diode D_3 until it becomes zero and the phase C current builds up.

From the above analysis, each of the upper switches is always chopped for one 120^0 interval and the corresponding lower switch is always turned on per interval. The freewheeling diodes provide the necessary paths for the currents to circulate when the switches are turned off and during the commutation intervals.

9.4.3 Sensors

There are two types of sensors for the BLDC drive system: a current sensor and a position sensor. Since the amplitude of the dc link current is always equal to that of the motor phase current in six-step current control, the dc link current is measured instead of the phase current. Thus, a shunt resistor, which is in series with the inverter, is usually used as the current sensor. Hall-effect position sensors typically provide the position information needed to synchronize the stator excitation with rotor position in order to produce constant torque. Hall-effect sensors detect the change in magnetic field. The rotor magnets are used as triggers for the Hall sensor. A signal conditioning circuit integrated within the Hall switch provides a TTL-compatible pulse with sharp edges and high noise immunity for connection to the controller.

For the six-step current control algorithm, rotor position needs to be detected at only six discrete points in each electrical cycle. The controller tracks these six points so that the proper switches are turned on or off for the correct intervals. Three Hall-effect sensors, spaced 120 electrical degrees apart, are mounted on the stator frame. The digital signals from the Hall sensors are then used to determine the rotor position and switch gating signals for the inverter switches.

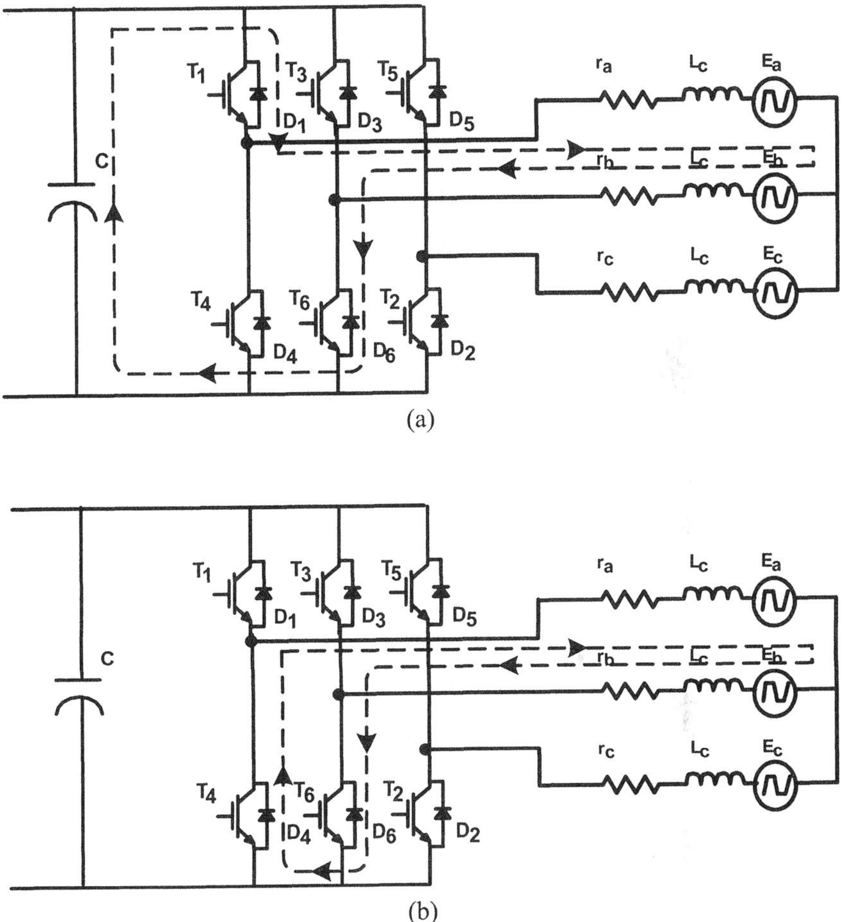

Figure 9.5 The current path when the switch T_1 turns on and turns off.

9.4.4 Controller

The controller of BLDC drive systems reads the current and position feedback, implements the speed or torque control algorithm, and finally generates the gate signals. Either analog controllers or digital signal processors serve well as controllers. In this chapter, the LF2407 DSP will be used as the controller.

The connectivity of the LF2407 in this application is illustrated in Fig. 9.6. Three capture units in the LF2407 are used to detect both the rising and falling edges of Hall-effect signals. Hence, every 60 electrical degrees of motor rotation, one capture unit interrupt is generated which ultimately causes a change in the gating signals and the motor to move to the next position. One input channel of the 10-bit Analog-to-Digital Converter reads the dc link current. The output pins PWM1-PWM6 are used to supply the gating signals to the inverter.

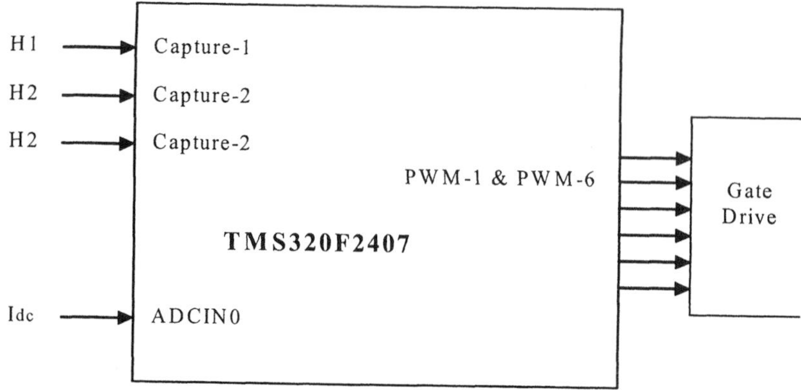

Figure 9.6 The interface of LF2407.

9.5 Implementation of the BLDC Motor Control System Using the LF2407

Since the LF2407 is used as the controller, much of the control algorithm is implemented in software. A block diagram of the BLDC motor control system is displayed in Fig. 9.7. The dashed line separates the software from the hardware components introduced in the previous section. It is necessary to choose hardware components carefully in order to ensure high processing speed and precision in the overall control system.

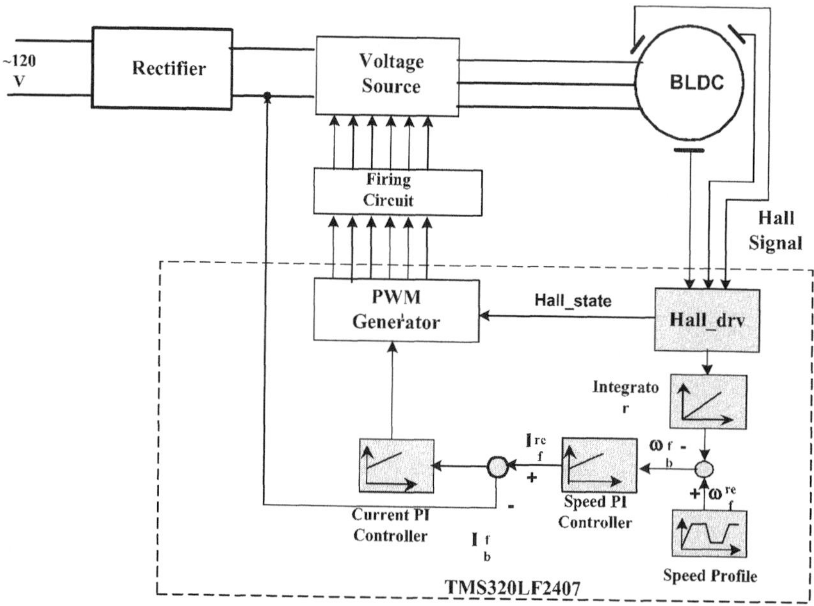

Figure 9.7 The block diagram of BLDC motor control system.

The overall control algorithm of the BLDC motor consists of nine modules:

1. Initialization procedure
2. Detection of Hall effect signals
3. Speed control subroutine
4. Measurement of current
5. Speed profiling
6. Calculation of actual speed
7. PID regulation
8. PWM generation
9. DAC output

The flowchart of the overall control algorithm is illustrated in Fig. 9.8.

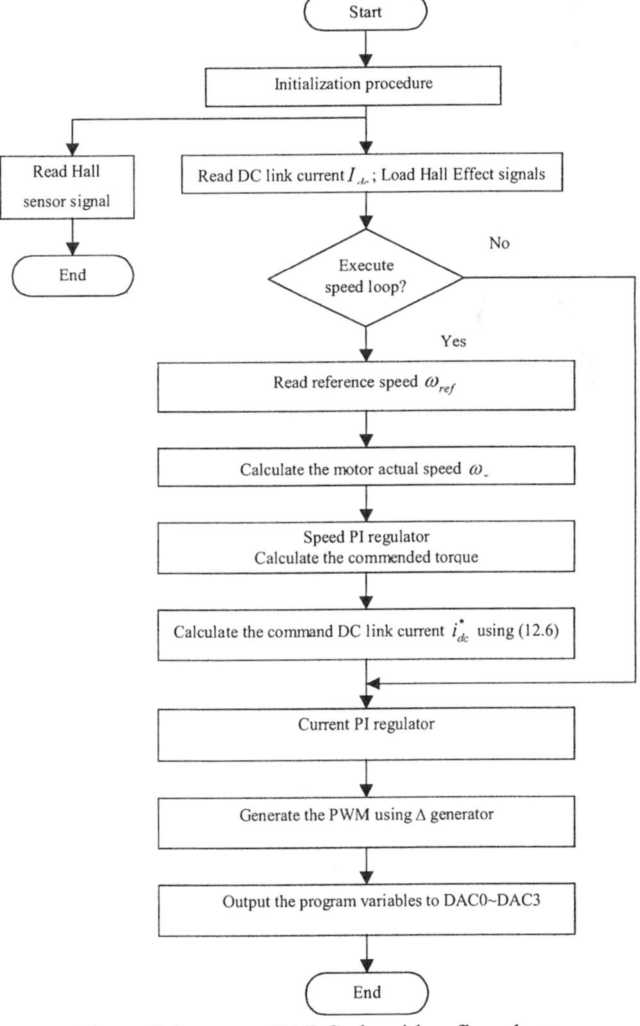

Figure 9.8 BLDC algorithm flowchart.

In the follwing sections, each block of the flow chart is dicussed in detail and the corresponding assembly code is given.

9.5.1 Initialization Procedure

The initialization procedures include the initialization of registers, memory allocations, and initializing constants and system variables. The TI website (www.ti.com) provides the standard linker command file for memory allocation on the LF2407. Readers can simply download it and then modify this file according to their own needs.

The need for and the initialization of system variables vary according to the application. The variables used in the BLDC control algorithm to generate the speed profile are initialized below:

```
POINT_B0
        SPLK    #0,       SPD_CNT
        SPLK    #0,       VTS_SEC
        SPLK    #0,       VTS_CNT
        SPLK    #0,       STEP_1
        SPLK    #5,       VTS_PRESCALE
        SPLK
#PSTEP_1, PROFILE_STEP_PTR
        SPLK    #04D0H,  SPD_SCALE
        SPLK    #0fffh,  SPD DESIRED
```

For BLDC motor control, the register initializations include four parts: system interrupt initialization, initialization of the ADC module, initialization of the Hall-effect signal detected, and initialization of the Event Manager. The assembly code for system interrupt initialization is given below:

```
;System Interrupt Init.
;Event Manager
POINT_EV
        SPLK    #0000001000000000b, EVIMRA
                                  ;Enable T1 Underflow Int (i.e. Period)
        SPLK    #0000000000000111b, EVIMRC
                                  ;Enable CAP1,2,3 ints
        SPLK    #0FFFFh, EVIFRA
                                  ;Clear all Group A interrupt flags
        SPLK    #0FFFFh, EVIFRB
                                  ;Clear all Group B interrupt flags
        SPLK    #0FFFFh,EVIFRC
                                  ;Clear all Group C interrupt flags
POINT_PG0
        SPLK    #0000000000001010b,IMR
                                  ;En Int lvl 2,4 (T2 & CAP ISR)
        SPLK    #0FFFFh, IFR
                                  ;Clear any pending Ints
```

9.5.2 The Detection of Hall-Effect Signals

Each edge of the Hall-effect sensor output signal generates a capture interrupt. The CPU responds to this interrupt and branches to the interrupt service subroutine to perform the following tasks: detect the Hall sensor sequences, decode the sequence, define the six states of the inverter, and record the time interval between the two nearest Hall-effect edges. The time between edges is used to calculate the rotor speed. The assembly code for the interrupt service subroutine is given below:

```
CAP_ISR:
;Context save regs
        MAR     *,      AR1             ;AR1 is stack pointer
        MAR     *+                      ;skip one position
        SST     #1,     *+              ;save ST1
        SST     #0,     *+              ;save ST0
        SACH    *+                      ;save acc high
        SACL    *                       ;save acc low
        CALL    HALL3_DRV
;Restore Context
END_ISR:
        MAR     *,      AR1             ;make stack pointer active
        LACL    *-                      ;Restore Acc low
        ADDH    *-                      ;Restore Acc high
        LST     #0,     *-              ;load ST0
        LST     #1,     *-              ;load ST1
        CLRC    INTM
        RET
```

The following code determines which one of the six switching states is needed:

```
HALL3_DRV:
                ...
                ...
Map_States:
        LDP     #hall_vars
        LACC    hall_seq, 2             ;x4 for jump table
        ADD     #STATE_TABLE
        BACC
STATE_TABLE:                    ;Map Hall connections and readings to
                                ;BLDC_PWM_DRV's states based on it's
                                ;state 0 alignment
        SPLK    #1, hall_state_next ;seq=0, BLDC_PWM_DRV next state 1
        B       HALL_END
        SPLK    #3, hall_state_next ;seq=1, BLDC_PWM_DRV next state 3
        B       HALL_END
        SPLK    #2, hall_state_next ;seq=2, BLDC_PWM_DRV next state 2
        B       HALL_END
        SPLK    #5, hall_state_next ;seq=3, BLDC_PWM_DRV next state 5
        B       HALL_END
        SPLK    #0, hall_state_next ;seq=4, BLDC_PWM_DRV next state 0
        B       HALL_END
        SPLK    #4, hall_state_next ;seq=5, BLDC_PWM_DRV next state 4
HALL_END:   RET
```

9.5.3 The Subroutine of Speed Control Algorithm

The Timer 1 underflow interrupt is used for the speed control subroutine. The speed control subroutine performs the task of reading the current, loading the inverter state obtained from capture interrupt, generating the commanded speed profile, calculating the actual motor speed, regulating speed and current, and finally generating the PWM signals to drive the inverter. The PWM frequency is determined by the time interval of this interrupt; the duty cycle is recalculated in every interrupt. The speed control algorithm is implemented by the following assembly code:

```
T1_PERIOD_ISR:
                                                    ;Context save regs
          MAR      *,        AR1                    ;AR1 is stack pointer
          MAR      *+                               ;skip one position
          SST      #1,       *+                     ;save ST1
          SST      #0,       *+                     ;save ST0
          SACH     *+                               ;save acc high
          SACL     *                                ;save acc low
          POINT_EV
          SPLK     #0FFFFh,EVIFRA                   ;Clear all Group A interrupt
                                                    flags (T1 ;ISR)
READ_HALL
          LDP      #hall_vars
          Lacc     hall_state_next
          POINT_B0
          sacl     cmtn_ptr_bd                      ;Input to BLDC_PWM_DRV
CUR_READ
          CALL     AD_CONV
          POINT_B0
          LACC     CL_SPD_FLG
          BCND     CURRENT_CNTL,GT                  ;speed-loop?
                                                    ;speed control
SPEED_CNTL:         POINT_B0
          CALL     SPEED_PROFILE
          CALL     VTIMER_SEC
          CALL     SPEED_CAL
          CALL     D_PID_spd
          LACC     D_spd_out
          SACL     I_ref
                                                    ;current control
CURRENT_CNTL
          CALL     D_PID_cur
          LACC     D_cur_out
          SACL     D_func
PWM_GEN            CALL          BLDC_PWM_DRV
DA_CONV            CALL          DAC_VIEW_Q15I
                                                    ;Restore Context
END_ISR:
          MAR      *,        AR1                    ;make stack pointer active
          LACL     *-                               ;Restore Acc low
          ADDH     *-                               ;Restore Acc high
          LST      #0,       *-                     ;load ST0
          LST       #1,      *-                     ;load ST1
          CLRC     INTM
          RET
```

9.5.4 Measurement of the Current (ADC Module)

For the BLDC motor control algorithm, the ADC converter reads in the voltage across the shunt resistor on ADCIN0. This voltage is proportional to the dc link current because the resistor is in series with the flow of current. The code section below reads the result register and obtains the ADC conversion result of the voltage across the shunt resistor.

```
AD_CONV
          LDP      #ADCTRL1>>7
          LACC     ADC_RESULT0
          SFR
          AND      #7FFFh
          SACL     Idc
          ….
          ….
AD_EXIT   RET
```

9.5.5 *Profile of the Reference Speed*

The reference speed profile is used to control the dynamic response and steady state behavior of the motor. The speed profile is divided into different sections, such as the acceleration interval, constant speed interval, and deceleration interval. We can use the different intervals to make the rotor accelerate, run at constant speed, or decelerate. One example of speed profile is shown in Fig. 9.8.

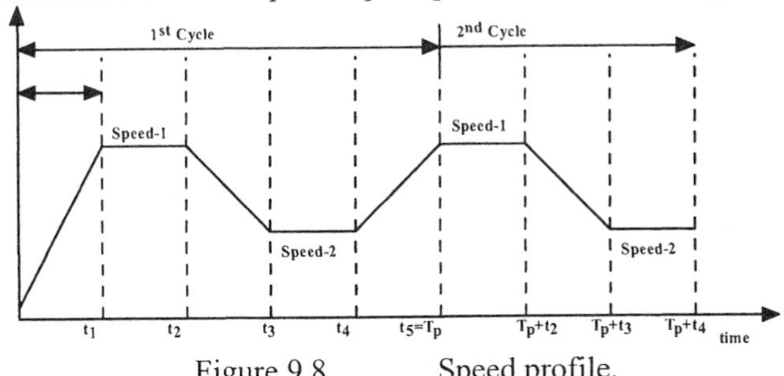

Figure 9.8 Speed profile.

In the speed profile given in Fig. 9.8, the interval from time 0 to t_1 represents a soft-start period where reference speed is slowly increased from zero to *speed-1*. For the time interval between t_1 to t_2, the speed reference is maintained constant at its value, *speed-1*. During the time interval from t_2 to t_3, the reference speed is slowly reduced to *speed-2*. The reference speed is then kept constant at *speed-2* for the time interval from t_3 to t_4. Finally, the speed is again increased to *speed-1* over the time interval t_4 to t_5. In our case, the sequence t_2 to t_5 is repeated continuously unless disabled by another routine. A sample of the assembly code used for such a speed profile is given below:

```
SPEED_PROFILE:
          ....
          ....
PSTEP_4   LACC  #SPEED_4
      AND   #0fFFH
      SACL  speed_ref
      LACC  SPD_CNT
      SUB   #03fFH
      BCND  GO_STEP5,GT
      LACC  VTS_SEC
      SUB   #TLENGTH_4
      BCND  SPR_END,LT
      SPLK  #0,VTS_SEC              ;RESET VIRTUAL TIMER
      SPLK  #0,VTS_CNT
      LACC  SPD_CNT
      ADD   #1
      SACL  SPD_CNT
      B             SPR_END
GO_STEP5    SPLK#0,VTS_SEC          ;RESET VIRTUAL TIMER
      SPLK  #0,VTS_CNT
      SPLK  #01ffH,STEP_3
      SPLK  #0,SPD_CNT
      LACC  #PSTEP_5
      SACL  PROFILE_STEP_PTR
      B     SPR_END
      .....
SPR_END    RET
```

9.5.6 The Calculation of the Actual Motor Speed

This module uses the value of the variable *Timestamp*, which represents the time interval between the two edges of the Hall-effect signal generated by the position interface module, to calculate the motor shaft speed. With a 30 MHz system clock as in the case of LF2407, *Timestamp* is related to the motor speed by

$$Timestamp = \left(\frac{60}{2 \times \omega_m \times prescalar \times 33 \times 10^{-9}} \right) \tag{9.7}$$

$$speed\_cal = \frac{32 \times 65536}{Timestamp} = 0.5906\ \omega_{\mathrm{m}} \tag{9.8}$$

where

ω_{m} = shaft speed in rpm.
prescalar = prescalar value for Timer-2 = 128
t_{cpu} = CPU period = 33 nsec
speed_cal = calculated speed in rpm

The speed calculation routine measures the time between two consecutive edge transitions of the position signal and cannot distinguish between the directions of rotation. A portion of the assembly code of the speed calculation routine is given below:

```
SPEED_CAL:
        ....
        ....
        LT      RES                     ;RES=1/Timestamp
        MPY     SPD_SCALE
        PAC
        SACH    speed_cal, 4
        RET
```

9.5.7 PID (Proportional, Integral, and Derivative) Regulation

PID controllers are used for both speed and current regulation. Both types of controllers have the same structure. The rectangular (trapezoidal) method of integration is used and depends upon the value of the parameters K_1, K_2, and K_3. Limits are set to limit the output of PI controller. This routine implements the following PI equation:

$$U(n) = U(n-2) + K_1\ e(n) + K_2\ e(n-1) + K_3\ e(n-2) \tag{9.9}$$

where

$U(n)$ is the current output of the PI controller (n^{th} sample)
$U(n-2)$ is the output of PI controller at $(n\text{-}2)^{th}$ sample
$e(n)$ is the error at n^{th} sample
$e(n-1)$ is the error at $(n\text{-}1)^{th}$ sample
$e(n-2)$ is the error at $(n\text{-}2)^{th}$ sample

The constants K_1, K_2, and K_3 for trapezoidal approximation are

$$K_1 = K_p + 2\frac{K_d}{T} + \frac{K_i T}{2} \tag{9.10}$$

$$K_2 = K_i T - 4\frac{K_d}{T} \tag{9.11}$$

$$K_3 = 2K_d - K_p + \frac{K_i T}{T} \tag{9.12}$$

and for rectangular approximation are

$$K_1 = K_p + \frac{K_d}{T} + K_i T \tag{9.13}$$

$$K_2 = K_i T - 2\frac{K_d}{T} \tag{9.14}$$

$$K_3 = \frac{K_d}{T} - K_p \tag{9.15}$$

In all of the above equations K_p, K_d, K_i are defined as in

$$u(t) = K_p e(t) + K_i \int dt + K_d \frac{de}{dt} \tag{9.16}$$

A portion of the assembly code implementing the PI controller is given below:

```
D_PID_spd:
        ....
        ....
        LACC     D_Un_H_0
        SUB      #MAX_POS_LIMIT
        BCND     D_PLUS_OK,LEQ          ;If maxed out, saturate at max -ve
        SPLK     #MAX_POS_LIMIT,D_Un_H_0
        SPLK     #0,D_Un_L_0
        B        D_EXIT

D_PLUS_OK:        LACC      D_Un_H_0        ;else keep current value
        SUB      #MAX_NEG_LIMIT
        BCND     D_NEG_OK,GEQ           ;if maxed out, saturate at max +ve
        SPLK     #MAX_NEG_LIMIT,D_Un_H_0          ;Saturation control
        SPLK     #0,D_Un_L_0
D_NEG_OK:
        .....
        .....
        RET
```

9.5.8 PWM Generation

The Compare Units have been used to generate the PWM signals. The PWM output signal is high when the output of current PI regulation matches the value of T1CNT and is set to low when the timer underflow occurs. The switch states are controlled by the ACTR register. As discussed earlier, in order to minimize the switching loss, the lower switches are always kept on and the upper switches are chopped on/off to regulate the phase current. From the implementation point of view, in using the LF2407, it is required that the ACTR register be reset for each interval. In other words, PWM1, PWM3, and PWM5 which gate the upper switches are set as *active low/high* and PWM2, PWM4, and PWM6 which trigger the lower switches are set as *force high*. The sample of code below illustrates this implementation.

```
BLDC_PWM_DRV
        ....
LACC    #COMMUTATION_TBL
        ADD     cmtn_ptr_bd
        TBLR    GPR0
        LACC    GPR0
        BACC
        STATE_ANB                       ;Input current path, Phase A
        POINT_EV                        ;Output current path, Phase B
        SPLK    #00C2H, ACTR            ;Non fed phase, Phase C
        B       STATE_END
        ....
STATE_END
        ....
        RET
```

9.5.9 *DAC Module*

The LF2407 evaluation board contains four channels DAC. In this application, the DAC on the evaluation board is used to display various system variables to be seen on an oscilloscope in real time. This feature is very useful during the development stage for real time debugging and verification of the software. The code below accepts the address pointers for four different system variables and then automatically updates the DAC channels to reflect the change in these variables.

```
;Convert Q15 input value to an absolute Q0 output to DAC0 channel
        POINT_B0
        SPM     1
        MAR     *, AR6
        LAR     AR6, DAC_IPTR0
        LT      *
        MPY     dac_hlf_rng             ;Normalize to half range of DAC
        PAC
        ADDH    dac_hlf_rng             ;offset by 1/2 DAC max value
        SACH    GPR0
        OUT     GPR0, PA0               ;DAC0 o/p
```

Chapter 10

CLARKE'S AND PARK'S TRANSFORMATIONS

10.1 Introduction

The performance of three-phase ac machines are described by their voltage equations and inductances. It is well known that some machine inductances are functions of rotor speed. The coefficients of the differential equations, which describe the behavior of these machines, are time varying except when the rotor is stalled. A change of variables is often used to reduce the complexity of these differential equations. There are several different methods to transform variables. In this chapter, the well-known Clarke and Park transformations are introduced, modeled, and implemented on the LF2407 DSP. Using these transformations, many properties of electric machines can be studied without complexities in the voltage equations. These transformations make it possible for control algorithms to be implemented on the DSP. By this approach, many of the basic concepts and interpretations of this general transformation are concisely established.

10.2 Clarke's Transformation

The transformation of stationary circuits to a stationary reference frame was developed by E. Clarke [2]. The stationary two-phase variables of Clarke's transformation are denoted as α and β. As shown in Fig. 10.1, α-axis and β-axis are orthogonal.

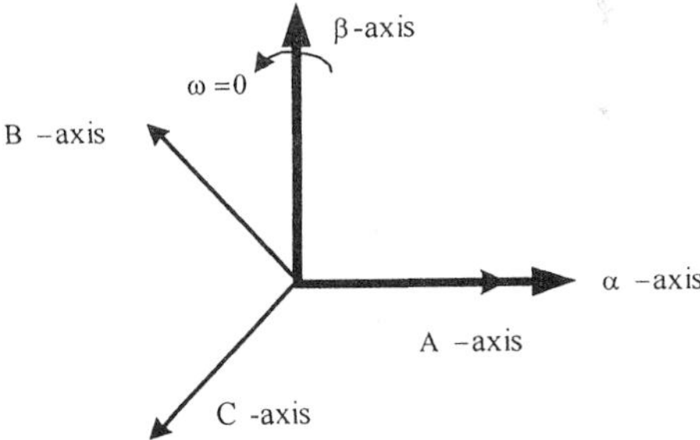

Figure 10.1 Clarke's transformation.

In order for the transformation to be invertible, a third variable, known as the zero-sequence component, is added. The resulting transformation is

$$\left[f_{\alpha\beta 0}\right] = T_{\alpha\beta 0}\left[f_{abc}\right] \tag{10.1}$$

where

$$\left[f_{\alpha\beta 0}\right] = \left[f_{\alpha} \quad f_{\beta} \quad f_{0}\right]^{T}$$

and

$$\left[f_{abc}\right] = \left[f_{a} \quad f_{b} \quad f_{c}\right]^{T}$$

where f represents voltage, current, flux linkages, or electric charge and the transformation matrix, $T_{\alpha\beta 0}$, is given by

$$T_{\alpha\beta 0} = \frac{2}{3}\begin{bmatrix} 1 & -\frac{1}{2} & -\frac{1}{2} \\ 0 & \frac{\sqrt{3}}{2} & -\frac{\sqrt{3}}{2} \\ \frac{1}{2} & \frac{1}{2} & \frac{1}{2} \end{bmatrix} \tag{10.2}$$

The inverse transformation is given by

$$[f_{abc}] = T_{\alpha\beta 0}{}^{-1}[f_{\alpha\beta 0}] \tag{10.3}$$

where the inverse transformation matrix is presented by

$$T_{\alpha\beta 0}{}^{-1} = \begin{bmatrix} 1 & 0 & 1 \\ -\frac{1}{2} & \frac{\sqrt{3}}{2} & 1 \\ -\frac{1}{2} & -\frac{\sqrt{3}}{2} & 1 \end{bmatrix} \tag{10.4}$$

10.3 Park's Transformation

In the late 1920s, R.H. Park [1] introduced a new approach to electric machine analysis. He formulated a change of variables which replaced variables such as voltages, currents, and flux linkages associated with fictitious windings rotating with the rotor. He referred the stator and rotor variables to a reference frame fixed on the rotor. From the rotor point of view, all the variables can be observed as constant values. Park's transformation, a revolution in machine analysis, has the unique property of eliminating all time varying inductances from the voltage equations of three-phase ac machines due to the rotor spinning.

Although changes of variables are used in the analysis of ac machines to eliminate time-varying inductances, changes of variables are also employed in the analysis of various static and constant parameters in power system components. Fortunately, all known real transformations for these components are also contained in the transformation to the arbitrary reference frame. The same general transformation used for the stator variables of ac machines serves the rotor variables of induction machines. Park's transformation is a well-known three-phase to two-phase transformation in synchronous machine analysis. Park's transformation is presented in Fig. 10.2.

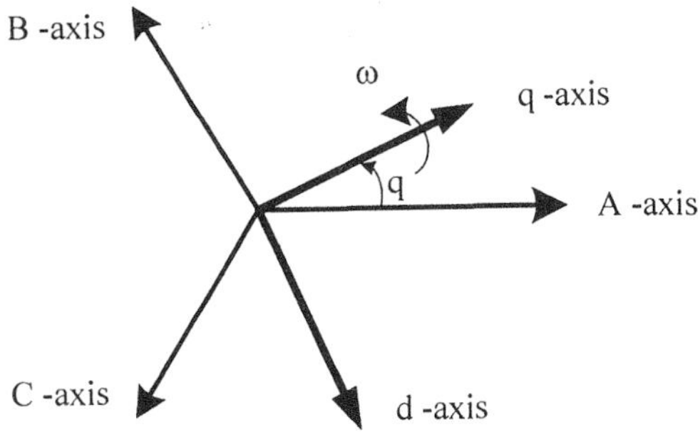

Figure 10.2 Park's transformation.

The transformation equation is of the form

$$\left[f_{qd0s}\right]=T_{qd0}(\theta)\left[f_{abcs}\right] \tag{10.5}$$

where

$$\left[f_{qd0s}\right]=\left[f_{qs} \quad f_{ds} \quad f_{0s}\right]^T$$

and $\left[f_{abcs}\right]=\left[f_{as} \quad f_{bs} \quad f_{cs}\right]^T$

and the dq0 transformation matrix is defined as

$$T_{qd0s}(\theta)=\frac{2}{3}\begin{bmatrix} \cos(\theta) & \cos(\theta-\frac{2\pi}{3}) & \cos(\theta+\frac{2\pi}{3}) \\ \sin(\theta) & \sin(\theta-\frac{2\pi}{3}) & \sin(\theta+\frac{2\pi}{3}) \\ \frac{1}{2} & \frac{1}{2} & \frac{1}{2} \end{bmatrix} \tag{10.6}$$

θ is the angular displacement of Park's reference frame and can be calculated by

$$\theta=\int_0^t \omega(\zeta)d\zeta+\theta(0) \tag{10.7}$$

where ζ is the dummy variable of integration. It can be shown that for the inverse transformation we can write

$$\left[f_{abcs}\right]=T_{qd0}(\theta)^{-1}\cdot\left[f_{qd0s}\right] \tag{10.8}$$

where the inverse of Park's transformation matrix is given by

$$T_{qd0}(\theta)^{-1} = \begin{bmatrix} \cos(\theta) & \sin(\theta) & 1 \\ \cos(\theta - \frac{2\pi}{3}) & \sin(\theta - \frac{2\pi}{3}) & 1 \\ \cos(\theta + \frac{2\pi}{3}) & \sin(\theta + \frac{2\pi}{3}) & 1 \end{bmatrix} \tag{10.9}$$

In the previous equations, the angular displacement θ must be continuous, but the angular velocity associated with the change of variables is unspecified. The frame of reference may rotate at any constant, varying angular velocity, or it may remain stationary. The angular velocity of the transformation can be chosen arbitrarily to best fit the system equation solution or to satisfy the system constraints. The change of variables may be applied to variables of any waveform and time sequence; however, we will find that the transformation given above is particularly appropriate for an a-b-c sequence.

10.4 Transformations Between Reference Frames

In order to reduce the complexity of some derivations, it is necessary to transform the variables from one reference frame to another one. To establish this transformation between any two reference frames, we can denote y as the new reference frame and x as the old reference frame. Both new and old reference frames are shown in Fig. 10.3.

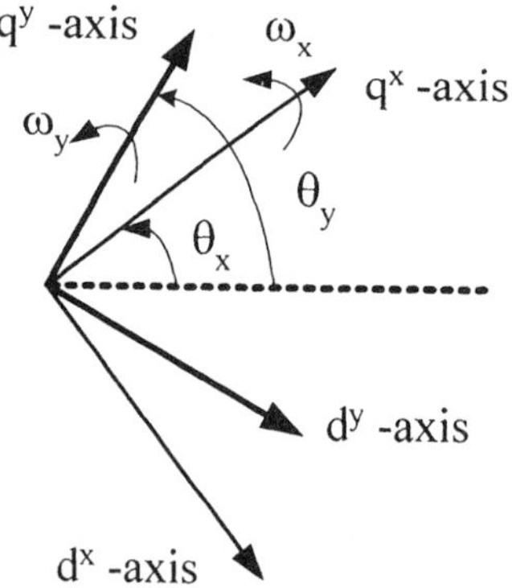

Figure 10.3 Transformation between two reference frames.

It is assumed that the reference frame x is rotating with angular velocity ω_x and the reference frame y is spinning with the angular velocity ω_y. θ_x and θ_y are angular displacements of reference frames x and y, respectively. In this regard, we can rewrite the transformation equation as

$$\left[f_{qd0s}^{y}\right] = T_{qd0s}^{x \to y} \cdot \left[f_{qd0s}\right] \tag{10.10}$$

But we have

$$\left[f^{x}_{qd0s}\right]=T^{x}_{qd0s}\cdot\left[f_{abcs}\right] \tag{10.11}$$

If we substitute (10.11) in (10.10) we get

$$\left[f^{y}_{qd0s}\right]=T^{x\to y}_{qd0s}\cdot T^{x}_{qd0s}\cdot\left[f_{abcs}\right] \tag{10.12}$$

In another way, we can find out that

$$\left[f^{y}_{qd0s}\right]=T^{y}_{qd0s}\cdot\left[f_{abcs}\right] \tag{10.13}$$

From (10.12) we obtain

$$T^{x\to y}_{qd0s}=T^{y}_{qd0s}\cdot {T^{x}_{qd0s}}^{-1} \tag{10.14}$$

Then, the desired transformation can be expressed by the following matrix:

$$T^{x\to y}_{qd0s}=\begin{bmatrix}\cos(\theta_y-\theta_x) & -\sin(\theta_y-\theta_x) & 0\\ \sin(\theta_y-\theta_x) & \cos(\theta_y-\theta_x) & 0\\ 1 & 1 & 1\end{bmatrix} \tag{10.15}$$

10.5 Field Oriented Control (FOC) Transformations

In the case of FOC of electric machines, control methods are performed in a two-phase reference frame fixed to the rotor *(q^r-d^r)* or fixed to the excitation reference frame *(q^e-d^e)*. We want to transform all the variables from the three-phase a-b-c system to the two-phase stationary reference frame and then retransform these variables from the stationary reference frame to a rotary reference frame with arbitrary angular velocity of ω. These transformations are usually cascaded. The block diagram of this procedure is shown in Fig. 10.4.

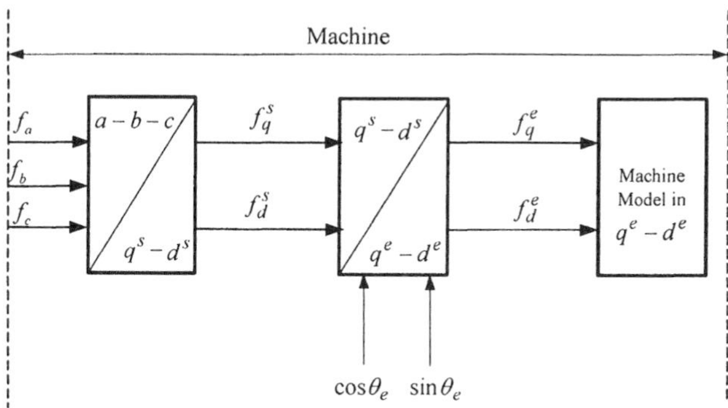

Figure 10.4 Machine side transformation in field oriented control.

In this figure, f denotes the currents or voltages and q^e-d^e represents the arbitrary rotating reference frame with angular velocity ω_e and q^s-d^s represents the stationary reference frame. In the vector control method, after applying field-oriented control it is necessary to transform variables to stationary a-b-c system. This can be achieved by taking the inverse transformation of variables from the arbitrary rotating reference frame to the stationary reference frame and then to the a-b-c system. The block diagram of this procedure is shown in Fig. 10.5. In this block diagram, * is a representation of commanded or desired values of variables.

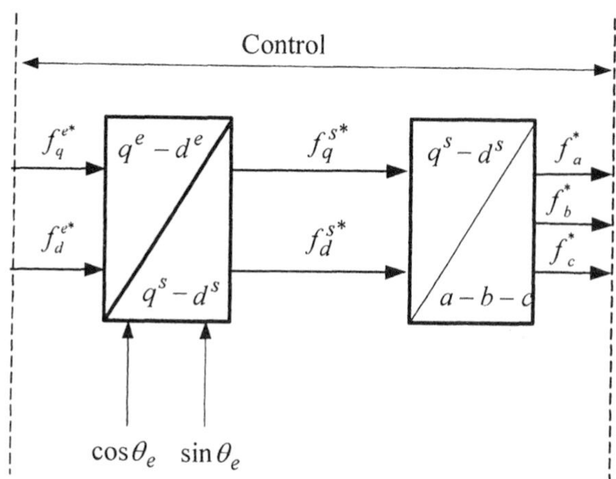

Figure 10.5 Variable transformation in the field oriented control.

10.6 Implementing Clarke's and Park's Transformations on the LF240X

10.6.1 Implementing Clarke's Transformation

It is desired to transfer the three-phase stationary parameters f_a, f_b, and f_c from the a-b-c system to the two-phase stationary reference frame. It is assumed that the system is balanced and we have

$$f_a + f_b + f_c = 0 \tag{10.16}$$

We can rewrite (10.1) as follows:

$$f_\alpha = \frac{2}{3} f_a - \frac{1}{3} f_b - \frac{1}{3} f_c \tag{10.17}$$

$$f_\beta = \frac{1}{\sqrt{3}} (f_b - f_c) \tag{10.18}$$

Substituting f_c from (10.16) into (10.17) and (10.18) results in

$$f_\alpha = f_a \tag{10.19}$$

$$f_\beta = \frac{1}{\sqrt{3}}(f_a + 2f_b) \quad (10.20)$$

10.6.1.1 Inputs and Outputs of Clarke's Transformation Block

The inputs and outputs of Clarke's transformation are shown in Fig. 10.6. As it is shown in this figure, f_a and f_b are inputs and f_α and f_β are outputs of this transformation.

Figure 10.6 Clarke transformation.

To enjoy better resolution of the variables in fixed point DSP, we transfer all variables to the Q15-based format. With this consideration, the maximum value of inputs and outputs can be $(2^{15}\text{-}1)$ or in hexadecimal, the format shall be $7FFF_h$. In this base, the variables can vary in the range 8000_h-$7FFF_h$. This transformation converts balanced three-phase quantities into balanced two-phase quadrature quantities as shown in Fig. 10.7.

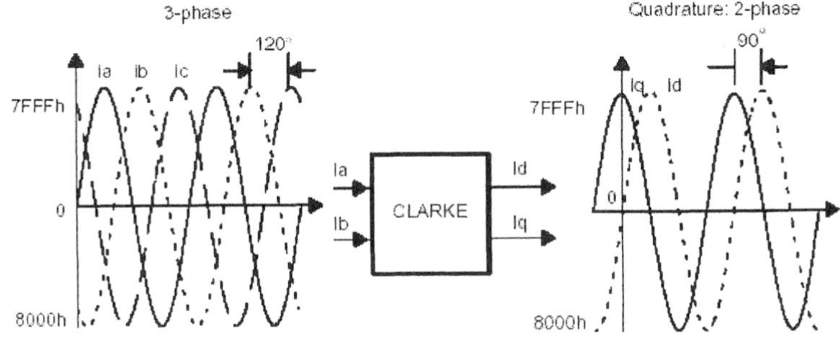

Figure 10.7 Quantities in Clarke's transformation. (Courtesy of Texas Instruments)

As we previously noted, our calculations are based on the Q15 format . So all the coeficients are present in this representation. Then 1/√3 is represented by

```
LDP     #sqrt3inv            ;sqrt3inv=(1/sqrt(3))
                             ;=0.577350269
SPLK    #018830,sqrt3inv     ;1/sqrt(3) (Q15)
```

Clarke's transformation is implemented as follows:

```
SETC    SXM             ;Sign extension mode on
LDP     #clark_a        ;clark_alfa = clark_a
LACC    clark_a         ;ACC = clark_a
SACL    clark_alfa      ;clark_d = clark_a
                        ;clark_beta=(2*clark_b+clark_a)/
                        ;sqrt(3)
SFR                     ;ACC = clark_a/2
ADD     clark_b         ;ACC = clark_a/2 + clark_b
SACL    clk_temp        ;clk_temp = clark_a/2 + clark_b
LT      clk_temp        ;TREG = clark_a/2 + clark_b
MPY     sqrt3inv        ;PREG=(clark_a/2+clark_b)*
                        ;(1/sqrt(3))
PAC                     ;ACC=(clark_a/2+clark_b)*
                        ;(1/sqrt(3))
SFL                     ;ACC=(clark_a+clark_b*2)*
                        ;(1/sqrt(3))
SACH    clark_beta      ;clark_beta=(clark_a+clark_b*2
                        ;(1/sqrt(3))
SPM     0               ;SPM reset
RET
```

10.6.2 Inverse Clarke's Transformation

From (10.3), the inverse Clarke functions for a balanced system can be obtained as

$$\begin{aligned} f_a &= f_\alpha \\ f_b &= \frac{-f_\alpha + \sqrt{3} * f_\beta}{2} \\ f_c &= \frac{-f_\alpha - \sqrt{3} * f_\beta}{2} \end{aligned} \tag{10.21}$$

This transformation converts balanced two-phase quadrature quantities into balanced three-phase quantities. The block diagram of the inverse Clarke transformation is shown in Fig. 10.8.

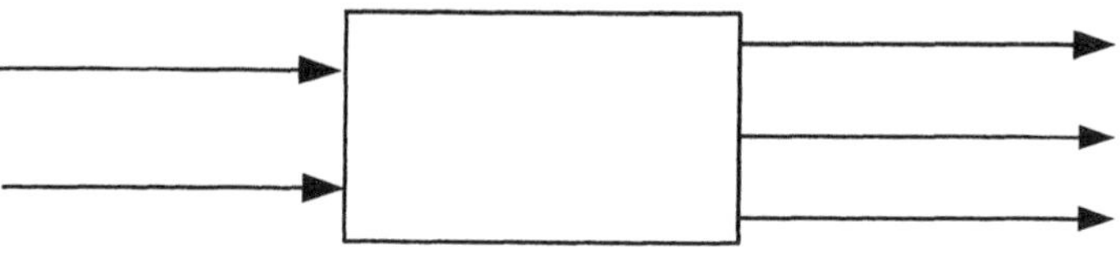

Figure 10.8 Inverse Clark transformation block.

In this block diagram, f_α and f_β are inputs and f_a, f_b, and f_c are outputs. Inputs and outputs are represented in Q15 format. Variation of quantities in the inverse Clark transformation is shown in Fig. 10.9.

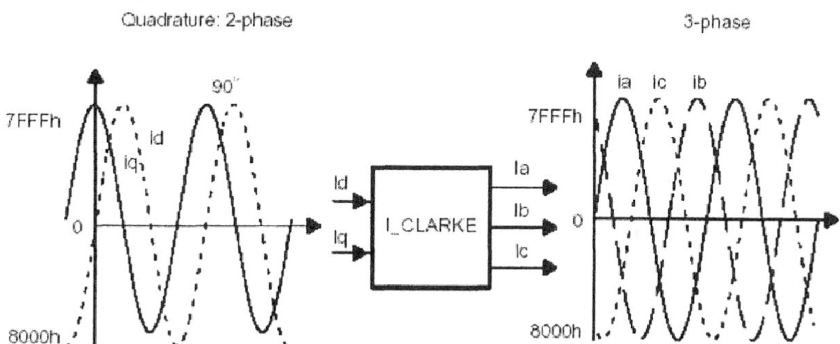

Figure 10.9 Quantities in inverse Clarke's transformation. (Courtesy of Texas Instruments)

Implementation of the inverse Clarke transformation via assembly code is as follows

```
I_CLARKE_INIT:
        LDP     #half_sqrt3             ;Variables data page
        SPLK    #28377,half_sqrt3       ;Set constant sqrt(3)*0.5 in Q15
                                        ;format
RET
I_CLARKE:
        LDP     #f_clark_alpha          ;Variables data page
        SPM     1                       ;SPM set for Q15 multiplication
        SETC    SXM                     ;Sign extension mode on
        LACC    f_clark_alpha           ;ACC = f_alpha
        SACL    f_clark_a               ;f_a = f_alpha
        LT      f_clark_beta            ;TREG = f_clark_beta
        MPY     half_sqrt3              ;PREG=f_clark_beta * half_sqrt3
        PAC                             ;ACC= f_clark_beta * half_sqrt3
SUB     f_clark_alpha,15                ;ACC=f_beta*half_sqrt3-f_alpha/2
        SACH    f_clark_b
        PAC                             ;ACC high = f_beta*half_sqrt3
        NEG                             ;ACC high = - f_beta*half_sqrt3
        SUB     f_clark_alpha,15        ;ACC high=-f_beta*half_sqrt3-
                                        ;f_alpha/2
        SACH    f_clark_c               ;f_c = - f_beta * half_sqrt3 -
                                        ;f_alpha/2
        SPM     0                       ;SPM reset
        CLRC    SXM                     ;Sign extension mode off
RET
```

10.6.3 Calculation of Sine/Cosine with Fast Table Direct Look-Up and Linear Interpolation

To implement the Park and the inverse Park transforms, the sine and cosine functions need to be implemented. This method realizes the sine/cosine functions with a look-up table of 256 values for 360° of sine and cosine functions. The method includes linear interpolation with a fixed step table to provide a minimum harmonic distortion. This table is loaded in program memory. The sine value is presented in Q15 format with the range of -1<value<1. The first few rows of the look-up sine table are presented as follows:

```
        ;SINVALUE                     ;        Index     Angle      Sin(Angle)
        ----------------                       --------  --------   ------------
SINTAB_360
        .word           0             ;        0         0          0.0000
        .word           804           ;        1         1.41       0.0245
        .word           1608          ;        2         2.81       0.0491
        .word           2410          ;        3         4.22       0.0736
        .word           3212          ;        4         5.63       0.0980
```

The following assembly code is written to read values of sine from the sine Table in Q15 format:

```
LACC    theta_p, 9                  ;Input angle in Q15 format and
                                    ;left shifted by 15
SACH    t_ptr                       ;Save high ACC to t_ptr (table
                                    ;pointer)
LACC    #SINTAB_360
ADD     t_ptr
TBLR    sin_theta                   ;sin_theta = Sin(theta_p) in Q15
```

Note that 0 < theta_p < 7FFFh (i.e., equivalent to 0 < theta_p < 360 deg). The TBLR instruction transfers a word from a location in program memory to a data-memory location specified by the instruction. The program-memory address is defined by the low-order 16 bits of the accumulator. For this operation, a read from program memory is performed, followed by a write to data memory.

To calculate the cosine values from the sine Table in Q15 format, we write the following code:

```
LACC    theta_p
ADD     #8192                       ;add 90 deg, cos(A)=sin(A+90°)
AND     #07FFFh                     ;Force positive wrap-around
SACL    GPR0_park                   ;here 90 deg = 7FFFh/4
LACC    GPR0_park,9
SACH    t_ptr
LACC    #SINTAB_360
```

10.6.4 Implementation of Park's Transformation on LF2407

As discussed in Section 10.5, with field-oriented control of motors, it is necessary to transform variables, i.e., currents and voltages, from a-b-c system to two-phase stationary reference frame, q^s-d^s, and from two-phase stationary reference frame q^s-d^s to arbitrary rotating reference frame with angular velocity of ω (q-d reference frame). The first transformation is dual to Clarke's transformation but the q^s axis is in the direction of α–axis, and d^s axis is in negative direction of β –axis. These two transformations are explained in the following sections.

10.6.4.1 Transformation from 3-phase to 2-phase Stationary Reference Frame $(a-b-c) \rightarrow (q^s - d^s)$

This transformation transfers the three-phase stationary parameters, f_a, f_b, and f_c from an a-b-c system to a two-phase orthogonal stationary reference frame. If we substitute θ=0 in (10.6) and assuming that the system is balanced, we get:

$$f_q^s = f_a \tag{10.23}$$

$$f_d^s = -\frac{1}{\sqrt{3}}(2f_b + f_a) \tag{10.24}$$

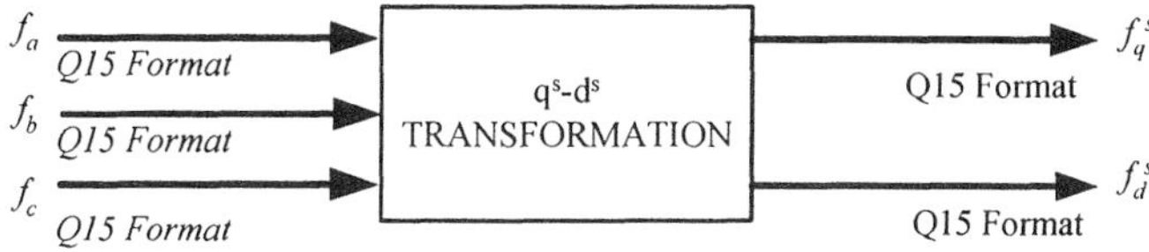

Figure 10.11 Two-phase stationary transformation.

Both input and output are represented in Q15 format with a block diagram of the transformation being shown in Fig. 10.11. The developed code is similar to what was mentioned in Section 10.6.1.1.

10.6.4.2 Transformation from the Stationary Reference Frame to the Arbitrary Rotary Reference Frame $(q^s - d^s) \rightarrow (q - d)$

This transformation converts vectors in a balanced two-phase orthogonal stationary system into an orthogonal rotary reference frame. The inputs are f_q^s, f_d^s, and θ, and the outputs are f_q and f_d. This is the transformation between the stationary reference frame and the arbitrary reference frame rotating with the angular velocity of ω. If we substitute $\theta_x = 0$ and $\theta_y = \theta$ we obtain:

$$\begin{aligned} f_q &= \cos\theta . f_q^s - \sin\theta \cdot f_d^s \\ f_d &= \sin\theta . f_q^s + \cos\theta \cdot f_d^s \end{aligned} \tag{10.25}$$

where θ is the angular displacement.

In this transformation, it is necessary to calculate $\sin\theta$ and $\cos\theta$, where the method to calculate them was presented in a previous section. In Fig. 10.12, the input and output of the Park transformation block has been shown. All the input and outputs are in the Q15 format and in the range of 8000_h-$7FFF_h$.

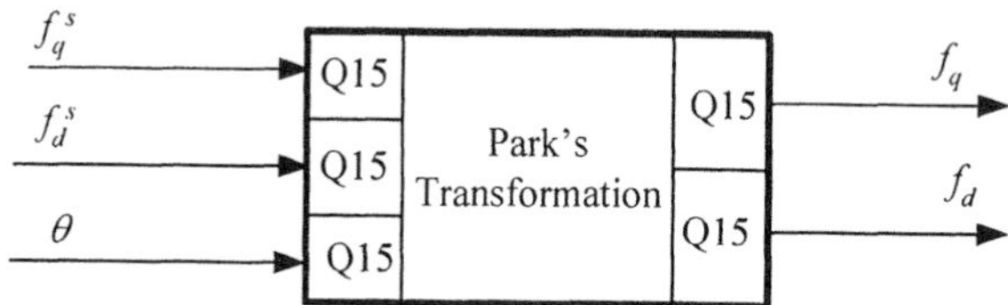

Figure 10.12 Park transformation block.

The following code is written to implement Park's transformation:

```
        SPM     1               ;SPM set for Q15 multiplication
        ZAC                     ;Reset accumulator
        LT      f_q_s           ;TREG = f_q_s
        MPY     sin_theta       ;PREG = f_q_s * sin(theta)
        LTA     f_d             ;ACC = f_q_s * sin(theta) and
                                ;TREG =f_q_s
        MPY     cos_theta       ;PREG = f_d_s* cos_teta
        MPYA    sin_theta       ;ACC=f_q_s*sin_teta+f_d_s*
                                ;cos_teta andPREG=f_q_s*sin_teta
        SACH    park_D          ;f_d =f_q_s * cos_teta + f_d_s*
                                ;sin(theta)
        LACC    #0              ;Clear ACC
        LT      f_d_s           ;TREG = f_d_s
        MPYS    cos_theta       ;ACC=- f_d_s* *sin(theta) and
                                ;PREG = f_q_s * cos(theta)
        APAC                    ;ACC=- f_d_s*sin(theta) +f_q_s*
                                ;cos(theta)
        SACH    f_q             ;fq = -f_d_s*sin(theta) +f_q_s*
                                ;cos(theta)
        SPM     0               ;SPM reset
RET
```

10.6.5 Transformation of the Arbitrary Rotating Reference Frame to the Stationary Reference Frame $(q-d)\rightarrow(q^s-d^s)$

This transformation projects vectors in an orthogonal rotating reference frame into a two-phase orthogonal stationary frame. From (10.15) we get:

$$
\begin{aligned}
f_q^s &= \cos\theta\cdot f_q + \sin\theta\cdot f_d \\
f_d^s &= -\sin\theta\cdot f_q + \cos\theta\cdot d_d
\end{aligned}
\tag{10.26}
$$

In this transformation, θ is the angular displacement. To transform variables to Park's reference frame, it is necessary to calculate $\sin\theta$ and $\cos\theta$. Use the method presented in the previous section. In Fig. 10.13, inputs and outputs of the inverse

Park transformation block are shown. The inputs are f_d, f_q, and θ, and the outputs are f_α and f_β. All the inputs and outputs are in the Q15 format and in the range of 8000_h-$7FFF_h$.

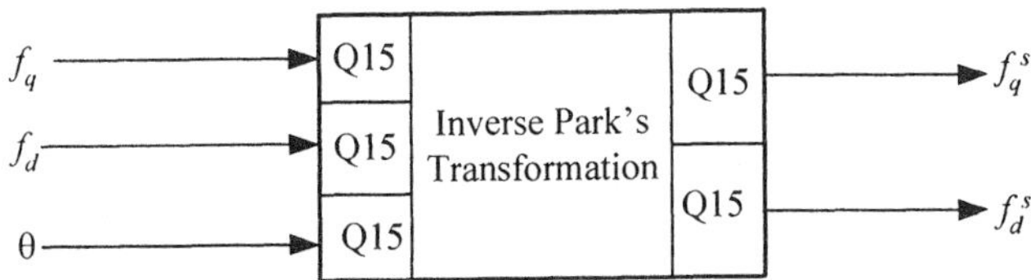

Figure 10.13 Inverse Park's transformation block.

The following code is written to implement this transformation:

```
SPM     1                       ;SPM set for Q15 multiplication
        ZAC                     ;Reset accumulator
        LT      f_q             ;TREG = fq
        MPY     cos_theta       ;PREG = fq * cos(theta)
        LTA     f_d             ;ACC=fq*cos(theta) and TREG =fd
        MPY     sin_theta       ;PREG = fd * sin(theta)
        MPYA    sin_theta       ;ACC=fq*cos(theta)+fd*sin(theta)
                                ;and PREG=fd*sin(theta)
        SACH    f_q_s           ;fd=fq*cos(theta)+fd*sin(theta)
        LACC    #0              ;Clear ACC
        LT      f_d             ;TREG = fd
        MPYS    cos_theta       ;ACC = -fd*sin_theta and
                                ;PREG = fd*cos_theta

        APAC
        SACH    f_d_s
        SPM     0               ;SPM reset

RET
```

10.6.6 The 2-Phase to 3-Phase Transformation $(q^s - d^s) \rightarrow (a - b - c)$

This transformation transforms the variables from the stationary two-phase q^s-d^s frame to the stationary a-b-c system. This system is also dual to the inverse Clarke transformation where the q^s-axis is in the direction of the α axis and the d^s-axis is in the negative direction of β–axis.

If we substitute θ=0 in (10.9) and assume a balanced system we get:

$$f_a = f_q^s$$
$$f_b = \frac{-f_q^s - \sqrt{3} f_d^s}{2} \tag{10.27}$$
$$f_c = \frac{-f_q^s - \sqrt{3} f_d^s}{2}$$

The implemented code is similar to the inverse Clarke transformation which will not be repeated in here.

10.7 Conclusion

With FOC of synchronous and induction machines, it is desirable to reduce the complexity of the electric machine voltage equations. The transformation of machine variables to an orthogonal reference frame is beneficial for this purpose. Park's and Clarke's transformations, two revolutions in the field of electrical machines, were studied in depth in this chapter. These transformations and their inverses were implemented on the fixed point LF2407 DSP.

References

1. R. H. Park, "Two-reaction theory of synchronous machines – Generalized method of analysis- Part I," *AIEE Trans.*, Vol. 48, July 1929, pp.716-727
2. E. Clarke, *Circuit Analysis of AC Power Systems*, Vol. I- Symmetrical and Related Components, John Wiley and Sons, New York, 1943.
3. P. Krause, O. Wasynczuk, S. D. Sudhoff, *Analysis of Electric Machinery*, IEEE Press, New York, 1995.
4. C. Ong, *Dynamic Simulation of Electric Machinery*, Prentice Hall, Englewood Cliffs, NJ, 1998.
5. H.C Stanley, "Analysis of the Induction Motor," *AIEE Trans.*, Vol. 57, 1938, pp. 751-755.

Chapter 11

SPACE VECTOR PULSE WIDTH MODULATION

11.1 Introduction

In this chapter, the concept of space vector pulse width modulation (SVPWM) as applied to an induction motor will be introduced. An explanation of the DSP assembly code is needed to implement the control algorithm. Several key functional parts of the DSP code will be discussed.

Of all motors, the squirrel cage induction motor is the most widely used motor in the industry. This leading position results mainly from certain excellent features of the squirrel cage motor such as:

- Uncomplicated, rugged construction; this means low initial cost and high reliability.
- Good efficiency coupled with low maintenance costs, resulting in low overall operating costs.

Squirrel cage motors, like all induction machines, are asynchronous machines with speed depending upon applied frequency, pole number, and load torque. In order to use the poly-phase ac motor as an adjustable speed device, it is necessary to control and adjust the frequency of the three-phase voltages applied to its terminals. The operating speed of the motor is determined by the following relationship

$$N = \frac{120 \cdot f}{P}(1-s) \tag{11.1}$$

where N is the shaft speed in rpm, f is the supplied frequency in Hz, P is the number of poles, and s is the operating slip.

A switching power converter can be used to control both the supplied voltage and frequency. Consequently, higher efficiency and performance can be achieved. The most common control principle for induction motors is the constant volts per hertz (V/Hz) principle, which will be explained in the next section.

11.2 Principle of Constant V/Hz Control for Induction Motors

For us to understand the V/Hz control, we will first assume that the voltage applied to a three-phase ac induction motor is sinusoidal, and neglect the voltage drop across the stator resistor. At steady state the machine terminal voltage is given by

$$\hat{V} \approx j\varpi\hat{\Lambda} \tag{11.2}$$

or

$$\hat{V} \approx \varpi\hat{\Lambda}$$

where $\hat{V}$ and $\hat{\Lambda}$ are the phasors of stator voltage and stator flux, and V and Λ are their respective magnitudes.

$$\Lambda \approx \frac{V}{\varpi} = \frac{1}{2\pi}\frac{V}{f} \tag{11.3}$$

It follows that if the ratio V/f remains constant with the change of f, then Λ also remains constant and the torque is independent of the supply frequency.

In actual implementation, the ratio between the magnitude and frequency of the stator voltage is usually based on the rated values of these variables, also known as motor ratings. However, when the frequency and voltage are low, the voltage drop across the stator resistance cannot be neglected. At frequencies higher than the rated value, to avoid insulation break, the constant V/f principle has to be violated. The realistic control limits that are placed on the applied voltage and frequency are illustrated in Fig. 11.1.

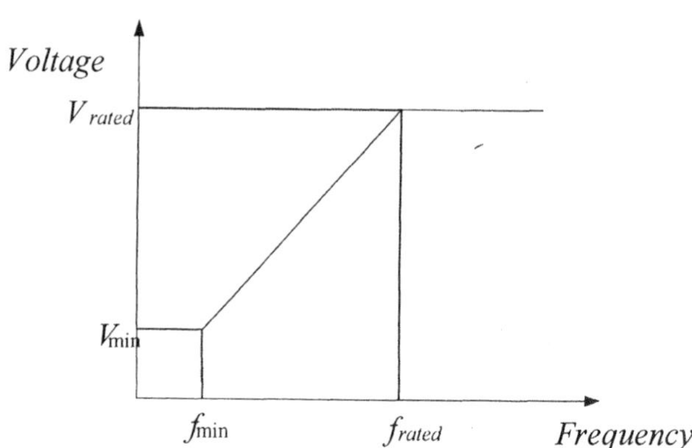

Figure 11.1 V/f limits on frequency and voltage.

11.3 Space Vector PWM Technique

Space Vector PWM (SVPWM) refers to a special technique of determining the switching sequence of the upper three power transistors of a three-phase voltage source inverter (VSI). It has been shown to generate less harmonic distortion in the output voltages or current in the windings of the motor load. SVPWM provides more efficient use of the dc bus voltage, in comparison with the direct sinusoidal modulation technique.

The structure of a typical three-phase voltage source inverter is shown in Fig. 11.2. The voltages, V_a , V_b, and V_c are the output voltages applied to the windings of a motor. Q1 through Q6 are the six power transistors which are controlled by a, a', b, b', c and c' gating signals and shape the output voltages. When an upper transistor is switched on, i.e., when a, b, and c are 1, the corresponding lower transistor is switched off, i.e., the corresponding a', b' or c' is 0. The on and off states of the upper transistors Q1, Q3, and Q5, or the states of a, b, and c are sufficient to evaluate the output voltage.

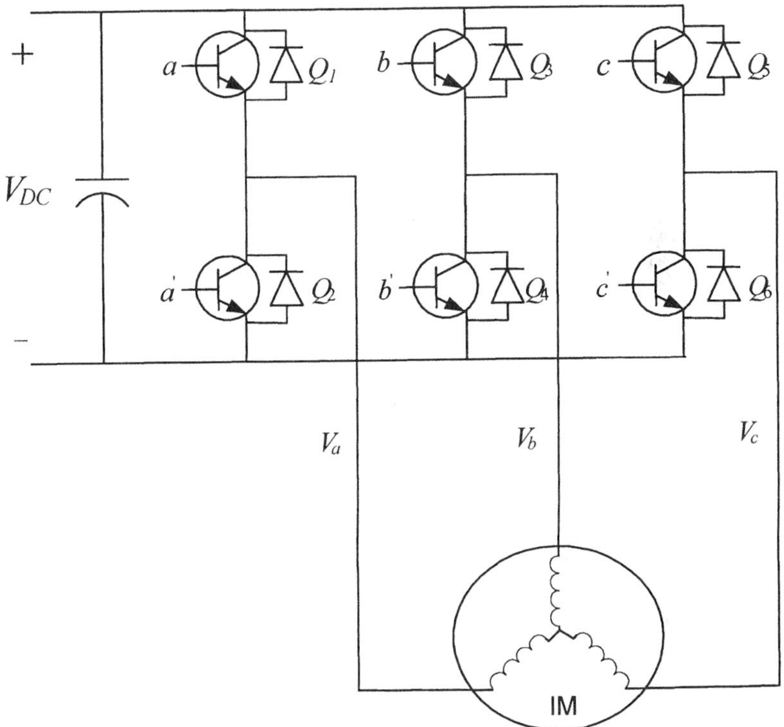

Figure 11.2 Three-phase power inverter supplying an induction motor.

11.3.1 Switching Patterns and the Basic Space Vectors

There are eight possible combinations of on and off states for the three upper power transistors. The on and off states of the lower power transistors are opposite to the upper ones, so they are determined once the states of the upper transistors are known. The eight combinations are the derived output line-to-line and phase voltages in terms of DC supply voltage, V_{dc}, according to (11.4) and (11.5), which are shown in Table 11.1.

The relationship between the switching variable vector $[a, b, c]^T$ and the line-to-line voltage vector $[V_{ab}, V_{bc}, V_{ca}]^T$ is given by the following:

$$\begin{bmatrix} V_{ab} \\ V_{bc} \\ V_{ca} \end{bmatrix} = V_{dc} \begin{bmatrix} 1 & -1 & 0 \\ 0 & 1 & -1 \\ -1 & 0 & 1 \end{bmatrix} \begin{bmatrix} a \\ b \\ c \end{bmatrix} \tag{11.4}$$

In addition, phase (line-to-neutral) output voltage vector $[V_a, V_b, V_c]^T$ is given by (11.5)

$$\begin{bmatrix} V_a \\ V_b \\ V_c \end{bmatrix} = \frac{1}{3} V_{dc} \begin{bmatrix} 2 & -1 & -1 \\ -1 & 2 & -1 \\ -1 & -1 & 2 \end{bmatrix} \begin{bmatrix} a \\ b \\ c \end{bmatrix} \tag{11.5}$$

Table 11.1 Switching patterns and output voltages of a three-phase power inverter

a	**b**	**c**	V_a	V_b	V_c	V_{ab}	V_{bc}	V_{ca}
0	0	0	0	0	0	0	0	0
1	0	0	2/3	−1/3	−1/3	1	0	-1
1	1	0	1/3	1/3	−2/3	0	1	-1
0	1	0	−1/3	2/3	−1/3	-1	1	0
0	1	1	−2/3	1/3	1/3	-1	0	1
0	0	1	−1/3	−1/3	2/3	0	-1	1
1	0	1	1/3	−2/3	1/3	1	-1	0
1	1	1	0	0	0	0	0	0

11.3.2 Expression of the Stator Voltages in the (d-q) Frame

Assuming *q* and *d* are the horizontal and vertical axes of the stator coordinate frame, the *d-q* transformation given in (11.6) can transform a three-phase voltage vector into a vector in the *d-q* coordinate frame. This vector represents the spatial vector sum of the three-phase voltage. The phase voltages corresponding to the eight combinations of switching patterns can be mapped into the *d-q* plane by the same *d-q* transformation as shown in Table 11.2. This mapping results in 6 non-zero vectors and 2 zero vectors. The non-zero vectors form the axes of a hexagonal as shown in Fig. 11.3. The angle between any two adjacent non-zero vectors is 60^0. The 2 zero vectors are positioned at the origin and apply zero voltage to a motor. The group of the 8 vectors are referred to as the basic space vectors and are denoted by V_0, through V_7. The *d-q* transformation can be applied to the reference a, b, and c voltages to obtain the reference V_{out} in the *d-q* plane as shown in Fig. 11.3.

$$\begin{bmatrix} V_q \\ V_d \end{bmatrix} = \frac{2}{3} \begin{bmatrix} 1 & -\frac{1}{2} & -\frac{1}{2} \\ 0 & \frac{\sqrt{3}}{3} & -\frac{\sqrt{3}}{2} \end{bmatrix} \begin{bmatrix} V_a \\ V_b \\ V_c \end{bmatrix} \tag{11.6}$$

Table 11.2 The eight switching states and corresponding d-q voltages.

a	b	c	V_q	V_d	V_{dq}
0	0	0	0	0	$V_0 = 0$
0	0	1	$-\frac{1}{3}V_{dc}$	$\frac{1}{\sqrt{3}}V_{dc}$	$V_1 = \frac{2}{3}V_{dc}$
0	1	0	$-\frac{1}{3}V_{dc}$	$-\frac{1}{\sqrt{3}}V_{dc}$	$V_2 = \frac{2}{3}V_{dc}$
0	1	1	$-\frac{2}{3}V_{dc}$	0	$V_3 = \frac{2}{3}V_{dc}$
1	0	0	$\frac{2}{3}V_{dc}$	0	$V_4 = \frac{2}{3}V_{dc}$
1	0	1	$\frac{1}{3}V_{dc}$	$\frac{1}{\sqrt{3}}V_{dc}$	$V_5 = \frac{2}{3}V_{dc}$
1	1	0	$\frac{1}{3}V_{dc}$	$-\frac{1}{\sqrt{3}}V_{dc}$	$V_6 = \frac{2}{3}V_{dc}$
1	1	1	0	0	$V_7 = 0$

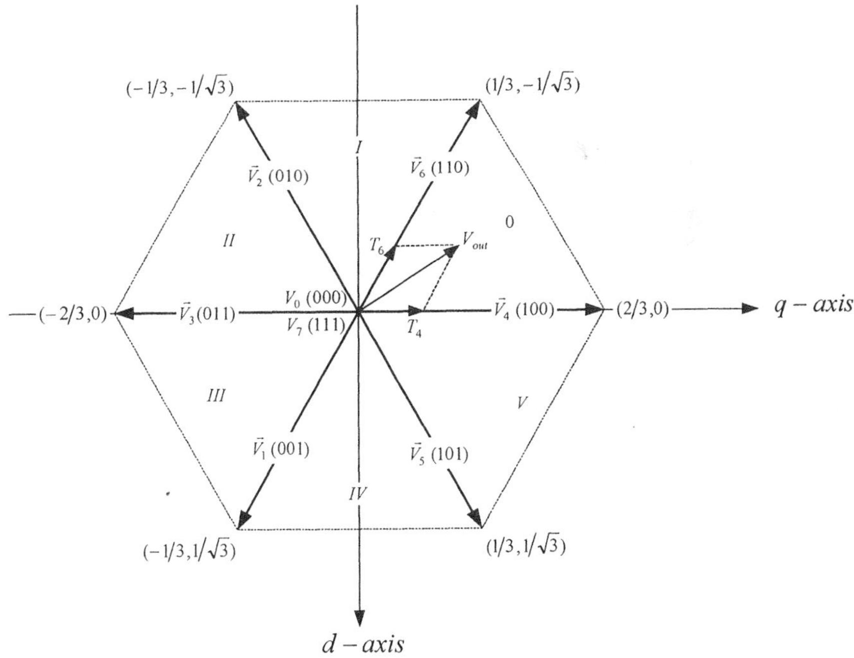

Figure 11.3 Space vector diagram.

11.3.3 Approximation of Output with Basic Space Vectors

The objective of the space vector PWM technique is to approximate the reference voltage vector V_{out} by a combination of the eight switching patterns. One simple means of approximation is to require the average output voltage of the inverter (in small period *T*) to be the same as the average of V_{out} in the same period. This is shown in (11.7) for the output voltage in the Sector 0, where T_4 and T_6 are the respective durations in time for which switching patterns are V_4 and V_6.

$$\frac{1}{T}\int_{nT}^{(n+1)T} V_{out}\,dt = \frac{1}{T}(T_4V_4 + T_6V_6) \qquad n = 0,1,2,\cdots,\ where\quad T_4 + T_6 \le T \qquad (11.7)$$

Assuming the PWM period, T_{pwm}, is small and the change of V_{out} is relatively slow, from (11.7), we obtain

$$\int_{nT_{PWM}}^{(n+1)T_{PWM}} V_{out}\,dt = T_{PWM}V_{out} = (T_4V_4 + T_6V_6) \quad n = 0,1,2,\cdots,\ where\ T_4 + T_6 \le T_{PWM}$$

(11.8)

Equation (11.8) shows that for every PWM period, the desired reference voltage V_{out} can be approximated by having the power inverter in a switching pattern of V_4 and V_6 for T_4 and T_6 periods of time, respectively. Since the sum of T_4 and T_6 is less than or equal to T_{pwm}, the inverter needs to have a 0 ((000) V_0 or (111) V_7) pattern for the rest of the period. Therefore, (11.8) will then become

$$T_{PWM}V_{out} = T_4V_4 + T_6V_6 + T_0(V_0\ or\ V_7) \qquad (11.9)$$

where

$$T_1 + T_2 + T_0 = T_{pwm}.$$

The reference voltage vector V_{out} is obtained by mapping the desired three-phase output voltages to the d-q plane through the d-q transform. When the desired output voltages are in the form of three sinusoidal voltages with a 120^0 phase shift between them, V_{out} becomes a vector rotating around the origin of the *d-q* plane with a frequency corresponding to that of the desired three-phase voltages. The envelope of the hexagon formed by the basic space vectors, as shown in Fig. 11.3, is the locus of maximum V_{out}. Therefore, the magnitude of V_{out} must be limited to the shortest radius of this envelope because V_{out} is a rotating vector. This gives a maximum magnitude of $V_{dc}/\sqrt{2}$ for V_{out}. The maximum root mean square (rms) values of the fundamental line-to-line and line-to-neutral output voltages are $V_{dc}/\sqrt{2}$ and $V_{dc}/\sqrt{6}$. Notice that these values are $2/\sqrt{3}$ times higher than what a standard sinusoidal PWM technique can generate.

An example of a symmetric space vector PWM waveform is shown in Fig. 11.4. It is assumed that the reference voltage V_{out} lies in Sector 0, which is bordered by vectors V_4 and V_6.

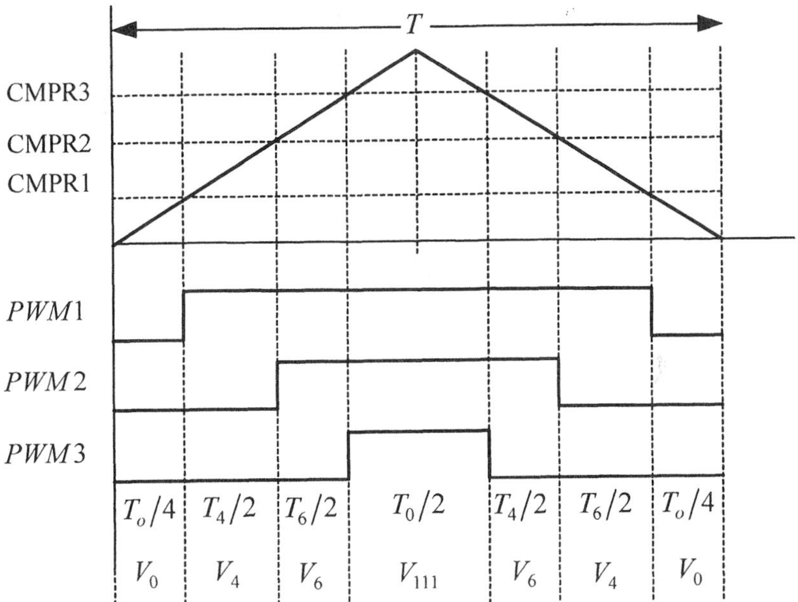

Figure 11.4 A symmetric space vector PWM switching pattern.

11.3.4 Calculating the Time Periods of the Switching States

The output voltage V_{out} can be in any one of Sector 0 to Sector 5. Equation (11.10) shows that for every PWM period, V_{out} is approximated by switching between the two non-zero basic vectors that border the sector of the current output voltage V_{out}. For instance, if V_{out} is in Sector 1, it can be approximated by switching the inverter between states V_2 and V_6 for periods of time T_2 and T_6, respectively. Because the sum of T_2 and T_6 should be less than or equal to T_{pwm}, the inverter should remain in T_0 or T_7 for the rest of the period.

From (11.10), we can calculate the time durations T_4 and T_6.

$$\begin{bmatrix} T_4 \\ T_6 \end{bmatrix} = T_{PWM} \begin{bmatrix} V_{4q} & V_{6q} \\ V_{4d} & V_{6d} \end{bmatrix}^{-1} \begin{bmatrix} V_{outq} \\ V_{outd} \end{bmatrix} \tag{11.10}$$

or

$$\begin{bmatrix} T_4 \\ T_6 \end{bmatrix} = T_{PWM} M_0 \begin{bmatrix} V_{outq} \\ V_{outd} \end{bmatrix}$$

where M_0 is the normalized decomposition matrix for sector 0. By substituting the values of V_{4q}, V_{4d}, V_{6q}, and V_{6d}, we obtain

$$\begin{bmatrix} T_4 \\ T_6 \end{bmatrix} = T_{PWM} \begin{bmatrix} 2/3 & 1/3 \\ 0 & -1/\sqrt{3} \end{bmatrix}^{-1} \begin{bmatrix} V_{outq} \\ V_{outd} \end{bmatrix} \tag{11.11}$$

The matrix inverse can be calculated before program execution for each sector and then obtained via a look-up table during execution. Doing so ensures smooth operation because the calculation load on the DSP is reduced. This approach is useful when V_{out} is given in the form of the vector $\begin{bmatrix} V_{outq} & V_{outd} \end{bmatrix}^T$. Table 11.3 shows the sector numbers and the associated normalized decomposition matrix.

Table 11.3 Normalized decomposition matrix vs. sector.

Sector	Durations Calculated	Decomposition Matrix
0	T_4 *and* T_6	$M_0 = \begin{bmatrix} \sqrt{3}/2 & 1/2 \\ 0 & -1 \end{bmatrix}$
1	T_2 *and* T_6	$M_1 = \begin{bmatrix} -\sqrt{3}/2 & -1/2 \\ \sqrt{3}/3 & -1/2 \end{bmatrix}$
2	T_2 *and* T_3	$M_2 = \begin{bmatrix} 0 & -1 \\ -\sqrt{3}/3 & 1/2 \end{bmatrix}$
3	T_1 *and* T_3	$M_2 = \begin{bmatrix} 0 & 1 \\ -\sqrt{3}/3 & -1/2 \end{bmatrix}$
4	T_1 *and* T_5	$M_4 = \begin{bmatrix} -\sqrt{3}/2 & 1/2 \\ \sqrt{3}/3 & 1/2 \end{bmatrix}$
5	T_5 *and* T_4	$M_5 = \begin{bmatrix} -\sqrt{3}/2 & -1/2 \\ 0 & 1 \end{bmatrix}$

11.3.5 Finding the Sector Number

It is necessary to know in which sector the output voltage is located to determine the switching time periods and switching sequence. The following algorithm can be used if the reference output voltage is in the *a-b-c* plane. If the output voltage is given in the *d-q* plane, we must transform the vector to the a-b-c plane before using the algorithm. In order to perform the transformation, first calculate the values of A, B, and C by using the following equations:

$$\begin{aligned} A &= sig(ref_1 - ref_2) \\ B &= sig(ref_2 - ref_3) \\ C &= sig(ref_3 - ref_1) \end{aligned} \tag{11.12}$$

where *sig* is the sign function, which is defined as

$$sig(x) = \begin{cases} 1 & x > 0 \\ undef & x = 0 \\ -1 & x < 0 \end{cases}$$

and ref_1, ref_2, and ref_3 are the output a, b, and c voltages. Then, find the value of N from the following relationship

$$N = |A + 2B + 4C| \tag{11.13}$$

Finally, we refer to Table 11.4 to map N to the sector of V_{out}.

Table 11.4 N vs. sector

N	1	2	3	4	5	6
Sector	1	5	0	3	2	4

11.3.6 SVPWM Switching Pattern

The order of the non-zero vectors and the zero vectors in each PWM period must be determined. Different switching orders result in different waveform patterns. Figure 11.5 shows the waveform produced for each sector of a symmetric switching scheme. Each waveform and sector has the following properties:

- Each PWM channel switches twice per PWM period except when the duty cycle is 0 or 100%.
- There is a fixed switching order among the three PWM channels for each sector.
- Every PWM period starts and ends with V_0.
- The amount of V_{000} inserted is the same as that of V_{111} in each PWM period.

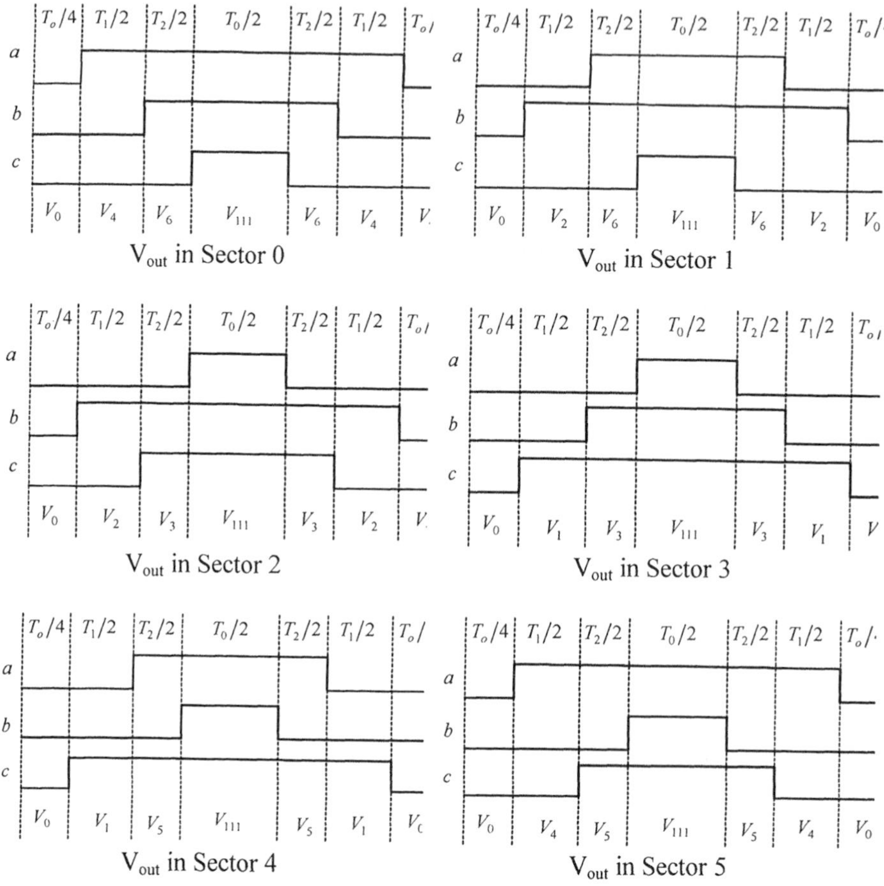

Figure 11.5 A symmetric space vector PWM switching pattern.

11.4 DSP Implementation

In this section, the space vector switching scheme discussed previously is implemented on a LF2407 DSP processor. The DSP-based algorithm is interrupt driven, meaning that the functionality of the code depends on a hardware interrupt, in this case the Timer 1 underflow interrupt. Figure 11.6 is a flowchart depicting the algorithm implemented on the LF2407 DSP processor.

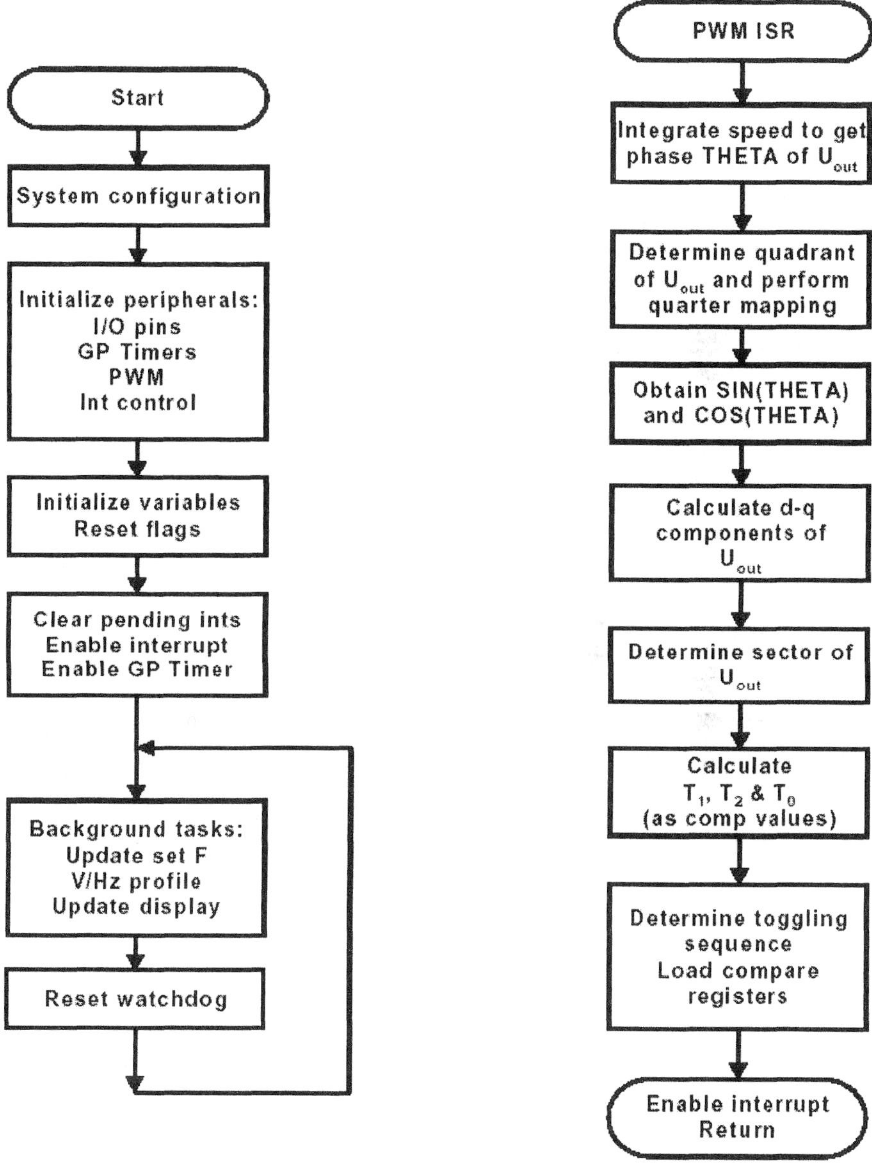

Figure 11.6 Space vector PWM algorithm flowchart.

The major features of this DSP implementation are:

- 32-Bit integration to obtain the phase of the reference voltage vector
- Quarter mapping to calculate sine and cosine functions
- Sector-based look-up table for the decomposition matrix

- Sector-based look-up table for the channel toggling order or Action Control Register reload pattern

11.4.1 Algorithm Subroutines

As shown in Fig. 11.6, while the DSP algorithm waits for an interrupt to occur, the DSP will continue to execute the code in the *main_loop* routine until a Timer 1 underflow interrupt is generated by the event manager. The task of *main_loop* is to first obtain the magnitude of reference voltage vector Vout based on the constant V/Hz profile. After the reference voltage vector is determined, the watchdog timer is reset and the DSP is instructed to branch back to the beginning of *main_loop*, repeating the above process, provided that an interrupt has not occurred yet. The *main_loop* algorithm can be seen below.

```
;==============================================================
; Start of background loop
;----------------------------------------------------------------
main_loop    LDP  #4
             SPLK  #debug_data,set_f ; Replace with debug data

f2omega LT          set_f                ; set f -> omega: D0
        MPY         f_omega              ; D0*D10=D(10+1)
        PAC                              ; product -> ACC: D11
        SACH        omega,1              ; -> set angular speed: D10
        lacc        omega
        sub         #min_omega_          ; compare W with its lower limit
        BGZ         winlimit             ; continue if within limit
        splk        #min_omega_,omega    ; saturate if not winlimit

; Note the following implies constant v/f

omega2v LT          omega                ; set angular speed -> T: D10
        MPY         omega_v              ; D10*D-9=D(1+1)
        PAC                              ; product -> ACC: D2
        SACH        set_v,1              ; -> mag of ref voltage and -> D1
        lacc              set_v
        sub               #max_v_        ; compare Uout w/ its upper limit
        BLEZ        uinuplim             ; continue if within limit
        splk              #min_v_,set_v  ; saturate if not
        B           reset_wd

uinuplim LACC       set_v
         SUB        #min_v_              ; compare Uout with its lower limit
         BGEZ       reset_wd             ; continue if within limit
         splk             #min_v_,set_v  ; saturate if not

reset_wd LDP        #WDKEY>>7            ; Reset WD timer
         SPLK       #wd_rst_1,WDKEY      ;
         SPLK       #wd_rst_2,WDKEY
         SPLK       #0000000001101111b,WDCR

     B    main_loop                      ; End of background loop
```

When a Timer underflow interrupt occurs, the DSP finishes its current instruction and branches to the interrupt service routine. In the interrupt service routine, tasks 1 through 5 are performed. Each task along with the corresponding code is shown below:

Obtain the phase (q) of Vout by integrating the command speed.

```
;------------------------------------------------------------------
; Generate revolving voltage vector Uout=trans(Ud Uq)
;------------------------------------------------------------------
            ldp         #omega          ; Integrate speed to get phase
            LT          omega           ; set W -> T: D10
            MPY         t_sample        ; D10*D-9=D(1+1)
            PAC                         ; product -> ACC: D2
            SFR                         ; -> D3
            ADDH        theta_h         ; D3+D3=D3 (32 bit)
            ADDS        theta_l
            SACH        theta_h         ; save
            SACL        theta_l

chk_lolim   bcnd        chk_uplim,GEQ ; check upper limit if positive
            ADDH        theta_360       ; D3+D3=D3, rollover if not
            SACH        theta_h         ; save
            B           rnd_theta

chk_uplim   SUBH        theta_360       ; D3-D3=D3 compare with 2*pi
            bcnd        rest_theta,LEQ          ; resume theta_h if within limit
            SACH        theta_h         ; rollover if not
            B           rnd_theta

rest_theta  ADDH        theta_360       ; resume theta high
rnd_theta   ADD         #1,15           ; round up to upper 16 bits
            SACH        theta_r
;------------------------------------------------------------------
; Quadrant mapping
;------------------------------------------------------------------
            LACC        one             ; assume theta (theta_h) is in
            SPLK        #-1,SS          ; quadrant 1
            SACL        SC              ; 1=>SC, sign of COS(theta)
            LACC        theta_r
            SACL        theta_m         ; theta=>theta_m
            SUB         theta_90
            BLEZ        E_Q             ; jump to end if 90>=theta

; assume theta (theta_h) is in quadrant 2
            SPLK        #-1,SC          ; -1=>SC
            LACC        theta_180       ;
            SUB         theta_r         ; 180-theta
            SACL        theta_m         ; =>theta_m
            BGEZ        E_Q             ; jump to end if 180>=theta

; assume theta (theta_h) is in quadrant 3
            SPLK        #1,SS           ; -1=>SS
            LACC        theta_r
            SUB         theta_180       ; theta-180
            SACL        theta_m         ; =>theta_m
            LACC        theta_270
```

```
        SUB     theta_r
        BGEZ    E_Q             ; jump to end if 270>=theta

; theta (theta_h) is in quadrant 4
        SPLK    #1,SC           ; 1=>SC
        LACC    theta_360
        SUB     theta_r
        SACL    theta_m         ; 360-theta_h=>theta_m
```

Obtain the sine and cosine of q with quarter mapping and table look-up, and calculate the d-q component of Vout.

```
;----------------------------------------------------------------
; sin(theta), cos(theta)
;----------------------------------------------------------------
E_Q     LT      theta_m         ; D3. Find index
        MPY     theta_i         ; D3*D6=D(9+1)
        PAC                     ; D10
        SACH    sin_indx        ; D10
        LACC    sin_indx,11     ; r/s 5 by l/s 11 -> integer (D15)
        SACH    sin_indx        ; right shift 5 bits => D15

        LACC    sin_entry       ; Look up sin
        ADD     sin_indx
        TBLR    sin_theta
        LACC    sin_end
        SUB     sin_indx
        TBLR    cos_theta

        LT      SS              ; Look up cos
        MPY     sin_theta       ; modify sign: D15*D1=D(16+1)
        PAC
        SACL    sin_theta       ; left shift 16 bits and save: D1
        LT      SC
        MPY     cos_theta       ; modify sin: D15*D1=D(16+1)
        PAC                     ;
        SACL    cos_theta       ; left shift 16 bits and save: D1
;----------------------------------------------------------------
; Calcualte Vd & Vq
;----------------------------------------------------------------
        LT      set_v           ; set v -> T: D1
        MPY     cos_theta       ; set v*cos(theta): D1*D1=D(2+1)
        PAC                     ; product -> ACC: D3
        SACH    Ud,1            ; d component of ref Uout: D2
        MPY     sin_theta       ; set v*sin(theta): D1*D1=D(2+1)
        PAC                     ; product -> ACC: D3
        SACH    Uq,1            ; q component of ref Uout: D2
```

Determine which sector Vout is in.

```
;----------------------------------------------------------------
; Determine sector
;----------------------------------------------------------------
        LT      theta_r         ; D3
        MPY     theta_s         ; D3*D0=D4
        PAC
        SACH    sector
        LACC    sector,5        ; r/s 11 by l/s 5 -> integer (D15)
```

```
        SACH        sector          ; right shift 11 bits
```

Decompose Vout to obtain T1, T2 and T0 as compare values.

```
;-----------------------------------------------------------------------------
; Calculate T1&T2 based on:     Tpwn Uout=V1*T1+V2*T2
;
; i.e.     [T1 T2]=Tpwn*inverse[V1 V2]*Uout
; i.e.     [0.5*T1 0.5*T2]=Tp*inverse[V1 V2]*Uout
; i.e.     [0.5*C1 0.5*C2]=inverse[V1 V2]*Uout=M(sector)*Uout
;
; where     C1=T1/Tp, C2=T2/Tp, are normalized wrt Tp
;          M(sector)=inverse of [V1 V2] = decomposition matrix
;             obtained through table lookup
;          Uout=Transpose of [Ud Uq]
;          Tp=Timer 1 period = 0.5*Tpwm
;          Tpwm=PWM period Tpwm
;-----------------------------------------------------------------------------
        LACC        #dec_ms
        ADD         sector,2
        SACL        temp            ; get the pointer
        LAR         AR0,temp        ; point to parameter table

; Calculate 0.5*C1 based on 0.5*C1=Ud*M(1,1)+Uq*M(1,2)
        LT          Ud              ; D2
        MPY         *+              ; M(1,1) Ud: D2*D1=D(3+1)
        PAC                         ; D4
        LT          Uq              ; D4
        MPY         *+              ; M(1,2) Uq: D2*D1=D(3+1)
        APAC                        ; 0.5*C1: D4+D4=D4
        BGEZ        cmp1_big0       ; continue if bigger than zero
        ZAC                         ; set to 0 if less than 0

cmp1_big0
        SACH        temp            ; 0.5*C1: D4
        LT          temp            ; D4
        MPY         t1_periods      ; D4*D10 = D(14+1)
        PAC                         ; D15
        SACH        cmp_1           ; 0.5*C1*Tp: D15

; Calculate 0.5*C2 based on 0.5*C2=Ud*M(2,1)+Uq*M(2,2)
        LT          Ud              ; D2
        MPY         *+              ; M(2,1) Ud: D2*D1=D(3+1)
        PAC                         ; D4
        LT          Uq              ; D2
        MPY         *+              ; M(2,2) Uq: D2*D1=D(3+1)
        APAC                        ; 0.5*C2: D4+D4=D4

        BGEZ        cmp2_big0       ; continue if bigger than zero
        ZAC                         ; zero it if less than zero

cmp2_big0
        SACH        temp            ; 0.5*C2: D4
        LT          temp            ; D4
        MPY         t1_periods      ; D4*D10 = D(14+1)
        PAC                         ; D15
        SACH        cmp_2           ; 0.5*C2*Tp: D15
```

```
; Calculate 0.5*C0 based on 0.5*C3*Tp=Tp*(1-0.5*C1-0.5*C2)
        LACC    #t1_period_
        SUB     cmp_1           ;
        SUB     cmp_2           ; D15
        BGEZ    cmp0_big0       ; continue if bigger than zero
        ZAC                     ; zero it if less than zero

cmp0_big0
        SACL    cmp_0           ;
        LACC    cmp_0,15        ; right shift 1b (by l/s 15b)
        SACH    cmp_0           ; 0.25*C0*Tp
```

Determine the switching sequence and load the obtained compare values into corresponding compare registers.

```
;-------------------------------------------------------------------
; Determine channel toggling sequence and load compare registers
;-------------------------------------------------------------------
        LACC    #first_         ;
        ADD     sector          ; point to entry in look up table
        TBLR    first_tog       ; get 1st-to-toggle channel
        LAR     AR0,first_tog   ; point to the channel
        LACC    cmp_0

        SACL    *               ; cmp_0 => the channel

        LACC    #second_        ;
        ADD     sector          ; point to entry in look up table
        TBLR    sec_tog         ; get 2nd-to-toggle channel
        LAR     AR0,sec_tog     ; point to the channel
        LACC    cmp_0           ;
        ADD     cmp_1           ; cmp_0+cmp_1
        SACL    *               ; => the channel

        LACC    #CMPR3
        SUB     first_tog
        ADD     #CMPR2
        SUB     sec_tog
        ADD     #CMPR1
        SACL    temp            ; get 3rd-to-toggle channel
        LAR     AR0,temp        ; point to the channel
        LACC    cmp_0
        ADD     cmp_1
        ADD     cmp_2           ; cmp_0+cmp_1+cmp_2
        SACL    *               ; =>the channel

        RET                     ; return
```

The code shown above composes the functional parts of the LF2407 assembly code which implements the SVPWM switching scheme.

11.4.2 Verification of the SVPWM Algorithm and Conclusions

The space vector PWM algorithm can be verified by probing the filtered PWM outputs of LF2407 using a very simple low-pass filter as shown in Fig. 11.7 and by viewing the resultant signal on a oscilloscope.

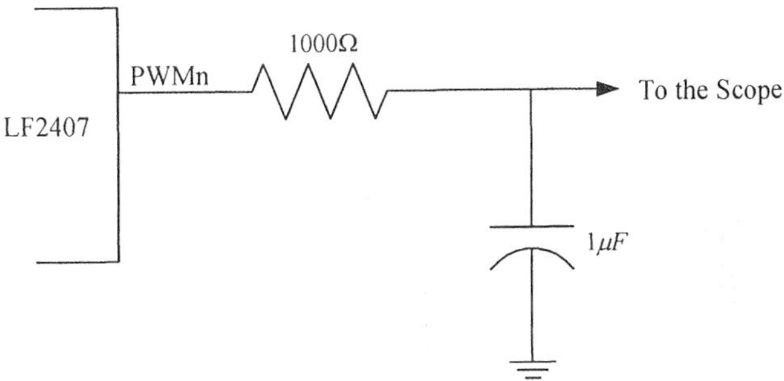

Figure 11.7 Low pass filter for filtering the LF2407 PWM outputs.

The output of the low-pass filter is illustrated by the oscilloscope screenshot in Fig. 11.8. It shows the three-phase voltages and the corresponding line-to-line voltage for an 11Hz waveform. The fundamental frequency and the third harmonic, which is inherently generated by the space vector method, are clearly shown. As expected, the three-phase wave forms are shifted from one another by 120 degrees.

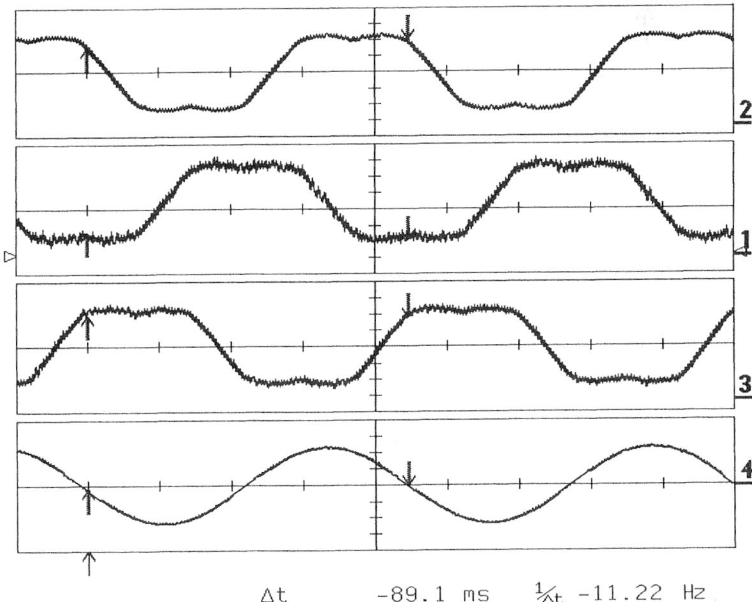

Figure 11.8 Top to bottom: the waveforms of filtered SVPWM outputs, phase voltages and line-to-line voltage (frequency = 11Hz).

This chapter presented the concept of constant V/Hz control of induction motors using the SVPWM. The theory of both the V/Hz control and the space vector PWM was discussed. The theoretical analysis first discussed has been supported by the implementation of the SVPWM algorithm via the LF2407 DSP. The output results verify the validity of both the theory and the DSP implementation.

References

1. H.A. Toliyat, ELEN689 Class Notes, Spring 2002.
2. TI application note, "AC Induction Motor Control Using Constant V/Hz Principle/Space Vector PWM- 'C240 (Rev. A) (SPRA284A).

Chapter 12

DSP-BASED CONTROL OF PERMANENT MAGNET SYNCHRONOUS MACHINES

12.1 Introduction

As described in Chapter 9, the permanent magnet synchronous motor (PMSM) is a PM motor with a sinusoidal back-EMF. Compared to the BLDC motor, it has less torque ripple because the torque pulsations associated with current commutation do not exist. A carefully designed machine in combination with a good control technique can yield a very low level of torque ripple (<2% rated), which is attractive for high-performance motor control applications such as machine tool and servo applications.

In this chapter, following the same procedures used in Chapter 9, the principles of the PMSM drive system will be introduced. Later, the control implementation using the LF2407 DSP will be described in detail.

12.2 The Principle of the PMSM

12.2.1 Mathematical Model of PMSM in the abc Stationary Reference Frame

Figure 12.1 depicts a cross-section of the simplified three-phase surface mounted PMSM motor for our discussion. The stator windings, as-as', bs-bs', and cs-cs', are shown as lumped windings for simplicity, but are actually distributed about the stator. The rotor has two poles. Mechanical rotor speed and position are denoted as ω_{rm} and θ_{rm}, respectively. Electrical rotor speed and position, ω_r and θ_r, are defined as P/2 times the corresponding mechanical quantities, where P is the number of poles.

Based on the above motor definition, the voltage equation in the abc stationary reference frame is given by

$$V_{abcs} = R_s i_{abcs} + \frac{d}{dt}\lambda_{abcs} \tag{12.1}$$

where

$$f_{abcs} = [f_{as} \quad f_{bs} \quad f_{cs}]^T \tag{12.2}$$

and the stator resistance matrix is given by

$$R_s = diag[r_s \quad r_s \quad r_s] \tag{12.3}$$

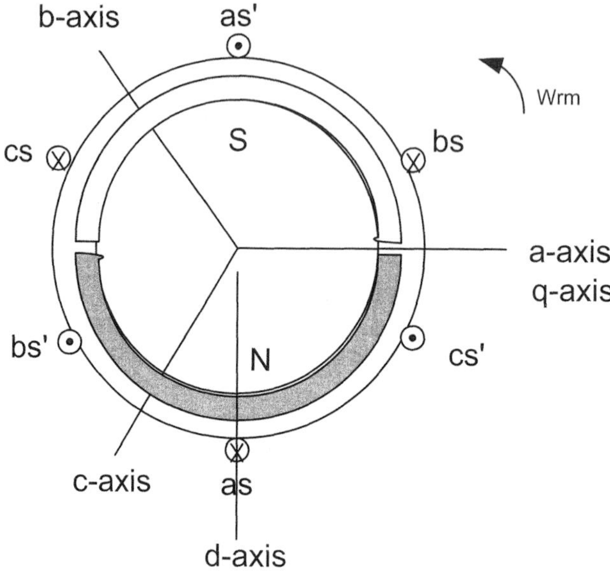

Figure 12.1 The cross-section of PMSM.

The flux linkages equation can be expressed by

$$\lambda_{abcs} = L_s i_{abcs} + \lambda'_m \begin{bmatrix} \sin\vartheta_r \\ \sin(\vartheta_r - \frac{2\pi}{3}) \\ \sin(\vartheta_r - \frac{4\pi}{3}) \end{bmatrix} \tag{12.4}$$

where λ_m' denotes the amplitude of the flux linkages established by the permanent magnet as viewed from the stator phase windings. Note that in (12.4) the back-EMFs are sinusoidal waveforms that are 120^0 apart from each other. The stator self inductance matrix, L_s, is given as

$$L_s = \begin{bmatrix} L_{ls} + L_A - L_B\cos 2\theta_r & -\frac{1}{2}L_A - L_B\cos 2(\theta_r - \pi/3) & -\frac{1}{2}L_A - L_B\cos 2(\theta_r + \pi/3) \\ -\frac{1}{2}L_A - L_B\cos 2(\theta_r - \pi/3) & L_{ls} + L_A - L_B\cos 2(\theta_r - 2\pi/3) & -\frac{1}{2}L_A - L_B\cos 2(\theta_r + \pi) \\ -\frac{1}{2}L_A - L_B\cos 2(\theta_r + \pi/3) & -\frac{1}{2}L_A - L_B\cos 2(\theta_r + \pi) & L_{ls} + L_A - L_B\cos 2(\theta_r + 2\pi/3) \end{bmatrix} \tag{12.5}$$

The electromagnetic torque may be written as

$$T_e = \frac{P}{2}\{\lambda_m'[(i_{as} - \frac{1}{2}i_{bs} - \frac{1}{2}i_{cs})\cos\theta_r - \frac{\sqrt{3}}{2}(i_{bs} - i_{cs})\sin\vartheta_r] + \frac{L_{md} - L_{mq}}{3}[(i_{as}^2 - \frac{1}{2}i_{bs}^2 - \frac{1}{2}i_{cs}^2 - i_{as}i_{bs} - i_{as}i_{cs} + 2i_{bs}i_{cs})\sin 2\theta_r + \frac{\sqrt{3}}{2}(i_{bs}^2 i_{cs}^2 - 2i_{as}i_{bs} + 2i_{as}i_{cs})\cos 2\theta_r]\} + T_{cog}(\theta_r) \tag{12.6}$$

In (12.6), $T_{cog}(\theta_r)$ represents the cogging torque and the d- and q-axes magnetizing inductances are defined by

$$L_{md} = \frac{3}{2}(L_A - L_B)$$

and

$$L_{md} = \frac{3}{2}(L_A + L_B) \tag{12.7}$$

The torque and speed are related by the electromechanical motion equation

$$J\frac{d}{dt}\omega_{rm} = \frac{P}{2}(T_e - T_L) - B_m\omega_{rm} \tag{12.8}$$

where J is the rotational inertia, B_m is the approximated mechanical damping due to friction, and T_L is the load torque.

12.2.2 Mathematical Model of PMSM in Rotor Reference Frame

The voltage and torque equations can be expressed in the rotor reference frame in order to transform the time-varying variables into steady state constants. Since the stator has two poles and the rotor has four poles, the transformation of the three-phase variables in the stationary frame to the rotor reference frame is defined as

$$f_{qd0r} = K_r f_{abcs} \tag{12.9}$$

where

$$K_r = \frac{2}{3}\begin{bmatrix} \cos\theta_r & \cos(\theta_r - \frac{2\pi}{3}) & \cos(\theta_r + \frac{2\pi}{3}) \\ \sin\theta_r & \sin(\theta_r - \frac{2\pi}{3}) & \sin(\theta_r + \frac{2\pi}{3}) \\ \frac{1}{2} & \frac{1}{2} & \frac{1}{2} \end{bmatrix}$$

If the applied stator voltages are given by

$$\begin{cases} V_{as} = \sqrt{2}V_s \cos\theta_{ev} \\ V_{bs} = \sqrt{2}V_s \cos(\theta_{ev} - \frac{2\pi}{3}) \\ V_{cs} = \sqrt{2}V_s \cos(\theta_{ev} + \frac{2\pi}{3}) \end{cases} \tag{12.10}$$

Then, applying (12.9) to (12.1), (12.4) and (12.10) yields

$$v_{qs}^r = r_s i_{qs}^r + \omega_r \lambda_{ds}^r + \frac{d}{dt}\lambda_{qs}^r \tag{12.11}$$

$$v_{ds}^r = r_s i_{ds}^r - \omega_r \lambda_{qs}^r + \frac{d}{dt}\lambda_{ds}^r \tag{12.12}$$

$$\lambda_{qs}^r = L_{qs} i_{qs}^r \tag{12.13}$$

$$\lambda_{ds}^r = L_{ds} i_{ds}^r + \lambda_m^{'r} \tag{12.14}$$

where the q- and d-axes self inductances are given by $L_{qs} = L_{ls} + L_{mq}$ and $L_{ds} = L_{ls} + L_{md}$, respectively.

The electromagnetic torque can be written as

$$T_e = \frac{3}{2}\frac{P}{2}[\lambda_m^{'r} i_{qs}^r + (L_{ds} - L_{qs}) i_{qs} i_{ds}] \tag{12.15}$$

From (12.15), it can be seen that torque is related only to the d- and q-axes currents. Since $L_q \geq L_d$ (for surface mount PMSM, both of inductances are equal), the second item contributes a negative torque if the flux weakening control has been used. In order to achieve the maximum torque/current ratio, the d-axis current is set to zero during the constant torque control so that the torque is proportional only to q-axis current. Hence, this results in the control of q-axis current for regulating the torque in rotor reference frame.

12.3 PMSM Control System

Based on the above analysis, a PMSM drive system is developed as shown in Fig. 12.2. The total drive system looks similar to that of the BLDC motor and consists of a PMSM, power electronics converter, sensors, and controller. These components are discussed in detail in the following sections.

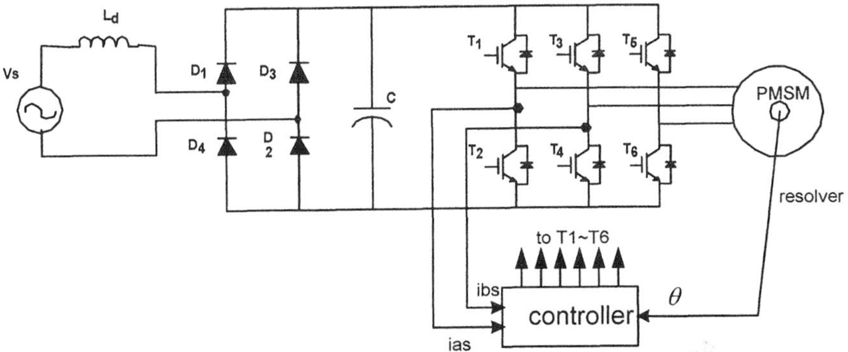

Figure 12.2 The PMSM speed control system.

12.3.1 PMSM Machine

The design consideration of the PMSM is to first generate the sinusoidal back-EMF. Unlike the BLDC, which needs concentrated windings to produce the trapezoidal back-EMF, the stator windings of PMSM are distributed in as many slots per pole as deemed practical to approximate a sinusoidal distribution. To reduce the torque ripple, standard techniques such as skewing and chorded windings are applied to the PMSM. With the sinusoidally excited stator, the rotor design of the PMSM becomes more flexible than the BLDC motor where the surface mount permanent magnet is a favorite choice. Besides the common surface mount non-salient pole PM rotor, the salient pole rotor, like inset and buried magnet rotors, are often used because they offer appealing performance characteristics during the flux weakening region. A typical PMSM with 36 stator slots in stator and four poles on the rotor is shown in Fig. 12.3.

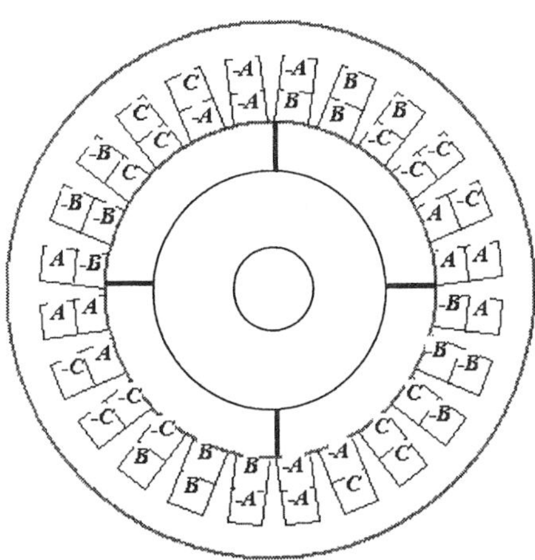

Figure 12.3 A four-pole 24-slot PMSM.

12.3.2 Power Electronic Converter

The PMSM shares the same topology of the power electronics converter as the BLDC motor drive system. The converter is the standard two-stage configuration with a dc link capacitor between a front-end rectifier and a three-phase full-bridge inverter as the output. The rectifier is either a full-bridge diode or power switch rectifier.

Due to the sinusoidal nature of the PMSM, control algorithms such as V/f and vector control, developed for other AC motors, can be directly applied to the PMSM control system. If the motor windings are Y-connected without a neutral connection, three phase currents can flow through the inverter at any moment. With respect to the inverter switches, three switches, one upper and two lower in three different legs conduct at any moment as shown in Fig. 12.4. PWM current control is still used to regulate the actual machine current. Either a hysteresis current controller, a PI controller with sine-triangle, or a SVPWM strategy is employed for this purpose. Unlike the BLDC motor, the three switches are switched at any time.

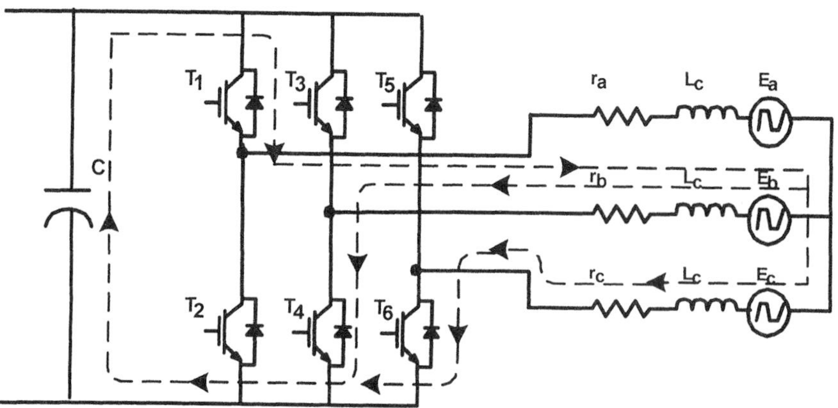

Figure 12.4 The current path when the three phases are chopped.

12.3.3 Sensors

There are two types of sensors used in the PMSM drive system: the current sensor, which measures the phase currents, and the position sensor which is used to sense the rotor position and speed. The resistances in series with the power switches as shown in Fig. 12.2 are usually used as shunt resistor phase current sensors. Either an encoder or resolver serves as the position sensor. Rotor position is needed in order to synchronize the stator excitation of the PMSM with the rotor speed and position.

Figure 12.5 shows the structure of an optical encoder. It consists of a light source, slotted disk, and photo sensors. The disk rotates with the rotor. The two photo sensors output a logic "1" when they detect light. When the light is blocked, a logic "0" is generated by the sensors. When the light passes through the slots of the disk and strikes the sensor, a logic "1" is produced. These logic signals are shown in Fig. 12.5. By counting the number of pulses, the motor speed can be calculated. The direction of rotation can be determined by detecting the leading edge between signal A and signal B.

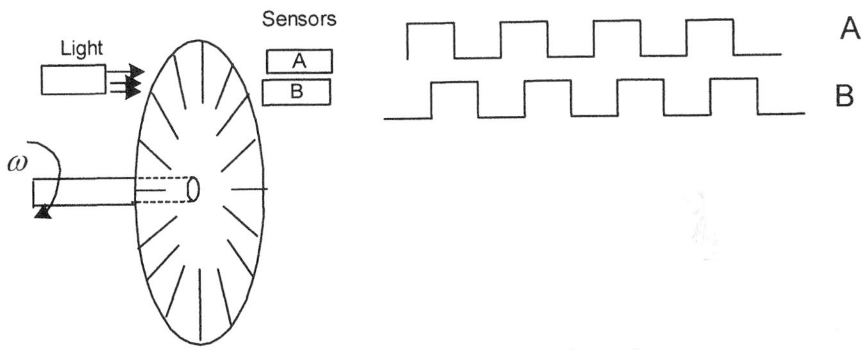

Figure 12.5 The structure of encoder.

A resolver is a rotary electromechanical transformer. It outputs to sinusoidal signals such that one wave is a sinusoidal function of the rotor angle θ, while the other signal is a cosinusoidal function of θ. The difference between these two waveforms reveals the position of the rotor. Integrated circuits such as the AD2S80 can be used to decode the signals. The resolver output waveform and the corresponding rotor position are given in Fig. 12.6.

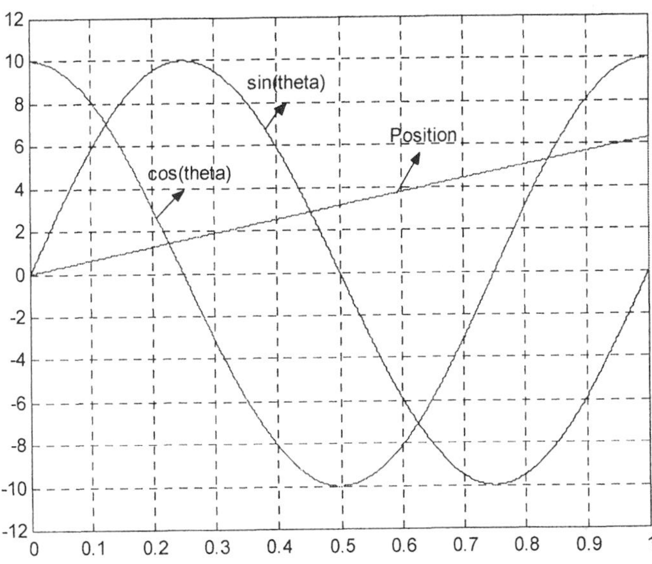

Figure 12.6 The resolver output and the corresponding rotor position.

12.3.4 Controller

The LF2407 is used as the controller to implement speed control of the PMSM system. The interface of the LF2407 is illustrated in Fig. 12.7. Similar to the BLDC motor control system, three input channels are selected to read the two phase currents and resolver signal. Because a resolver is used in one case, the QEP inputs are not used. QEP inputs work only with a QEP signal that a rotary encoder supplies. The DSP output pins PWM1-PWM6 used to supply the gating signals to the switches and form the output of the control part of the system.

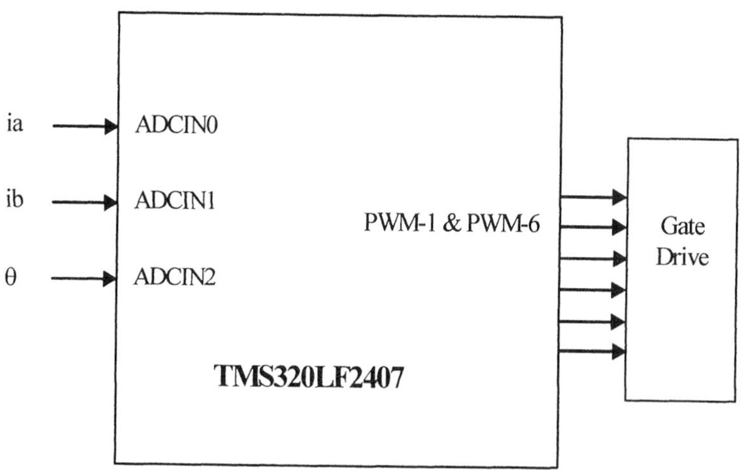

Figure 12.7 The interface of LF2407.

12.4 Implementation of the PMSM System Using the LF2407

A block diagram of the PMSM drive system is displayed in Fig. 12.8. An assembly code algorithm was written for the LF2407 to implement the control system shown inside the dashed line in Fig. 12.8.

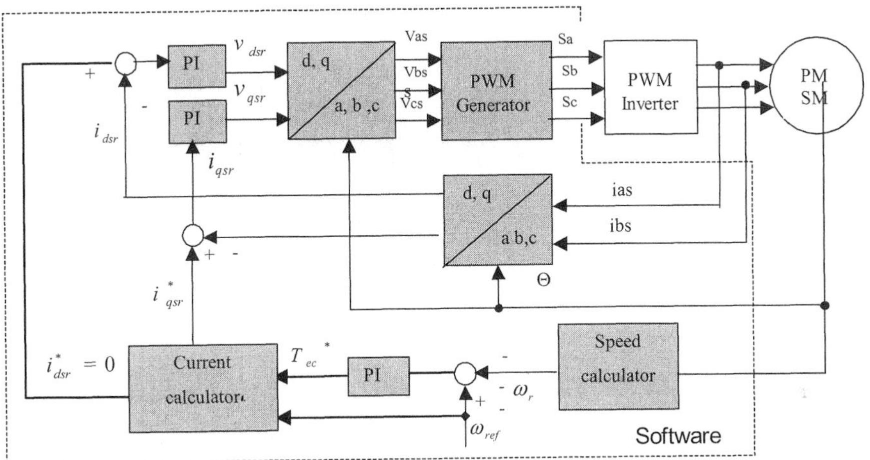

Figure 12.8 Block diagram of PMSM speed control system.

The flowchart of the developed software is shown in Fig. 12.9. The control program of the PMSM has one main routine and includes four modules:

1. Initialization procedure
2. DAC module
3. ADC module
4. Speed control module

The first three items introduced in Chapter 9. Hence, in the following section, only the speed control module is discussed in detail, with the corresponding assembly code given.

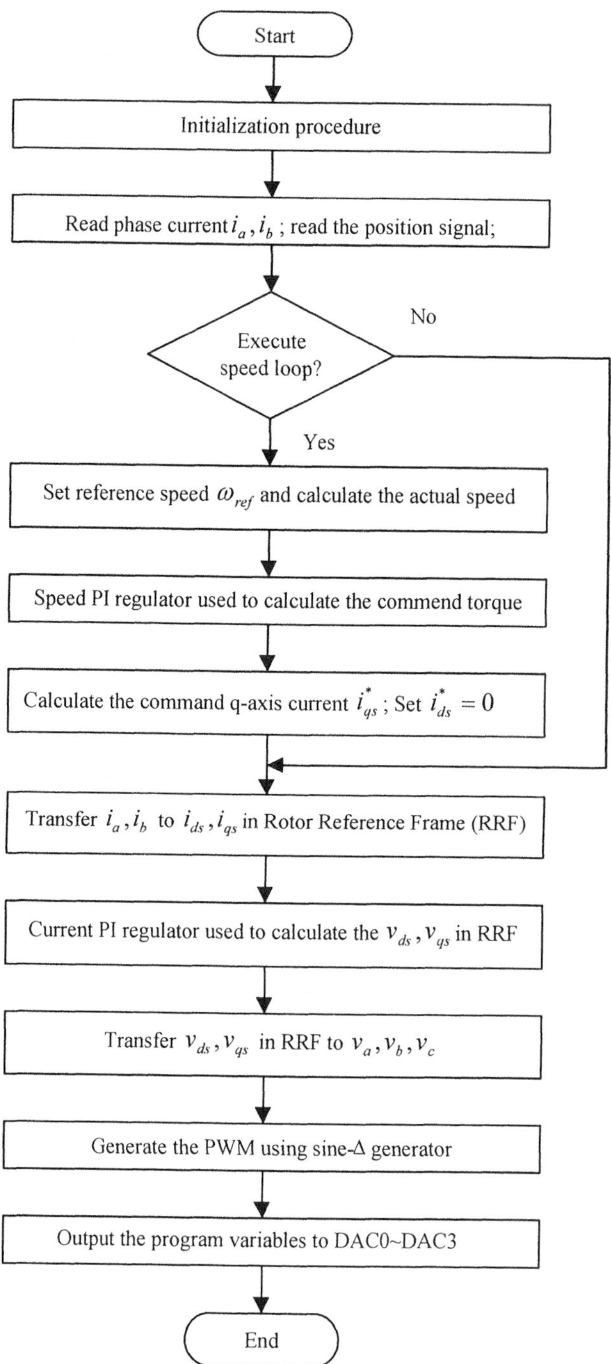

Figure 12.9 The flow chart of PMSM control system.

12.4.1 The Speed Control Algorithm

In the BLDC motor control system the Timer 1 underflow interrupt is used for the subroutine of speed control. This routine performs the tasks of:

- Reading the current and position signal, then generating the commanded speed profile.
- Calculating the actual motor speed, transferring the variables in the *abc* model to the *d-q* model and reverse.
- Regulating the motor speed and currents using the vector control strategy.
- Generating the PWM signal based on the calculated motor phase voltages.

The PWM frequency is determined by the time interval of the interrupt, with the controlled phase voltages being recalculated every interrupt. The modules of this routine are detailed in the following section. The code below shows this routine.

```
T1_PERIOD_ISR:
;Context save regs
        MAR     *,AR1                       ;AR1 is stack pointer
        MAR     *+                          ;skip one position
        SST     #1, *+                      ;save ST1
        SST     #0, *+                      ;save ST0
        SACH    *+                          ;save acc high
        SACL    *                           ;save acc low
        POINT_EV
        SPLK    #0FFFFh,EVIFRA              ;Clear all Group A interrupt
                                            ;flags (T1 ISR)
READ_SIG
        CALL    ADC_CONV
        CALL    CAL_TRIANGLE
        CALL    ADC_DQ
        POINT_B0
        LACC    CL_SPD_FLG
        BCND    CURRENT_CNTL,GT             ;speed-loop?
; speed control
SPEED_CNTL:     POINT_B0
        CALL    SPEED_PROFILE
        CALL    VTIMER_SEC
        CALL    SPEED_CAL
        CALL    D_PID_spd
        BLDD    #D_PID_out                  ;iqsr
        SPLK    #0, idsr_ref

; current control
CURRENT_CNTL
        CALL    D_PID_cur
        BLDD    #D_out_iq, Vqr
        BLDD    #D_out_id, Vdr
        CALL    DQ_ABC
        BLDD    #a_out, Va
        BLDD    #b_out, Vb
        BLDD    #c_out, Vc
PWM_GEN       CALL    PWM_DRV
DA_CONV       CALL    DAC_VIEW_Q15I
;Restore Context
END_ISR:
        MAR     *, AR1                      ;make stack pointer active
```

```
        LACL    *-                  ;Restore Acc low
        ADDH    *-                  ;Restore Acc high
        LST     #0, *-              ;load ST0
        LST     #1, *-              ;load ST1
        CLRC    INTM
        RET
```

12.4.1.1 The Calculation of sinθ and cosθ

A lookup table is used to calculate the sine and cosine values of the rotor position θ. The rotor electrical angle depends only on its sine value in lookup table. The cosine value is calculated by shifting the sine value 90 degrees. The sine and cosine values, which are used in the transformation, can be obtained by simply knowing the rotor angle. The code below shows how to read the 1:1 look-up table with the LF2407.

```
TRI_CAL
        ...
        LACC    TRI_INT             ;load accumulator
        AND     #0ffh               ;get lower bits
        ADD     #SINTAB             ;table read
        TBLR    sine_a
        ...
        RET
```

The block of code below shows a portion of the sine value lookup table.

```
        ;SINVAL                         Index     Angle     Sin(Angle)
SINTAB
        ....
        .word       12539       ;       16        22.50      0.3827
        .word       13279       ;       17        23.91      0.4052
        .word       14010       ;       18        25.31      0.4276
        .word       14732       ;       19        26.72      0.4496
        .word       15446       ;       20        28.13      0.4714
        .word       16151       ;       21        29.53      0.4929
        .word       16846       ;       22        30.94      0.5141
        .word       17530       ;       23        32.34      0.5350
        .word       18204       ;       24        33.75      0.5556
        ....
        RET
```

12.4.1.2 The abc-to-dq Transformation

The abc-to-dq transformation is defined in (12.9). It transfers the three-phase stationary motor model to a two-phase rotational motor model. In other words, under the restriction of the same motor performance, three phase stationary stator windings with 120^0 separation can be replaced by a two-phase rotational winding with the q-phase 90^0 ahead of d-phase. The two-phase currents are related to the three-phase currents as defined by the transformation in (12.9). After this transformation, a significant simplification is achieved. The d and q-axis variables are decoupled and independent with time and rotor position, which implies that these variables become constant in steady state. It is possible to control the d and q variables independently. Since the d-axis variables are associated with the field variable and q-axis variables are related to the torque, this feature enables us to control the ac motor similar to a dc motor. For more detailed information on this topic we can refer to vector control theory. A portion of the abc-to-dq transformation using the assembly code is given in the code below:

```
ABC_DQ:
        ...
        ...
        LACC    #0
        LT       ABC_ain
        MPY     sone_a
        LTA     ABC_bin
        MPY     sone_b
        LTA     ABC_cin
        MPY     sone_c
        LTA     ABC_ain
        SACH    ABC_D_out
        ...
        ...
        RET
```

12.4.1.3 The d-q to a-b-c Transformation

After the commanded d and q-axes variables are calculated, these two variables are transferred to the a-b-c stationary frame to drive the motor. This reverse transform is defined as follows:

$$f_{abcs} = K_r^{'} f_{qd0r} \tag{12.16}$$

where

$$K_r^{'} = \begin{bmatrix} \cos\theta_r & \sin\theta & \frac{1}{2} \\ \cos(\theta_r - \frac{2\pi}{3})_r & \sin(\theta_r - \frac{2\pi}{3}) & \frac{1}{2} \\ \cos(\theta_r + \frac{2\pi}{3}) & \sin(\theta_r + \frac{2\pi}{3}) & \frac{1}{2} \end{bmatrix} \tag{12.17}$$

An example of the assembly code to implement the above equation is given in the code below:

```
DQ_ABC
        ...
        ...
        LACC    #0
        LT       DQ_D_ref
        MPY     sone_a
        LTa     DQ_Q_ref
        MPY     cosone_a
        MPYA    cosone_b
        SACH    DQ_aout
        ...
        ...
        RET
```

12.4.1.4 PWM Generation

The PWM circuits of the 2407 Event Manager are used to generate the gating signals. Figure 12.10 displays the principle of this method. The control signal with frequency f1 is constantly compared with a triangle signal which has a high-frequency f2 (usually f2/f1>21). If the controlled signal is larger than the triangle signal, a PWM output signal becomes a logic "1". Otherwise, a "0" is given.

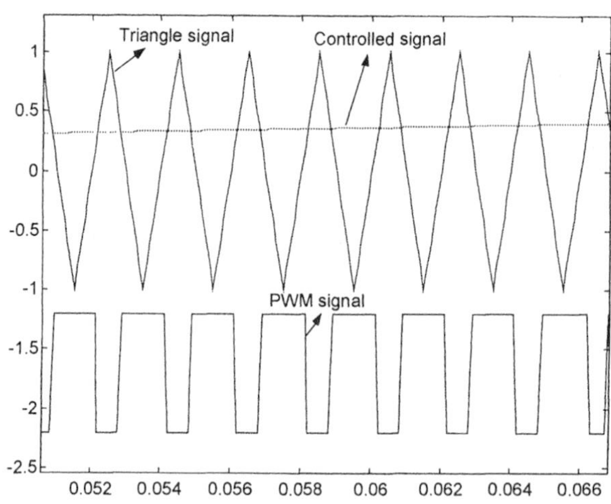

Figure 12.10 The principle of sine-triangle PWM generation.

The full-compare units have been used to generate the PWM outputs. The PWM signal is high when the output of current PI regulation matches the value of T1CNT and set low when the Timer underflow occurs. The switch states are controlled by the ACTR register. As discussed in Section 3.2, the lower switches should always be on and the upper switches should be chopped. From the point of implementation on the LF2407, this requires that the ACTR register is reset for each interval. Therefore, PWM1, PWM3, and PWM5, which trigger the upper switches, are set as *active low/high* and PWM2, PWM4, and PWM6, which trigger the lower switches are set as *force high*. The code below illustrates this implementation.

```
SINE_PWM:
        ....
        ....
POINT_B0
        MPY     Ub
        PAC
        ADD     PERIOD,15
        POINT_EV
        SACH    CMPR2
        ....
        RET
```

Chapter 13

DSP-BASED VECTOR CONTROL OF INDUCTION MOTORS

13.1 Introduction

For many years, induction motors have been preferred for a variety of industrial applications because of their robust and rugged construction. Until a few years ago, the induction motor could either be plugged directly into the grid (uncontrolled) or controlled by means of the well-known scalar volts per Hertz (V/f) method. In variable speed drives, both methods have serious drawbacks in the areas of efficiency, reliability, and electromagnetic interference (EMI). With the uncontrolled method, even a simple change in the reference speed is not possible. Additionally, its system integration depends highly on the motor design (i.e., starting torque vs. maximum torque, torque vs. inertia, number of pole pairs, etc).

The scalar V/f method is able to provide speed variation, but this method cannot provide real-time control. In other words, the system response is only satisfactory at steady state and not during transient conditions. This results in excessive current and over-heating, which necessitate the drive to be oversized. This over-design no longer makes the motor cost effective due to the high cost of the drive circuitry. By using real-time processors such as the LF2407 DSP controller, and with an accurate induction motor model, the development of highly reliable and accurate variable speed motor drives becomes possible.

With the advent of field-oriented control (FOC) schemes, induction motors can be made to operate similar to separately excited dc motors. The indirect field oriented controls, or vector control, for speed and torque controlled AC drives are becoming the industry standard in order to obtain high dynamic motor performance.

The control algorithm explained in this chapter is a rotor flux field-orientated control strategy. In this chapter, we will go through not only the implementation of the control software, but also the theoretical and practical aspects of the vector control. In the end, the reader will be familiar with the different parts of the FOC strategy of the induction motor as well as the developmental steps involved. The reader should also be able to apply this induction motor drive solution to other desired systems. This chapter deals with the structure of an induction motor and develops its model followed by its FOC schemes. Finally, hardware and software development procedures covered.

13.2 Three-Phase Induction Motor Basic Theory

13.2.1 Three-Phase Induction Motor

Three-phase induction machines are asynchronous machines that operate below the synchronous speed when motoring and above the synchronous speed when generating. They are the most popular machine used in industry today and are rugged and require very little maintenance. Compared to dc motors, induction motors are not as easy to control. They typically draw large starting currents, about six to eight times their full load values, and operate with lagging power factor when loaded. However, with the advent of the vector control concept for motor control, it is possible to decouple the torque and the flux, thus making the control of the induction motor very similar to that of the dc motor.

13.2.2 Induction Motor Construction

The dc motor can be called a conduction motor because the electric power is conducted directly to the armature through the brushes and commutator. In the case of induction motors, the rotor receives power by induction; the same way a secondary of a two-winding transformer receives power from the primary. This is why the induction motor can be treated as a rotating transformer, where the primary winding is stationary, but the secondary is free to rotate. We use this concept to develop the equivalent circuit for induction motors.

The most popular type of induction motor used is the squirrel cage induction motor shown in Fig. 13.1. The rotor consists of a laminated core with parallel slots for carrying the rotor conductors, which are usually heavy bars of copper, aluminum, or alloys. One bar is placed in each slot; or rather, the bars are inserted from the end when the semi-closed slots are used. The rotor bars are brazed, electrically welded, or bolted to two heavy and stout short-circuiting end-rings, thus completing the squirrel cage construction. The rotor bars are permanently short-circuited on themselves. The rotor slots are usually not parallel to the shaft, but are given a slight angle, called a skew, which increases the rotor resistance due to increased length of rotor bars and an increase in the slip for a given torque. The skew is also advantageous because it reduces the magnetic hum while the motor is operating and reduces the locking tendency, or cogging, of the rotor teeth.

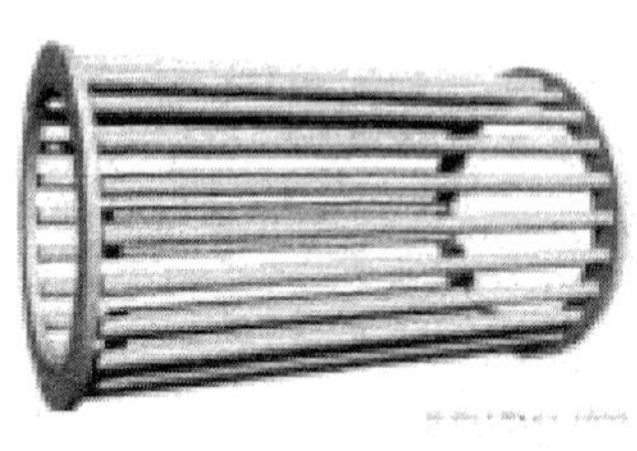

Figure 13.1 Short-circuited rotor bars of the squirrel cage induction motor.

13.2.3 Operation

When the three-phase stator windings are fed by a three-phase supply, a magnetic flux of a constant magnitude rotating at synchronous speed is created inside the motor. Due to the relative speed between the rotating flux and the stationary conductors, an electromagnetic force (EMF) is induced in the rotor in accordance with Faraday's laws of electromagnetic induction. The frequency of the induced EMF is the same as the supply frequency, and the magnitude is proportional to the relative velocity between the flux and the conductors. The direction of this EMF is given by Fleming's right-hand rule. Since the rotor bars form a closed path as shown in Fig. 13.1, a rotor current is produced which, according to Lenz's law, is opposite to that of the relative velocity between the rotating flux and the conductors. Therefore, the rotor current develops in the same direction as the flux and tries to catch up with the rotating flux.

13.2.4 Slip

The difference between the synchronous speed ω_e and the actual speed ω_r of the motor is called the slip.

$$s = \frac{\omega_e - \omega_r}{\omega_e} \tag{13.1}$$

13.3 Model of the Three-Phase Induction Motor in Simulink

13.3.1 Voltage Equations of the Idealized Motor Model

The idealized circuit model of the three-phase induction machine is shown in Fig. 13.2:

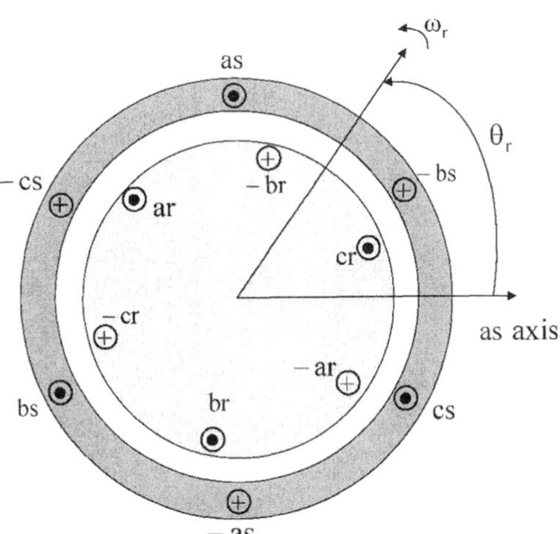

Figure 13.2 Idealized circuit model of the three-phase induction motor.

Stator voltage equations:

$$v_{as} = r_s i_{as} + \frac{d\lambda_{as}}{d_t} \tag{13.2}$$

$$v_{bs} = r_s i_{bs} + \frac{d\lambda_{bs}}{d_t} \tag{13.3}$$

$$v_{cs} = r_s i_{cs} + \frac{d\lambda_{cs}}{d_t} \tag{13.4}$$

Rotor voltage equations:

$$v_{ar} = i_{ar} r_r' + \frac{d\lambda_{ar}}{d_t} \tag{13.5}$$

$$v_{br} = i_{br} r_r' + \frac{d\lambda_{br}}{d_t} \tag{13.6}$$

$$v_{cr} = i_{cr} r_r' + \frac{d\lambda_{cr}}{d_t} \tag{13.7}$$

Flux linkage equations:

$$\begin{bmatrix} \lambda_s^{abc} \\ \lambda_r^{abc} \end{bmatrix} = \begin{bmatrix} L_{ss}^{abc} & L_{sr}^{abc} \\ L_{rs}^{abc} & L_{rr}^{abc} \end{bmatrix} \begin{bmatrix} i_s^{abc} \\ i_r^{abc} \end{bmatrix} \tag{13.8}$$

where:

$$\lambda_s^{abc} = \begin{bmatrix} \lambda_{as} \\ \lambda_{bs} \\ \lambda_{cs} \end{bmatrix}, \ \lambda_r^{abc} = \begin{bmatrix} \lambda_{ar} \\ \lambda_{br} \\ \lambda_{cr} \end{bmatrix}, \ i_s^{abc} = \begin{bmatrix} i_{as} \\ i_{bs} \\ i_{cs} \end{bmatrix}, \ i_r^{abc} = \begin{bmatrix} i_{ar} \\ i_{br} \\ i_{cr} \end{bmatrix} \tag{13.9}$$

The stator-to-stator and rotor-to-rotor winding inductances are:

$$L_{ss}^{abc} = \begin{bmatrix} L_{ls} + L_{ss} & L_{sm} & L_{sm} \\ L_{sm} & L_{ls} + L_{ss} & L_{sm} \\ L_{sm} & L_{sm} & L_{ls} + L_{ss} \end{bmatrix},$$

$$L_{rr}^{abc} = \begin{bmatrix} L_{lr} + L_{rr} & L_{rm} & L_{rm} \\ L_{rm} & L_{lr} + L_{rr} & L_{rm} \\ L_{rm} & L_{rm} & L_{lr} + L_{rr} \end{bmatrix} \tag{13.10}$$

The stator-to-rotor mutual inductances are dependent on the rotor angle:

$$L_{sr}^{abc} = \left[L_{rs}^{abc}\right]^T = L_{sr} \begin{bmatrix} \cos\theta_r & \cos(\theta_r + 2\pi/3) & \cos(\theta_r - 2\pi/3) \\ \cos(\theta_r - 2\pi/3) & \cos\theta_r & \cos(\theta_r + 2\pi/3) \\ \cos(\theta_r + 2\pi/3) & \cos(\theta_r - 2\pi/3) & \cos\theta_r \end{bmatrix} \tag{13.11}$$

where:

$L_{\ell s}$ = Stator winding leakage inductance per phase

L_{ss} = Self inductance of stator winding

L_{sm} = Peak value of stator to rotor mutual inductance

L_{sr} = Peak value of stator to rotor mutual inductance

If

P_g = air-gap permeance,

then

$$L_{ss} = N_s^{\,2} P_g, \quad L_{sr} = N_s N_r P_g, \quad L_{sm} = N_s^{\,2} P_g Cos(2\pi/3),$$

$$L_{rm} = N_r^{\,2} P_g Cos(2\pi/3), \quad L_{rr} = N_r^{\,2} P_g \tag{13.12}$$

We can see that the idealized machine is described by six first-order differential equations; one for each winding. These differential equations are coupled to one another by the mutual inductances between the windings. The stator-to-rotor coupling terms are a function of the rotor position, so when the rotor rotates, the coupling terms change with time. To solve this problem, induction motor equations are transferred to the quadrature rotating reference frame such that the mutual inductances are not time dependent.

13.4 Reference Frame Theory

Reference frame theory is an integral part of electric drives. Reference frames are powerful tools for the analysis and application of sophisticated control techniques, particularly in the case of the three-phase induction and synchronous machines. Using reference frame theory, it is possible to transform the machine phase variables to another reference frame. By judicious choice of the reference frames, it is possible to considerably reduce the complexity of the model machine. Reference frame theory has become especially important for digital motor control where the need for accurate but simple motor models is essential. Though the theory can be extended to any arbitrary reference frame, the two most commonly used reference frames are the Stationary Reference Frame and the Synchronous Reference Frame. The Clarke and Park transformations are used to transfer the induction motor equations to these frames. The transformations are discussed in Chapter 10 in detail and are repeated here for reference. Clarke's transformation is given by

$$\begin{bmatrix} f_{qs}^{s} \\ f_{ds}^{s} \\ f_{0s}^{s} \end{bmatrix} = T(0) \begin{bmatrix} f_{as} \\ f_{bs} \\ f_{cs} \end{bmatrix} \tag{13.12}$$

where

$$T(0) = 2/3\begin{bmatrix} 1 & -1/2 & -1/2 \\ 0 & -\sqrt{3}/2 & \sqrt{3}/2 \\ 1/2 & 1/2 & 1/2 \end{bmatrix} \tag{13.13}$$

Park's Transformation is represented by

$$\begin{bmatrix} f_{qs}^{e} \\ f_{ds}^{e} \end{bmatrix} = \begin{bmatrix} \cos\rho & -\sin\rho \\ \sin\rho & \cos\rho \end{bmatrix} \begin{bmatrix} f_{qs}^{s} \\ f_{ds}^{s} \end{bmatrix} \tag{13.14}$$

where the rotor position is given by

$$\rho = \int \omega_e dt \tag{13.15}$$

13.5 Induction Motor Model in the Arbitrary q-d-0 Reference Frame

As mentioned previously, the two most common reference frames chosen to represent the induction motor are the stationary and the synchronous reference frames. The stationary reference frame has the q-d-0 variables of the machine in the same frame as those normally used for the supply network. This choice of network is usually made when the supply network is large or complex. In the case of the synchronously rotating reference frame, the q-d-0 variables are constants at steady state.

Assuming that the induction motor is rotating at speed ω in the direction of rotor rotation, the machine equations in the stationary reference frame can be obtained by setting ω = 0. Likewise, the equations in the synchronous reference frame are obtained by setting ω = ω_e. Applying transformation to the stator windings a-b-c voltages, the stator winding q-d-0 voltages in the arbitrary reference frame are obtained.

$$v_{qd0}^{s} = \omega \begin{bmatrix} 0 & 1 & 0 \\ -1 & 0 & 0 \\ 0 & 0 & 0 \end{bmatrix} \lambda_{qd0}^{s} + p\lambda_{qd0}^{s} + r_s i_{qd0}^{s} \tag{13.16}$$

where $p = {}^{d}\!/_{dt}$. Applying the transformation to the rotor voltage equation, we get

$$v_{qd0}^{r} = (\omega - \omega_r) \begin{bmatrix} 0 & 1 & 0 \\ -1 & 0 & 0 \\ 0 & 0 & 0 \end{bmatrix} \lambda_{qd0}^{r} + p\lambda_{qd0}^{r} + r_r i_{qd0}^{r} \tag{13.17}$$

Stator and rotor flux linkage equations are given by

$$\begin{bmatrix} \lambda_{qs} \\ \lambda_{ds} \\ \lambda_{0s} \\ \lambda'_{qr} \\ \lambda'_{dr} \\ \lambda'_{0r} \end{bmatrix} = \begin{bmatrix} L_{ls}+L_m & 0 & 0 & L_m & 0 & 0 \\ 0 & L_{ls}+L_m & 0 & 0 & L_m & 0 \\ 0 & 0 & L_{ls} & 0 & 0 & 0 \\ L_m & 0 & 0 & L'_{lr}+L_m & 0 & 0 \\ 0 & L_m & 0 & 0 & L'_{lr}+L_m & 0 \\ 0 & 0 & 0 & 0 & 0 & L'_{lr} \end{bmatrix} \begin{bmatrix} i_{qs} \\ i_{ds} \\ i_{0s} \\ i'_{qr} \\ i'_{dr} \\ i'_{0r} \end{bmatrix} \quad (13.18)$$

where the primed values are referred values to the stator side according to the following relationships:

$$\lambda'_{qr} = \frac{N_s}{N_r}\lambda_{qr} \quad (13.19)$$

$$\lambda'_{dr} = \frac{N_s}{N_r}\lambda_{dr} \quad (13.20)$$

$$i'_{qr} = \frac{N_s}{N_r}i_{qr} \quad (13.21)$$

$$i'_{dr} = \frac{N_s}{N_r}i_{dr} \quad (13.22)$$

$$L'_{lr} = \left(\frac{N_s}{N_r}\right)^2 L_{lr} \quad (13.23)$$

Magnetizing inductance on the stator side is given by

$$L_m = \frac{3}{2}L_{ss} = \frac{3}{2}\frac{N_s}{N_r}L_{sr} = \frac{3}{2}\frac{N_s}{N_r}L_{rr} \quad (13.24)$$

The electromagnetic torque equation is given by

$$\begin{aligned} T_{em} &= \frac{3}{2}\frac{P}{2\omega_r}\left[\omega(\lambda_{ds}i_{qs} - \lambda_{qs}i_{ds}) + (\omega - \omega_r)(\lambda'_{dr}i'_{qr} - \lambda'_{qr}i'_{dr})\right] \\ &= \frac{3}{2}\frac{P}{2}\left[\lambda'_{qr}i'_{dr} - \lambda'_{dr}i'_{qr}\right] \\ &= \frac{3}{2}\frac{P}{2}\left[\lambda_{ds}i_{qs} - \lambda_{qs}i_{ds}\right] \\ &= \frac{3}{2}\frac{P}{2}L_m\left[i'_{dr}i_{qs} - i'_{qr}i_{ds}\right] \end{aligned} \quad (13.25)$$

13.6 Field Oriented Control

The term "vector" control refers to the control technique that controls both the amplitude and the phase of ac excitation voltage. Vector control therefore controls the spatial orientation of the electromagnetic fields in the machine. This has led to the coining of the term field oriented control (FOC), which is used for controllers that maintain a 90° spatial orientation between the critical field components. The term "field angle control" refers to the control strategy where the system is not at 90° of spatial orientation. In order to properly comprehend vector control, we must understand the principle of dc machine torque control on which FOC is based.

13.7 DC Machine Torque Control

The required 90° of spatial orientation between key field components can be compared to the dc motor, where the armature winding magnetic field and the filed winding magnetic filed are always in quadrature. The objective is to force the control of the induction machine to be similar to the control of a dc motor, i.e., torque control. In dc machines, the field and the armature winding axes are orthogonal to one another, making the MMFs established orthogonal. If the iron saturation is ignored, then the orthogonal fields can be considered to be completely decoupled.

For dc machines, the developed torque is

$$T_{em} = K_a \phi(I_f) I_a \tag{13.26}$$

where

k_a = Constant

$\phi(I_f)$ = Field flux

I_a = Armature current

Since the torque angle is always 90°, the flux and the torque can be controlled independently. The torque is controlled by adjusting the field current I_f, and the flux is directly controlled by adjusting the armature current I_a.

It is important to maintain a constant field flux for good torque control. It is also important to maintain an independently controlled armature current in order to overcome the effects of the resistance of the armature winding, leakage inductance, and the induced voltage is needed. A spatial angle of 90° between the flux and MMF axes has to be maintained in order to limit interaction between the MMF and the flux. If these conditions are met at every instant of time, the torque will always follow the current. In the case of dc machines, there is constant field flux and 90° torque angle due to the commutator and the separate field excitation system. In ac machines, these conditions have to be attained by using external controls, making the system more complex and difficult to understand.

13.8 Field Oriented Control, Direct and Indirect Approaches

With vector control, the mechanically robust induction motors can be used in high performance applications where dc motors were previously used. The key feature of the control scheme is the orientation of the synchronously rotating q-d-0 frame to the rotor flux vector. The d-axis component is aligned with the rotor flux vector and regarded as the flux-producing current component. On the other hand, the q-axis current, which is perpendicular to the d-axis, is solely responsible for torque production.

In order to apply a rotor flux field orientation condition, the rotor flux linkage is aligned with the d-axis so $\lambda_{qr}^{e}=0$ and $\lambda_{dr}^{e}=\hat{\lambda}_{r}$. By manipulating (13.16) and (13.17) in the rotating reference frame, $\omega=\omega_r$, we can obtain the field oriented condition.

$$p\lambda_{dr}^{e}+\frac{r_r}{L_r}\lambda_{dr}^{e}-\frac{r_r}{L_r}L_m i_{ds}^{e}=0 \xrightarrow{\text{steady state}} i_{ds}^{e}=\frac{\lambda_{dr}^{e}}{L_m} \quad (13.27)$$

$$\omega_{slip}=\frac{r_r}{\hat{\lambda}_r}\left(\frac{L_m}{L_r}\right)i_{qs}^{e}=\frac{L_m i_{qs}^{e}}{\tau_r\hat{\lambda}_r} \quad (13.28)$$

$$i_{qr}^{e}=-\frac{L_m}{L_m+L_{lr}}i_{qs}^{e}$$

$$T_e=\frac{3}{2}\frac{P}{2}\frac{L_m}{L_r}\hat{\lambda}_d i_{qs}^{e} \quad (13.29)$$

We can find out that in this case i_{ds}^{e} controls the rotor flux linkage and i_{qs}^{e} controls the electromagnetic torque. The reference currents of the q-d-0 axis $(i_{qs}^{e*}, i_{ds}^{e*})$ are converted to the reference phase voltages $(v_{ds}^{e*}, v_{qs}^{e*})$ as the commanded voltages for the control loop. Given the position of the rotor flux and two-phase currents, this generic algorithm implements the instantaneous direct torque and flux control by means of coordinate transformations and PI regulators, thereby achieving accurate and efficient motor control.

In asynchronous drives, the mechanical rotor angular speed is not, by definition, equal to the rotor flux angular speed. This implies that the necessary rotor flux position cannot be detected directly by the mechanical position sensor provided with the asynchronous motor explained here.

It is clear that for implementing vector control we have to determine the rotor flux position. Two basic approaches to determine the rotor flux position angle have evolved. The direct scheme shown in Fig. 13.3(a), electrically determines the rotor flux position from measurements using field angle sensors. The indirect scheme illustrated in Fig. 13.3(b), measures the rotor position and utilizes the slip relation to compute the angle of the rotor flux relative to the rotor axis. From the feasibility point of view, implementation of the direct method is difficult if not sometimes impossible. Therefore, in this chapter, the indirect method is considered as a solution for implementing FOC.

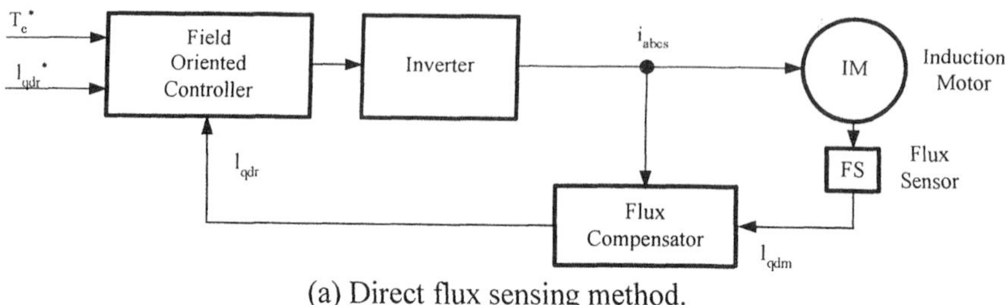

(a) Direct flux sensing method.

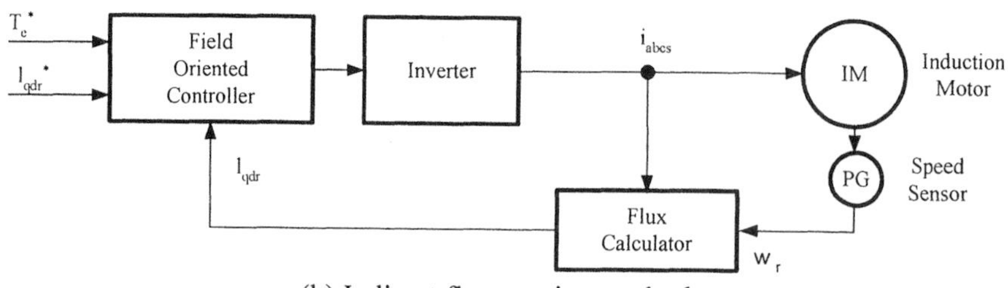

(b) Indirect flux sensing method.

Figure 13.3 Two generic types of induction motor vector control.

The indirect method is based on the calculation of the slip speed ω_{slip}, required for correct field orientation. Equations (13.27) and (13.28) show that we can control torque and field by i_{ds} and i_{qs} in the excitation frame. However, in the implementation of field-oriented control, we need to know i_{ds} and i_{qs} in the stationary reference frame. So, we have to know the angular position of the rotor flux to transform i_{ds} and i_{qs} from the excitation frame to the stationary frame. By using ω_{slip}, which is shown in (13.28) and using actual rotor speed, the rotor flux position is obtained.

$$\int_0^t \omega_{slip} dt + \int_0^t \omega_{re} dt = \theta_r(t) \tag{13.30}$$

or

$$\int_0^t \omega_{slip} dt + \theta_{re}(t) = \theta_r(t) \tag{13.31}$$

In literature, the algorithm of finding rotor flux position using the calculated ω_{slip} and measured ω_{re} or θ_{re} is called the Current Model Method. The Current Model takes i_{ds} and i_{qs} as inputs as well as the rotor mechanical speed and gives the rotor flux position as an output. Figure 13.4 shows the block diagram of the vector control strategy in which speed regulation is possible using a control loop.

The absence of the field angle sensors, along with the ease of operation at low speeds, has increased the popularity of the indirect vector control strategy. While the direct method is inherently the most desirable scheme, it suffers from the unreliability in measuring the flux. Although the indirect method can approach the performance of the direct measurement scheme, its major weakness is the accuracy of the control gain, which heavily depends on the motor parameters.

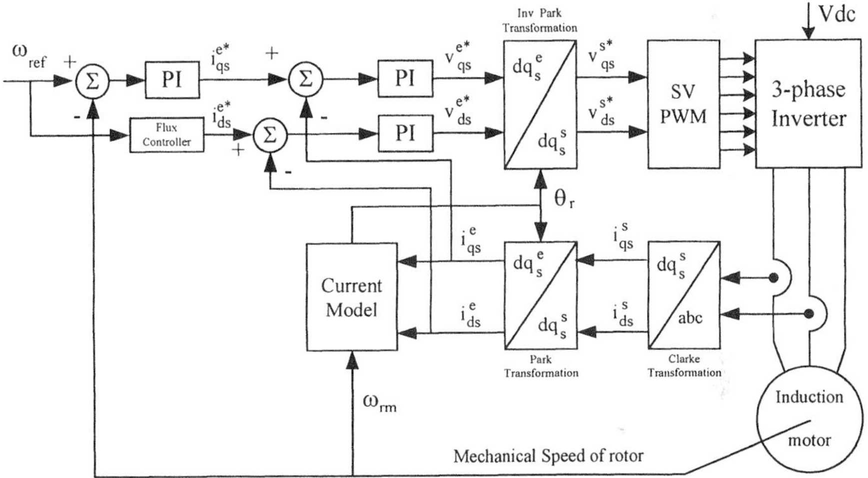

Figure 13.4 Vector control scheme for induction motor.

As shown in Fig. 13.4, two-phase current feeds the Clarke transformation block. These projection outputs are indicated as i_{ds}^{s} and i_{qs}^{s}. These two components of the current provide the inputs to Park's transformation, which gives the currents in the qds^{e} excitation reference frame. The i_{ds}^{e} and i_{qs}^{e} components, which are outputs of the Park transformation block, are compared to their reference values i_{ds}^{e*}, the flux reference, and i_{qs}^{e*}, the torque reference. The torque command, i_{qs}^{e*}, comes from the output of the speed controller. The flux command, i_{ds}^{e*}, is the output of the flux controller which indicates the right rotor flux command for every speed reference. For i_{ds}^{e*} we can use the fact that the magnetizing current is usually between 40 and 60% of the nominal current. For operating in speeds above the nominal speed, a field weakening section should be used in the flux controller section. The current regulator outputs, v_{ds}^{e*} and v_{qs}^{e*}, are applied to the inverse Park transformation. The outputs of this projection are v_{ds}^{s} and v_{qs}^{s}, which are the components of the stator voltage vector in the dqs^{s} orthogonal reference frame. They form the inputs of the SVPWM block. The outputs of this block are the signals that drive the inverter.

Note that both the Park and the inverse Park transformations require the exact rotor flux position, which is given by the current model block. This block needs the rotor resistance or rotor time constant as a parameter. Accurate knowledge of the rotor resistance is essential to achieve the highest possible efficiency from the control structure. Lack of this knowledge results in the detuning of the FOC. In Fig. 13.4, a SVPWM has been used to emulate v_{ds}^{s} and v_{qs}^{s} in order to implement current regulation. The reader can find more information about SVPWM in Chapter 11.

13.9 Simulation Results for the Induction Motor Control System

The drive system with the proposed control strategy has been simulated prior to laboratory experimentation. For simulation purposes, software packages such as Matlab/Simulink™ and Advanced Continuous Simulation Language (ACSL)™ can be used. In this section, Simulink™ has been used to model the induction motor, the vector control, and the SVPWM. The induction motor has been simulated with the dynamic q-d-0 model using the nominal parameters as given in Table 13.1. The dc link voltage in the simulation is equal to 100V. Maximum phase current has been limited to the rated value. Initially, the magnetizing current is set at 60% of the rated current. The simulation results of the control system to a command speed are shown in Fig. 13.5.

13.10 Induction Motor Speed Control System

Based on the previous analysis, an induction motor speed control system is developed as shown in Fig. 13.6. The total control system consists of the induction motor, the power electronics converter, the sensor, and the controller. These components are discussed in detail in the following section.

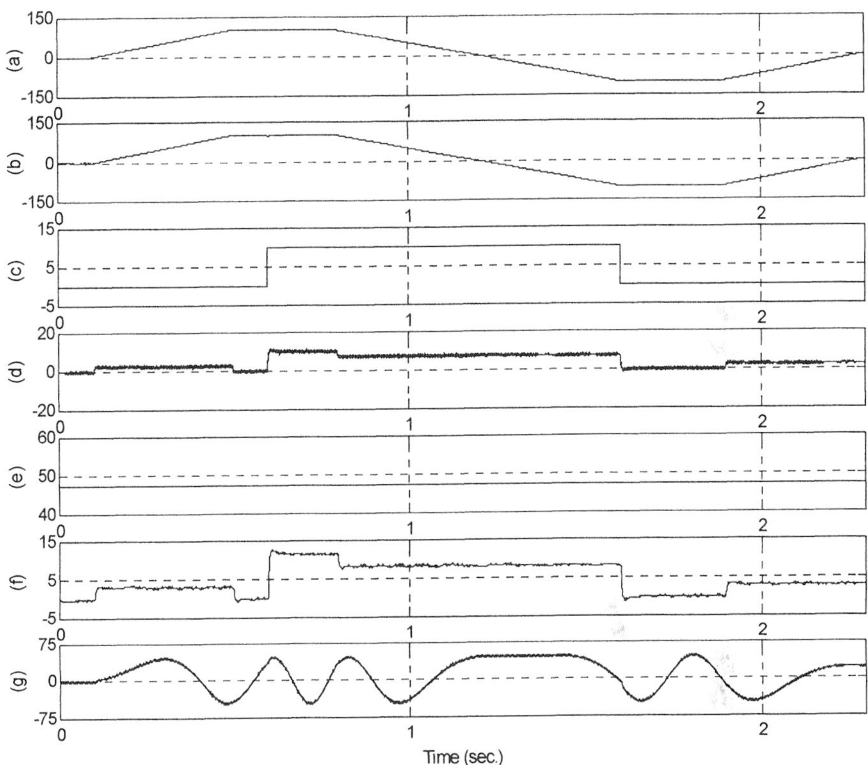

Figure 13.5 (a) reference speed, (b) actual speed, (c) load torque, (d) electromagnetic torque, (e) stator d-axis current in the rotating reference frame, (f) stator q-axis current in the rotating reference frame, (g) phase-A current.

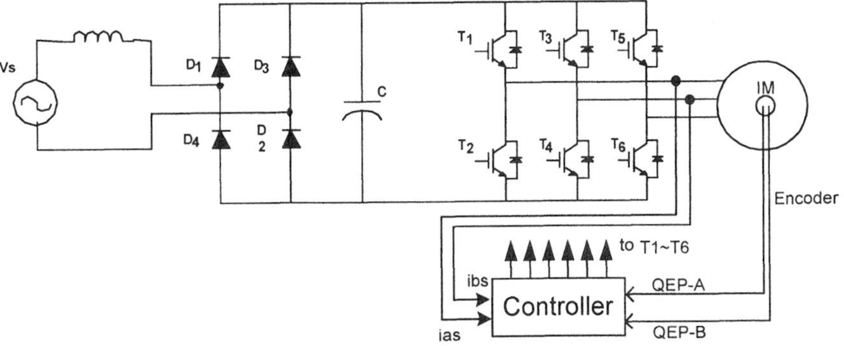

Figure 13.6 Induction motor speed control system.

13.11 System Components

13.11.1 Power Electronic Converter

As shown in Fig. 13.9, the power electronics converter in induction motor control system consists of two parts: a front-end rectifier and a three-phase full-bridge inverter in the right-hand side. The rectifier usually is a full-bridge diode. In case of a regenerative system, a power switch rectifier is used.

The inverter is usually responsible for both the electronic commutation and current regulation. Pulse-width-modulated current controllers are typically used to regulate the actual machine currents to match the sinusoidal current reference waveforms.

The power hardware used to implement and test the induction motor drive system can support an input voltage of 1200 V and a maximum current of 50 A. The hardware is based on six power IGBTs (SKM 50GB 123D), driven by the DSP controller via the integrated driver SKHI22. The power and the control components are insolated from one another by the use of opto-couplers in the gate drive signal path.

Table 13.1 Induction motor parameters

Motor Parameters	Value
Rated power	3.0 hp
Rated Voltage	230/460 Volt
Rated Current	7.6/3.8 Amp
Rated Speed	1760 rpm
Pole pairs	2
Rated frequency	60 Hz
Nominal efficiency	87.5%
Base impedance	23.64631 Ω
Stator resistance	0.044225 Ω
Magnetizing impedance	1.1178 Ω
Stator leakage impedance	0.05956 Ω
Rotor leakage impedance	0.05956 Ω
Rotor resistance	0.03078 Ω

13.11.2 Sensors

Two types of sensors for the induction motor control system are used. One is a current sensor and the other is a position sensor. The phase current sensing is performed via two current sensors supplied with ±15 V. Their maximum input currents can be changed by the number of turns in the primary winding, and the output is a bipolar voltage.

Encoders or resolvers serve as the position sensor because every point of the rotor position is needed to synchronize the rotor with the stator excitation. Figure 13.7 shows the structure of an optical encoder. It consists of a light source, a radially slotted disk and photoelectric sensors. The disk rotates with the rotor. The two photo sensors detect the light passing through the slots in the disk. When the light is hidden, a logic "0" is generated by the sensors. When the light passes through the slots of the disk, a logic "1" is produced. These logic signals are shown in Fig. 13.7. By counting the number of pulses, the motor speed can be calculated. The direction of rotation can be determined by detecting the leading signal between signal A and signal B.

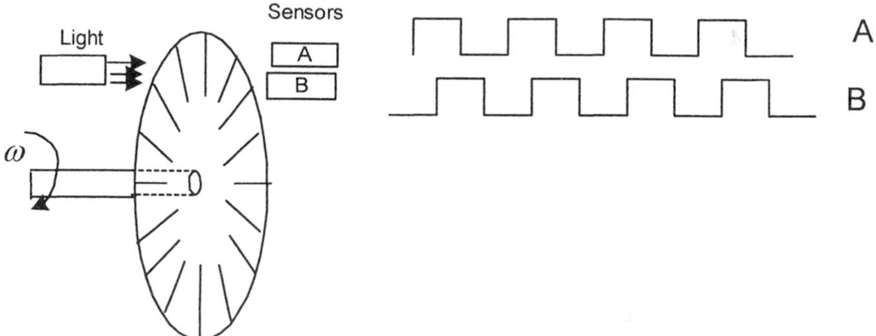

Figure 13.7 The structure of an encoder.

13.11.3 Controller

The controller of the induction motor control system is used to read the feedback current and position signals, to implement the speed or torque control algorithm, and to generate the gate signals based on the control signal. Analog controllers or digital signal processors can perform this task. We have used the LF2407 as a controller.

The interface of the LF2407 is illustrated in Fig. 13.8. Two quadrature counters detect the rising and falling edges of the encoder signals. Two input channels related to the 10-bit Analog-to-Digital Converter (ADC) are selected to read the two-phase currents. The pins PWM1 to PWM6 output the gating signals to the gate drive circuitry.

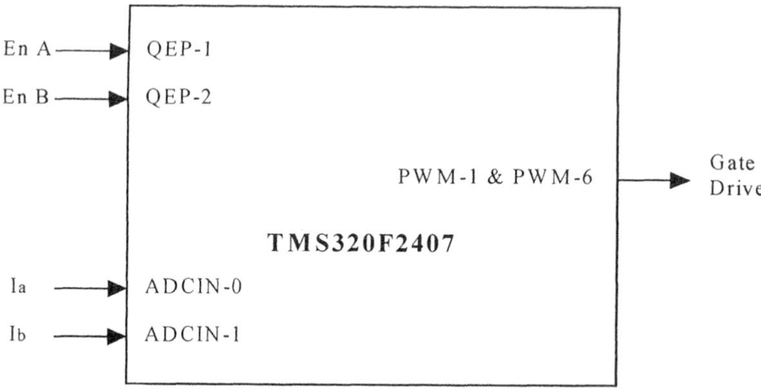

Figure 13.8 The interface of LF2407.

13.12 Implementation of Field-Oriented Speed Control of Induction Motor

Some practical aspects of implementing the block diagram of Fig. 13.8 are discussed in this section and subsections. The software organization, the utilization of different variables, and the handling of the DSP controller resources are described. In addition, the control structure for the per-unit model is presented. Next, some numerical considerations have been made in order to address the problems inherent within fixed-point calculation. As described, current model is one of the most important blocks in the block diagram depicted in Fig. 13.4. The inputs of this block are the currents and mechanical speed of rotor. Two sections of this chapter deal with technical points that should be considered during current and speed measurement, as well as their scaling. Also, there are some points to be noted during development of the current model in software; therefore, one section is dedicated to current model implementation. A PI controller is used in the field-oriented speed control of the induction motor as a regulator for current and speed control. The PI structure and block diagram are presented in another section.

13.12.1 Software Organization

The body of the software consists of two main modules: the initialization module and the PWM Interrupt Service Routine (ISR) module. The initialization model is executed only once at startup. The PWM ISR module interrupts the waiting infinite loop when the timer underflows. When the underflow interrupt flag is set, the corresponding ISR is served. Figure 13.9 shows the general structure of the software. The complete FOC algorithm is executed within the PWM ISR so that it runs at the same frequency as the switching frequency or at a fraction of it. The wait loop could be easily replaced with a user interface [1].

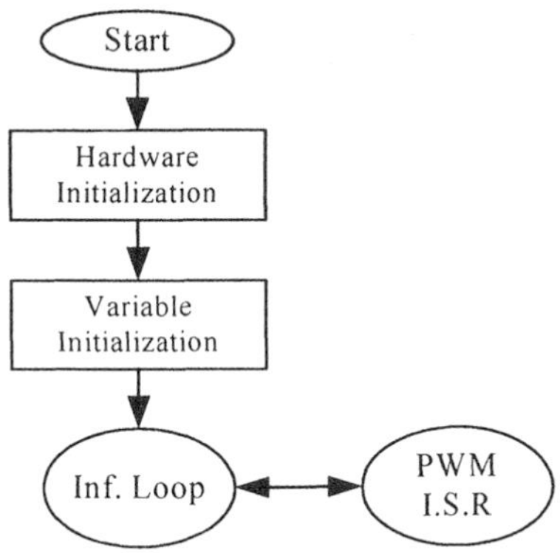

Figure 13.9 General structure of software.

13.12.2 Base Values and Per-Unit Model

It is often convenient to express machine parameters and variables of per-unit quantities. Moreover, the LF2407 is a fixed point DSP, so using a normalized per-unit model of the induction motor is easier than using real parameters. In this model, all quantities refer to the base values. Base power and base voltage are selected, and all parameters and variables are normalized using these base quantities. Although one might violate this convention from time to time when dealing with instantaneous quantities, the rms values of the rated phase voltage and current are generally selected as the base voltage for the a-b-c variables while the peak value is generally selected as the base voltage for d-q variables. The base values are determined from the nominal values by using (13.31), where I_n, V_n, f_n are the nominal phase current, the nominal phase to neutral voltage, and the nominal frequency in a star-connected induction motor, respectively. The base value definitions are as follows:

$$\begin{aligned} I_b &= \sqrt{2} I_n \\ V_b &= \sqrt{2} V_n \\ \omega_b &= 2\pi f_n \\ \psi_b &= \frac{V_b}{\omega_b} \end{aligned} \tag{13.31}$$

I_b and V_b are the maximum values of the nominal phase current and voltage, ω_b is the electrical nominal rotor flux speed, and ψ_b is the base flux.

13.12.3 Numerical Considerations

The per-unit model has been developed so that the software representation of speed, current, and flux is equal to 1.0 when the motor has reached its nominal speed under nominal load and magnetizing current. During transients, the current might reach higher values than the nominal current I_b in order to achieve a short response time. Also, the motor speed might exceed the nominal speed (ω_b), and then every per-unit value might be greater than 1.0. This fact necessitates foreseeing these situations and determines the most suitable numerical format used for the software.

13.12.4 The Numerical Format Determination

The numerical format used in the major part of this chapter is as follows: four bits are dedicated to the integer part, and twelve bits are dedicated to the fractional part. This numeric format is denoted by Q4.12. The resolution for this format is given by

$$\frac{1}{2^{12}} = 0.00024414$$

With the sign extension mode of the LF2407 set, the link between the real quantity and its Q4.12 format representation is illustrated in Fig. 13.10.

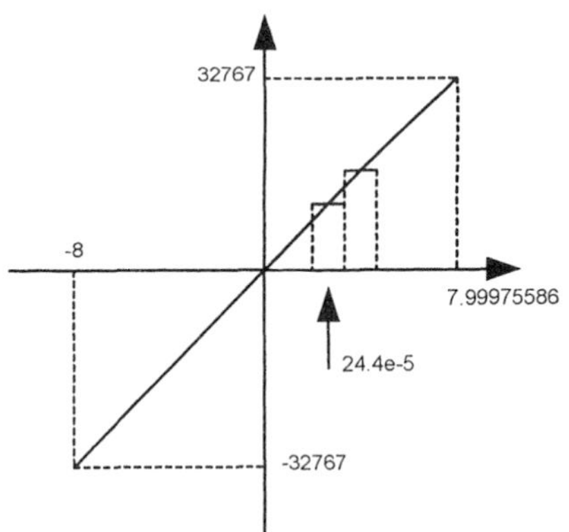

Figure 13.10 Representation of Q4.12 format.

The reason for this particular format is that the drive control quantities are, for the most part, not usually greater than four times their nominal values. In other words, not greater than four when the per-unit model is considered. Where this is not the case, a different format will be chosen. The selection of a range of [-8,8] ensures that the software values can handle each drive control quantity, not only during steady state operation but also during transient operation.

The Qx.y numeric format uses x bits for the integer part and y bits for the fractional part, so the resolution is 2^{-y}. If z is the per-unit value to implement, then its software value is $z \times 2^{y}$ in Qx.y format. Care must be taken when performing operations with a generic Qx.y format. Adding two Qx.y formatted numbers may result in numerical representation overflow. To avoid this kind of problem, one possible solution is to perform the addition in the high side of the Accumulator and set the saturation bit. Another option is to assume that the result will not be out of the maximum range.

The second solution can be used in this implementation if we know that the control quantities do not exceed half of the maximum value in the Q4.12 format. The result can still be represented in the Q4.12 format and directly considered as Q4.12 format, thereby allowing for a higher level of precision. As far as the multiplication is concerned, the result (in the 32-bit Accumulator) must either be shifted x positions to the left and the least significant word stored or be shifted y positions to the right with the last significant word being stored. The stored result is in Qx.y format. Figure 13.11 shows two Qx.y formatted 16-bit variables that are multiplied by one another.

The result of this multiplication in Qx.y format is represented in gray in the 32-bit Accumulator. Both solutions are depicted in Fig. 13.11.

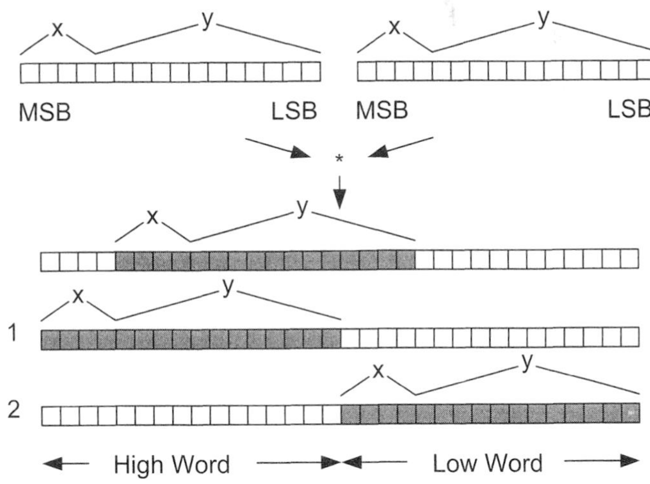

Figure 13.11 (1) Left shift and store high accumulator, (2) Right shift and store low accumulator.

Note that in this section there are also constants that cannot be represented by the Q4.12 format. Operations requiring different formats follow exactly the same process as that explained above.

13.12.5 Current Measurement

The field-oriented control structure requires two-phase current as inputs. Here, current transducers sense these two currents. The current sensor output therefore needs to be rearranged and scaled so that it may be used by the control software in Q4.12 format value. The complete process of acquiring the current is depicted in Fig. 13.12.

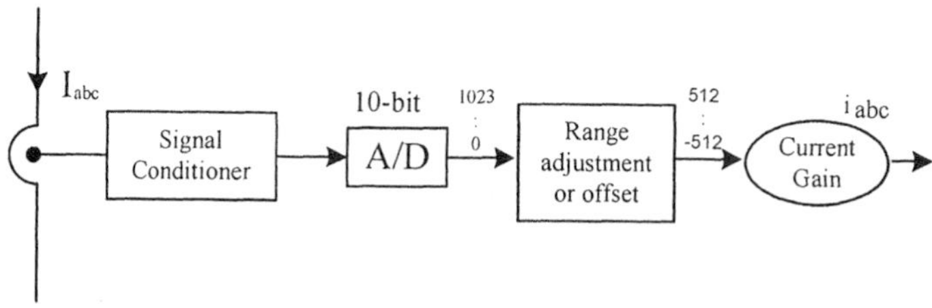

Figure 13.12 Current measurement block diagram.

The output signal of current transducer can be either positive or negative. This signal must be adjusted by the analog interface into a range of (0,3.3V) to allow the ADC module to read both positive and negative values. Figure 13.13 shows the inside of the signal conditioner.

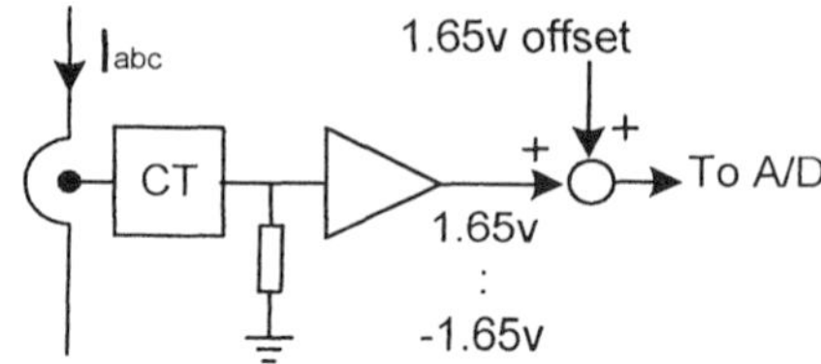

Figure 13.13 Current signal conditioner block diagram.

The amplifier gain is chosen such that sensing $I_{abc} = I_{\max}$ results in the absolute value of the amplifier output to be equal to 1.65V. Note that $I_{\max}$ represents the maximum measurable current, which is not necessarily equal to the maximum phase current. This information is useful at the point where current scaling becomes necessary. The ADC input voltage is now converted into a 10-bit digital value. The 1.65V analog offset is digitally subtracted from the converted result, thereby giving a signed integer value of the sensed current. The result of this process is shown in Fig. 13.14.

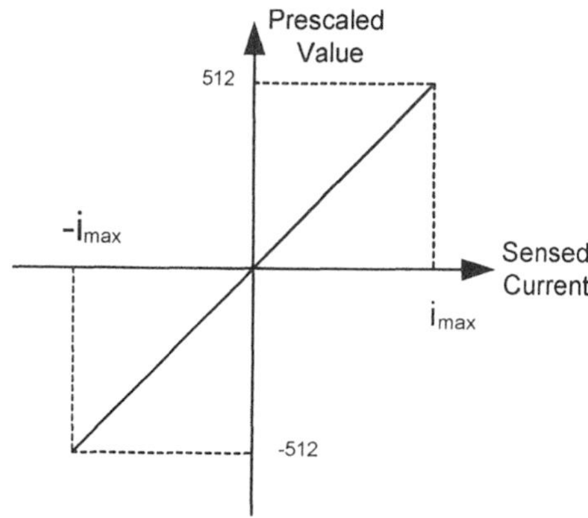

Figure 13.14 Sensed current values before scaling.

Because the variable format is Q4.12, the sensed phase currents must now be expressed with the per-unit model and then be converted into the Q4.12 format. Notice that the per-unit representation of the current is defined as the ratio between the measured current and the base current, and the maximum current handled by the hardware is represented by the value 512. The per-unit current conversion into the Q4.12 format is achieved by multiplying the sensed current by the following constant:

$$K_{cu} = \frac{4096}{\left(\dfrac{512 \times I_b}{I_{\max}}\right)} \tag{13.32}$$

In one single calculation, this constant performs not only the per-unit modeling but also the numerical conversion into Q4.12 format. When nominal current flows in a motor running at nominal speed, the current sensing and scaling block output is 1000h (equivalent to 1 per-unit).

The reader may change the numerical format by amending the numerator value and may adapt this constant to its own current sensing range by recalculating K_{cu} with its own $I_{\max}$ value. In this control system, maximum measurable current and base current are $I_{\max} = 12\text{A}$ and $I_b = 10.7\text{A}$, respectively. The constant value is:

$$K_{cu} = \frac{4096}{\left(\dfrac{512 \times 10.7}{12}\right)} = 8.97 \Leftrightarrow 08F8h \quad Q8.8$$

Note that K_{cu} is outside the Q4.12 format range. The most appropriate format to accommodate this constant is the Q8.8 format, which has a resolution of

$$\frac{1}{2^8} = 0.00390625$$

and the following correspondence to Fig. 13.15.

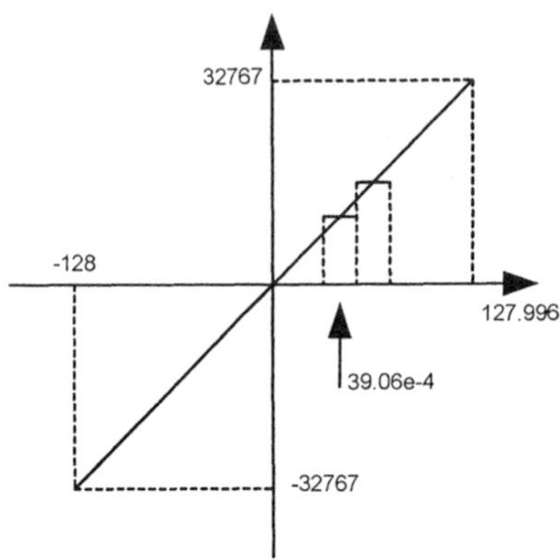

Figure 13.15 Representation of Q8.8 format.

The currents of two phases can be sampled by means of the DSP controller using two channels of the ADC module. The block of assembly code below reads and scales the current of phase A.

```
; Reading and scaling current value of phase A
        LDP #RESULT2>>7
        LACL RESULT2              ;Reading A/D result register
        RPT #5                    ;Shift to right 6 times
        SFR
        AND #0000001111111111b
        SUB #512                  ;Subtracting offset value
        LDP #IA
        SACL IA
        LAR AR0, #Kcur
        LT IA
        MPY *                     ;Multiplying by coefficient to scale the
                                  ;current value
        PAC
        SFL
        SACH IA,7                 ;Save current value in proper variable
```

13.12.6 Speed Measurement

As previously mentioned, for finding rotor flux position, it is necessary to measure the rotor mechanical speed. Usually an incremental encoder is used as a speed sensor. A 64 pulse per revolution incremental encoder is used to measure the motor speed. The software speed resolution is thus based on $4 \times 64 = 256$ increments per revolution. This sensor has two outputs and produces two pulse trains that are 90^0 out of phase with respect to each other. The periods of the pulses proportionally to the rotor speed. The two output channels A and B of speed sensor can be wired directly to the QEP input pins of the LF2407.

Because a low count encoder is used in the control system, and because this encoder does not have enough resolution at low speeds, the control system uses two methods in order to estimate the induction motor speed. One method has enough accuracy in the high speed region, above 200 rpm, and the other has appropriate resolution in the low speed region under 200 rpm. The first method, which is utilized during high speed, is based on counting the number of encoder pulses in a specific time interval. However, the second method is based on the measurement of time between two encoder pulses. Based on the motor speed, the developed program can utilize the advantages of both methods and switch between the two methods based on the actual speed of the motor.

13.12.7 Speed Estimation during High-Speed Region

As previously mentioned, this method is based on counting the number of encoder pulses in a specified time interval. The QEP assigned timer counts the number of pulses and records it in the timer counter register (TxCNT). As the mechanical time constant is much slower than the electrical one, the speed regulation loop frequency might be lower than the current loop frequency. The speed regulation loop frequency is obtained in this program by means of a software counter. This counter accepts the PWM interrupt as input clock and its period is the software variable called SPEEDSTEP. The counter variable is named *speedstep*. When *speedstep* is equal to SPEEDSTEP, the number of pulses counted is stored in another variable called n_p and thus the speed can be calculated. The scheme depicted in Fig. 13.16 shows the structure of the speed feedback generator.

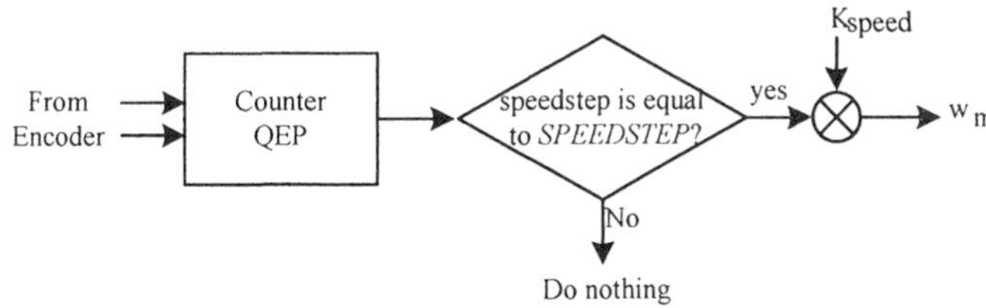

Figure 13.16 Block diagram of speed feedback calculator.

Assuming that n_p is the number of encoder pulses in one SPEEDSTEP period when the rotor turns at the nominal speed, a software constant K_{speed} should be chosen as follows:

$$01000h = K_{speed} \cdot n_p$$

The speed feedback can then be transformed into a Q4.12 format, which can be used in the control software. In the proposed control system, the nominal speed is 1800 rpm and SPEEDSTEP is set to 125. n_p can be calculated as follows:

$$n_p = \frac{1800 \times 64 \times 4}{60} \times SPEEDSTEP \times T_p = 288 \tag{13.33}$$

where $T_p = \dfrac{3}{f_{pwm}} = 3 \times 10^{-4}$ (PWM frequency is 10 kHz but the program is running at 3333 Hz) and hence K_{speed} is given by:

$$K_{speed} = \frac{4096}{288} = 14.22 \Leftrightarrow 0E38h \quad Q8.8$$

Note that K_{speed} is out of the Q4.12 format range. The most appropriate format to handle this constant is the Q8.8 format. The speed feedback in Q4.12 format is then obtained from the encoder by multiplying n_p by K_{speed}. The flow chart of speed measurement is presented in Fig. 13.17. A portion of the assembly code that measures and scales the rotor speed is given below.

```
; Start of speed calculation in high speed region
        LDP #T2CON>>7
        LACC T2CNT              ;Read counter register of encoder pulse
                                ;counter
        SPLK #7FFFh, T2CNT      ;Set counter value to 7FFFh
        SUB #7FFFh              ;Subtract 7FFFh from counter read value
                                ;to omit ;offset
        LDP #Speedtmp           ;Save this value in Speedtmp
        SACL Speedtmp
        LAR AR0, #Kspeed
        LT Speedtmp             ;Multiply Speedtmp by Kspeed to find out
                                ;scaled speed value
        MPY *
        PAC
        SACH N,4                ;Save speed value in proper variable in
                                ;Q4.12 format
```

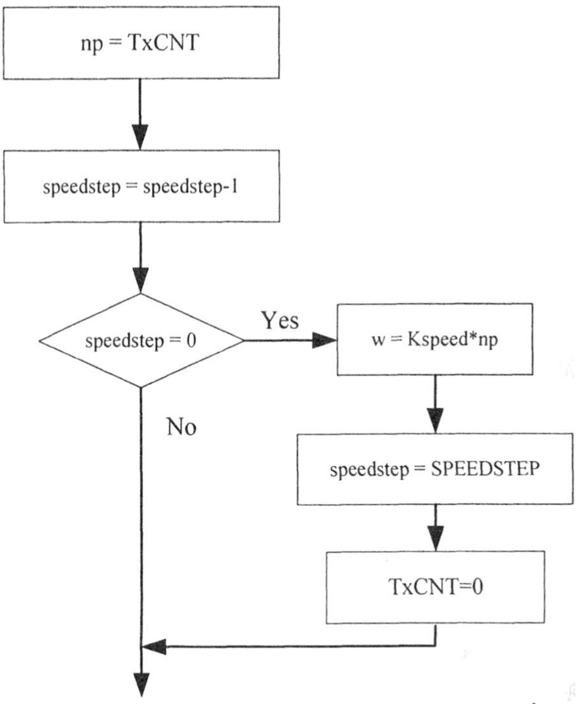

Figure 13.17 Complete flowchart of speed measurement block during high-speed region.

13.12.8 Speed Measurement during Low-Speed Region

To detect the edges of two successive encoder pulses, the developed program can use either the QEP counter or the capture unit input pins. The program has to measure the time between two successive pulses, so therefore it must utilize another GP timer. In this program, Timer 3 has been dedicated to the time measurement. During the interrupt service routine of the capture unit or counter QEP, speed can be calculated. To obtain the actual speed of the motor, the appropriate number is divided by the value in the count register of Timer 3.

As it can be inferred, at very low speeds an overflow may occur in Timer 3. The counter would then reset itself to zero and start counting up again. This event results in a large error in speed measurement. To avoid this event, Timer 3 will be disabled in the overflow interrupt service routine. However, this timer is enabled in the capture unit (counter QEP) interrupt.

The prescalar of Timer 3 is set to x/128, giving the input clock a 234375 Hz frequency. To obtain the speed value in Q4.12 format, 31238×4 (a constant number) is loaded in the accumulator. This number will be divided by the counter register of Timer 3. The flow-chart of this implementation is presented in Fig. 13.18.

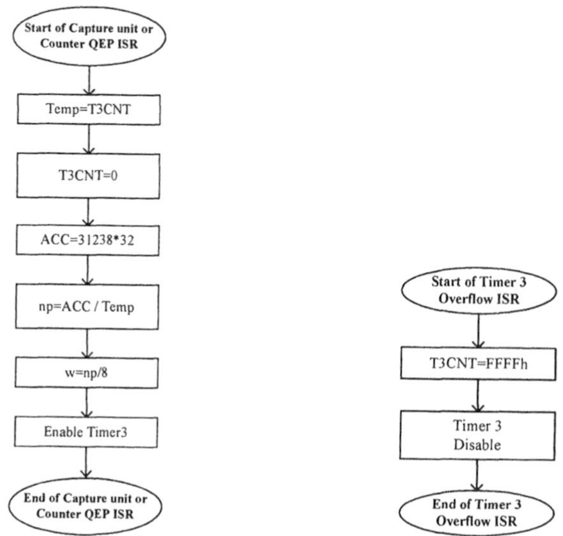

Figure 13.18 Flowchart of speed measurement at low speed.

13.12.9 The Current Model

In indirect FOC, the Current Model is used to find the rotor flux position. This module takes i_{ds} and i_{qs} as inputs plus the rotor electrical speed and then calculates the rotor flux position. The current model is based on (13.27) and (13.28). Equation (13.27) can be written as follows, in transient case:

$$\frac{L_r}{r_r L_m}\frac{d\lambda_{dr}}{dt}+\frac{\lambda_{dr}}{L_m}=i_{ds} \tag{13.34}$$

Assume $\frac{\lambda_{dr}}{L_m}=i_m$ where i_m is the magnetizing current, therefore (13.34) can be written as follows:

$$T_r\frac{d}{dt}i_m+i_m=i_{ds} \tag{13.35}$$

In order to find the rotor flux speed, we use (13.36) which has been inferred from (13.28) and (13.30) in a per-unit system.

$$f_s=\frac{1}{\omega_b}\frac{d\theta}{dt}=\omega_{re}+\frac{i_{qs}}{T_r i_m \omega_b} \tag{13.36}$$

where θ is the rotor flux position and $T_r = \frac{L_r}{r_r}$ and ω_{re} are the rotor time constant and rotor electrical speed, respectively. The rotor time constant is critical to the correct functionality of the Current Model. This system outputs the rotor flux speed, which in turn will be integrated to get the rotor flux position. Assuming that $i_{qs_{(k+1)}} \approx i_{qs_{(k)}}$, (13.35) and (13.36) can be discretized as follows:

$$\begin{aligned} i_{mr_{(k+1)}} &= i_{mr_{(k)}} + \frac{T_p}{T_r}\left(i_{ds_{(k)}} - i_{mr_{(k)}}\right) \\ f_{s_{(k+1)}} &= n_{(K+1)} + \frac{1}{T_r \omega_b} \frac{i_{qs_{(k)}}}{i_{mr_{(k+1)}}} \end{aligned} \tag{13.37}$$

For example, let the constants $\frac{T_p}{T_r}$ and $\frac{1}{T_r \omega_b}$ be renamed to K_t and K_r, respectively. Here $L_r = 73.8\,mH$, $r_r = 0.73\,\Omega$, and $f_n = 60Hz$. So for K_t and K_r we have:

$$K_r = \frac{T_p}{T_r} = \frac{(10000/3)^{-1}}{101.09 \times 10^{-3}} = 2.967 \times 10^{-3} \Leftrightarrow 000Ch \qquad Q4.12$$

$$K_t = \frac{1}{T_r \omega_b} = \frac{1}{30.232 \times 10^{-3} \times 377} = 26.237 \times 10^{-3} \Leftrightarrow 006Bh \qquad Q4.12$$

By knowing the rotor flux speed (f_s), the rotor flux position (θ_{cm}) is computed by the integration formula in the per-unit system.

$$\theta_{cm_{(k+1)}} = \theta_{cm_{(k)}} + \omega_b . f_{s_{(k)}} . T \tag{13.38}$$

As the rotor flux position range is $[0, 2\pi]$, 16-bit integer values have been used to achieve the maximum resolution. Figure 13.19 illustrates the relationship between the flux position and its numerical representation:

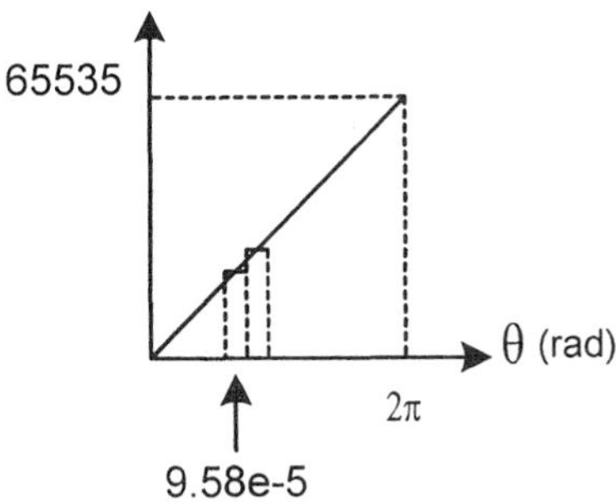

Figure 13.19 Relation between rotor flux position and its numerical representation.

In (13.38), let $\omega_b f_s T$ be called θ_{inc}. This variable is the rotor angle variation within one sampling period. At nominal operation (in other words when $f_s = 1$, the mechanical speed is 1800 rpm), θ_{inc} is equal to $\frac{2\pi \times 60}{10000/3} = 0.11309\,rad$. In one mechanical revolution performed at nominal speed, there are $\frac{2\pi}{0.11309} \approx 56$ increments of the rotor flux position. Let K be defined as the constant, which converts the [0, 2π] range into the [0, 65535] range. K is calculated as follows:

$$K = \frac{65536}{56} = 1170 \Leftrightarrow 0492h \quad Q1.15$$

Note that here we choose the Q1.15 format for this constant because the maximum value of θ which is 65535, represents 1 per-unit and the value of θ cannot be greater than 1 per-unit (2π). With the help of this constant, the rotor flux position computation and its formatting becomes:

$$\theta_{cm_{(k+1)}} = \theta_{cm_{(k)}} + Kf_{s_{(k)}}$$

Thus, the Current Model is a block, as depicted in Fig. 13.20, with three input variables i_{ds}, i_{ds}, ω_{re} (represented in Q4.12 format) and one output, which is the rotor flux position θ_{cm} represented as a 16-bit integer value. The code block below shows a portion of the assembly algorithm that determines the rotor flux position.

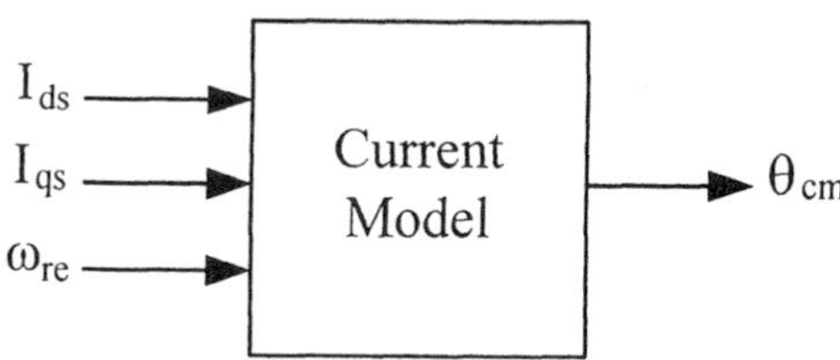

Figure 13.20 Inputs and output of the Current Model block.

```
;start of calculation rotor flux position
        LDP #IDS_R              ;start of calculation magnetizing
                                ;current
        LACC IDS_R
        SUB Imr
        SACL temp1
        MAR *, AR0
        LAR AR0, #Kr
        LT temp1
        MPY *
        PAC
        SACH temp1,4
        LACC temp1
        ADD Imr
        SACL Imr
```

```
        BCND Imr_Neqz, NEQ
        LACC #0
        SACL temp1
        B IQS_Rp

Imr_Neqz:                      ;if  Imr != 0 then  start of  slip
                               ;frequency calculation
        LACC Imr
        SACL temp2
        LACC IQS_R
        ABS
        SACL temp1
        LACC temp1,12
        RPT #15
        SUBC temp2
        SACL temp1
        LACC IQS_R
        BCND IQS_Rp, GT        ;if IQS is negative then change sign of
                               ;IQS/Imr
        LACC temp1
        NEG
        SACL temp1
IQS_Rp:

        LAR AR0, #Kt
        LT temp1
        MPY *
        PAC
        SACH temp1,4
        LACC temp1
        ADD N                  ;add rotor speed to slip frequency
        SACL fs                ;find rotor flux speed
;end of calculation of rotating flux speed
;Start of finding Rotor Flux position by using integral of fs
        LACC fs
        ABS
        SACL temp1
        LAR AR0, #Kfs          ;multiplying fs bu Kfs, a constant value
                               ;to find increment or decrement in rotor
                               ;flux position
        LT temp1
        MPY *
        PAC
        SACH teta_inc,4
        bit fs,0
        BCND fs_neg, TC        ;go to fs_neg if teta_inc is negative
        LACL teta_inc
        ADDS TETA
        SACL TETA              ;find new rotor flux position if
                               ;teta_inc is negative

        B fs_pos
fs_neg
        LACL TETA
        SUBS teta_inc
        SACL TETA              ;find new roto flux position if teta_inc
                               ;is positive
fs_pos
; end of finding Rotor flux position
```

13.12.10 The PI regulator

An electrical drive based on the Field Orientated Control needs two constants as control parameters: the torque component reference i_{qs}^{e*} and the flux component reference i_{ds}^{e*}. The classical PI regulator is well suited to regulate the torque and flux feedback to the desired values. This is because it is able to reach constant references by correctly setting both the proportional term (K_p) and the integral term (K_i), which are, respectively, responsible for the error sensibility and for the steady state error. The numerical expression of the PI regulator is as follows:

$$Y_{(k)} = K_p e_{(k)} + K_i e_{(k)} + \sum_{n=0}^{k-1} e_{(n)} \tag{13.39}$$

which is represented in Fig. 13.21.

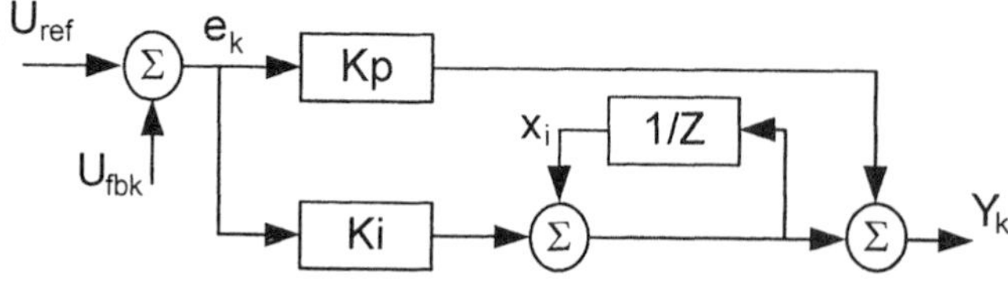

Figure 13.21 Classical PI regulator structure in discrete domain.

The limiting point is that during normal operation, large reference value variations or disturbances may occur, resulting in the saturation and overflow of the regulator variables and output. If they are not controlled, this non-linearity detriments the dynamic performance of the system. To solve this problem, one solution is to add to the previous structure a correction of the integral component as depicted in Fig. 13.22 [2].

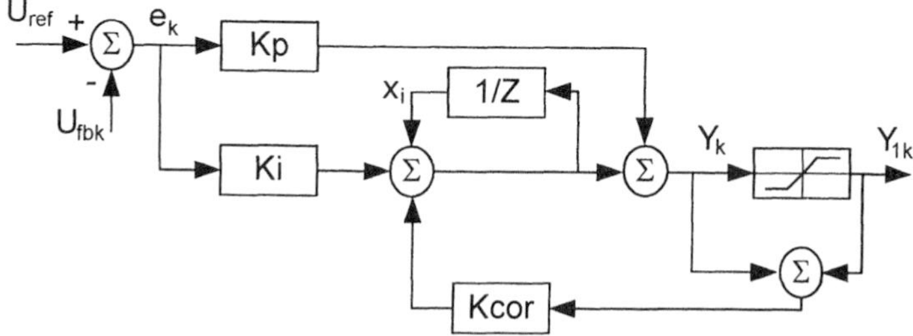

Figure 13.22 Numerical PI regulator with correction of the integral term.

The PI regulators are implemented with output saturation and integral component correction. The constants *Kpi, Ki, Kcor,* proportional, integral, and integral correction components, are selected based on the sampling period and on the motor parameters. These values should be changed based on the motor speed. These changes can be done automatically within a dummy loop in the program. To show the routine of the PI controller in assembly, the following section of code is given:

```
        LDP #N_ref              ;Start of PI procedure
        LACC N_ref              ;Load reference speed
        SUB N                   ;subtract motor speed from reference
                                ;speed to find error
        SACL err_N              ;put error in err_N
        LACC xi_N,12            ;load previous value of PI output
                                ;controller
        LAR AR0, #Kp_N          ;start of multiplication Kp*error
        LT err_N
        MPY *
        APAC                    ;Add previous output of controller with
                                ;new value Y(p)=Y(p-1)+Kp*error
        SACH Upi_N,4
        LACC Upi_N              ;start of positive saturation value <
                                ;(max)
        ABS
        SUB IQS_Rmax
        BCND N_sat, GT          ;if value is less than (max) go to
                                ;negative saturation
        LACL Upi_N
        B N_LIMIT
N_sat
        BIT Upi_N, 0            ;start of negative saturation (min)<
                                ;value
        BCND Upi_Ngz, NTC       ;if Upi_N is positive, then go to
                                ;Upi_Ngz
        LACL IQS_Rmin
        SACL Upi_N
        BN_LIMIT
Upi_Ngz
        LACL IQS_Rmax
        SACL Upi_N
N_LIMIT
        SACL IQS_R              ;start of correction procedure
        LAR AR0, #Ki_N
        LT err_N
        MPY *
        PAC
        ADD xi_N, 12
        SACH xi_N,4
        LACC xi_N               ;start of saturation on integrator
                                ;output
        ABS
        SUB IQS_Rmax
        BCND x_sat, GT
        LACL xi_N
        B x_LIMIT
x_sat
        BIT xi_N, 0
        BCND xi_Ngz, NTC        ;if xi_N is positive, then go to xi_Ngz
        LACL IQS_Rmin
        SACL xi_N
        B x_LIMIT
xi_Ngz
        LACL IQS_Rmax
        SACL xi_N
x_LIMIT
        SACL xi_N
; end of PI procedure
```

13.12.11 Calculation of Sine and Cosine Functions

In order to generate the sine and the cosine of an angle, a sine table and indirect addressing mode by auxiliary register AR has been used. This algorithm and code examples are presented in Chapter 11. The flow chart of the field-oriented speed control of induction motor is presented in Fig. 13.23. This routine is placed inside the PWM interrupt service routine.

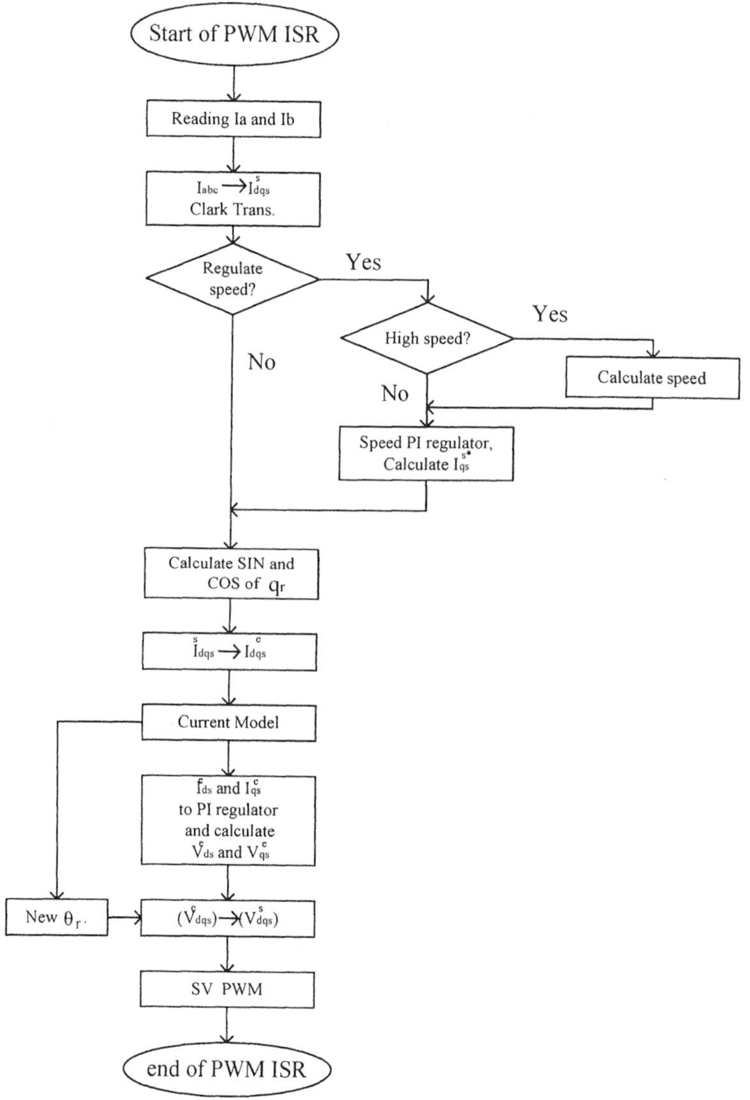

Figure 13.23 Flow chart of FOC software.

13.13 Experimental Results

In our experience, the motor has been coupled to a DC generator. The generator can be loaded with an adjustable resistor providing a variables load. As explained in the previous sections, the flux reference (i_{ds}^{e*}) in the normal speed range has been set at 0.4 per-unit. To avoid the maximum phase current being greater than 1 per-unit ($\left|i_{abc}\right| = \sqrt{i_{ds}^{e*2} + i_{qs}^{e*2}}$), i_{qs}^{e*} may not be higher than 0.8 per-unit. This torque reference limitation is integrated into the control software using the *IQS_ref_max* constant, which is set to 0CCDh (4.12 format). The following scope captures show the transient and steady state operations. Figure 13.24 shows the load torque, reference speed, motor speed, and phase current of the motor during transient operation. Before a change of the reference speed, a magnetizing current is applied to the motor to build the magnetizing flux. By increasing the load, a breaking torque is applied to the motor. In this figure, the reference speed is 100 rpm and the load torque is 45.5 (lb-in).

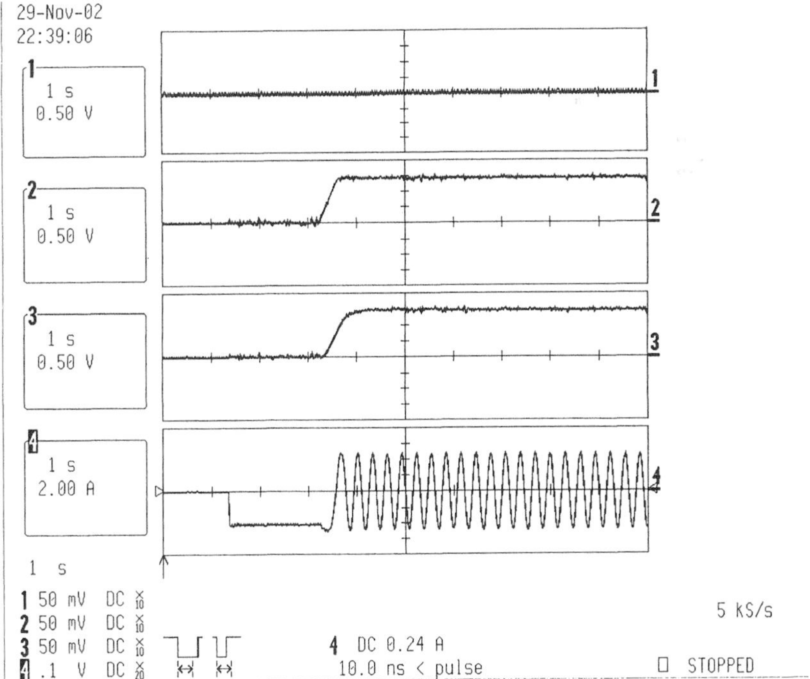

Figure 13.24 Start up, no-load condition, (1) load torque, (2) reference speed 100 rpm, (3) motor speed, (4) phase current.

13.14 Conclusion

In this chapter, the theory of field-oriented control of induction motors was described. The structure and organization of software written for the LF2407 DSP controller was also presented. Some technical points and tools were presented to assist in developing a working model for an induction motor drive. The modular structure of this presentation and guidelines allow the reader to quickly grasp the aspects of FOC, thereby assisting the reader in developing software for specific needs.

References

1. Texas Instruments, Implementation of a Speed Field Orientated Control of Three Phase AC Induction Motor using TMS320LF240, Literature Number: BPRA076, March 1998.
2. Texas Instruments, Field Orientated Control of Three Phase AC-motors, Literature Number: BPRA073, December 1997.

Chapter 14

DSP-BASED CONTROL OF SWITCHED RELUCTANCE MOTOR DRIVES

14.1 Introduction

Switched Reluctance Motor (SRM) drives are relatively new entities in the perpetually growing market of Adjustable Speed Motor Drives (ASMD). A rugged, modular structure and a relatively simple geometry are among the advantages of the SRM drive. In addition, the absence of magnetic sources, i.e., windings or permanent magnets on the rotor shown in Fig. 14.1, makes SRM relatively easy to cool and insensitive to high temperatures. The latter is of prime interest in applications that demand operation under harsh conditions such as automotive starter or alternator.

Figure 14.1 Rotor and stator of an 8/6 SRM.

As a singly excited synchronous machine, SRM generates its electromagnetic torque solely on the principle of reluctance. In most electric machines, an attraction and repletion force between the magnetic fields caused by the armature and field windings forms the dominant part of the torque. In a SRM, the tendency of a polarized rotor pole to align with an excited stator pole is the only source of torque. It must be noted that optimal performance is achieved by proper positioning of the current pulse with respect to the magnetic status of the machine. Therefore, the sensing of the rotor position becomes an integral part of the control in a SRM drive.

A unipolar power inverter is usually used to supply the SRM. The generation of the targeted current profile is performed using a hysteresis or PWM type current controller. Although a square-shaped current pulse is commonly used for excitation in a SRM, different optimal current profiles are sometimes used to mitigate the undesirable effects of excessive torque undulation and audible noise. In fact, SRM drives serve as an outstanding example of advanced motor drive systems where the focus is not on the complicated geometries of the motor, but on the development of a sophisticated control algorithm. The development of complex control algorithms is facilitated by the recent development of high-performance, cost-effective DSP-based controllers.

Optimal torque control is the major differentiating factor among various arts of electric drives. A detailed explanation of the torque generation process and optimal torque control in the four quadrants of operation of a SRM is given. Later, the development of speed controllers will be discussed.

14.2 Fundamentals of Operation

Switched reluctance machines can operate as either a motor or a generator. To explain the torque generation process, we investigate the mechanism of electromechanical energy conversion. As shown in Fig. 14.2, in order to establish a reluctance torque, a stator phase is excited at unaligned position displayed by a in the figure, viz., the position at which a pair of stator and rotor poles exhibits its largest air gap length. By magnetizing the stator pole, the closest rotor pole will be magnetically polarized and will experience an attractive force. The tangential component of this force substantiates an electromagnetic torque in the direction which reduces the air gap length. The shape of the current is usually controlled such that a maximum torque per ampere is generated. As the rotor approaches the aligned position, shown by u in the figure, the radial component of the attraction force becomes dominant and the tangential component reduces to zero. Therefore, it makes economic sense to remove the current before the aligned position. The shaded area in Fig. 14.2 depicts the magnetic energy converted into mechanical form, whereas the area denoted by "R" demonstrates the magnetic energy that has not been converted into useful work. Notably, the ratio between mechanical work and total converted energy into magnetic form is an indication of power quality in SRM drives. In Fig. 14.2, ψ and θ represent the flux linkages and rotor position, respectively.

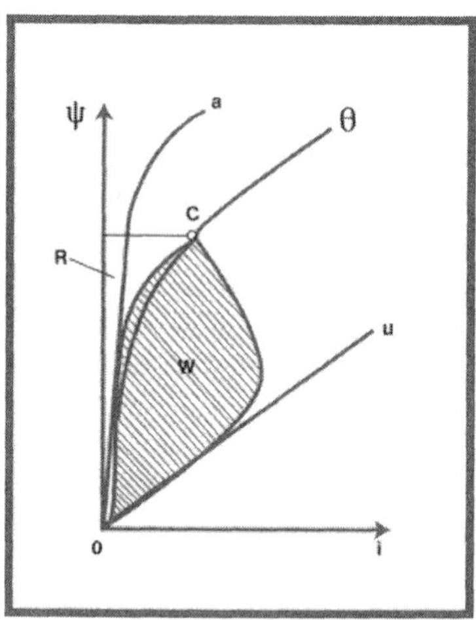

Figure 14.2 Electromechanical energy conversion in SRM.

In order to obtain motoring action, a stator phase is excited when the rotor is moving from the unaligned position toward the aligned position. Similarly, by exciting a stator phase when the rotor is moving from aligned toward unaligned position, a generating action will be achieved. By sequential excitation of the stator phases, a continuous rotation can be achieved. Figure 14.3 shows the distribution of the magnetic field during commutations in an 8/6 SRM drive. Notably, the direction of the rotation is opposite that of the stator excitation. A short flux path in the back-iron of the motor occurs in each electrical cycle. This, in turn, may cause asymmetry in the torque production process.

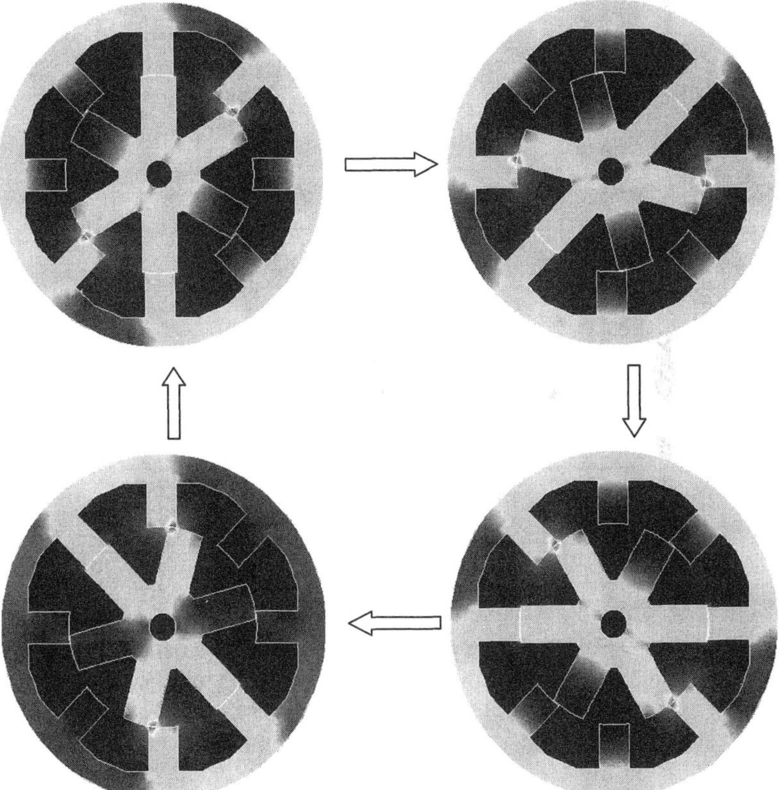

Figure 14.3 Illustration of short vs. long flux paths for a 8/6 SRM.

Proper synchronization of the stator excitation with the rotor position is a key step in the development of an optimal control strategy in SRM drives. Because the magnetic characteristics of the SRM, such as phase inductance or phase flux linkage, portray a one-to-one correspondence with the rotor position, they may be directly used for control purposes. In either case, direct or indirect detection of the rotor position forms an integral part of the control in the SRM drives.

The asymmetric bridge shown in Fig. 14.4 is the most commonly used power electronics inverter for a SRM drive. This topology features a unipolar architecture that allows for satisfactory operation in SRM drives. If both switches are closed, the available dc link voltage is applied to the winding. By opening the switches, the negative dc link voltage will be applied to the winding and freewheeling diodes guarantee a continuous current in the windings. Obviously, by keeping one of the switches closed while the other one is open, the respective freewheeling diode will provide a short-circuited path for the current. This topology can be used effectively to implement PWM-based or hysteresis-based current regulation as demanded by the control system. However, one should notice that at high speeds the induced EMF in the winding is dominant and does not allow effective control of the current waveform. Therefore, current regulation is an issue related only to the low speed mode of operation. During generation, the mechanical energy supplied by the prime mover will be converted into an electrical form manifested by the induced EMF. Unlike the motoring mode of operation, this voltage acts as a voltage source that increases the current in the stator phase, thereby resulting in the generation of electricity.

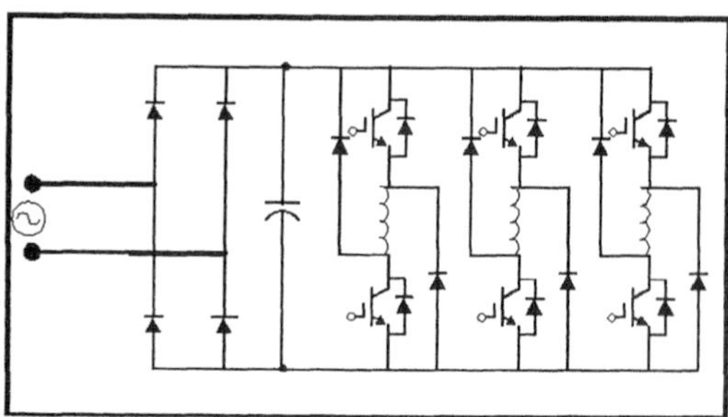

Figure 14.4 An asymmetric bridge with the front end rectifier for a 3-ϕ SRM drive.

14.3 Fundamentals of Control in SRM Drives

The control of electromagnetic torque is the main differentiating factor between various types of adjustable speed motor drives. In switched reluctance motor drives, tuning the commutation instant and profile of the phase current controls electromagnetic torque. Figure 14.5 depicts the basics of commutation in SRM drives. It can be seen that by properly positioning the current pulse, one can obtain positive (motoring) or negative (generating) modes of operation.

The induced EMF and electromagnetic torque generated by the SRM drive can be expressed in terms of co-energy as follows:

$$E = \frac{\partial^2 W_c}{\partial\theta\,\partial i}\omega \approx \frac{dL(\theta)}{d\theta} i\omega$$

$$T = \frac{\partial W_c}{\partial\theta} \approx \frac{1}{2}\frac{dL(\theta)}{d\theta} i^2 \tag{14.1}$$

where W_C, L, θ, i, and ω stand for co-energy, phase inductance, rotor position, phase current, and angular speed, respectively.

It must be noted that the nonlinear effects of magnetic saturation are neglected here. It is evident that a positive torque is achieved only if the current pulse is positioned in a region with an increasing inductance profile. Similarly, a generating mode of operation is achieved when the excitation is positioned in a region with a decreasing inductance profile. In order to enhance the productivity of the SRM drive, the commutation instants, (i.e., θ_{on}, θ_{off}) need to be tuned as a function of the angular speed and phase current. To fulfill this goal, the optimization of torque per Ampere is a meaningful objective. Therefore, exciting the motor phase when the inductance has a flat shape should be avoided. At the same time, the phase current needs to be removed well before the aligned position to avoid the generation of negative torque.

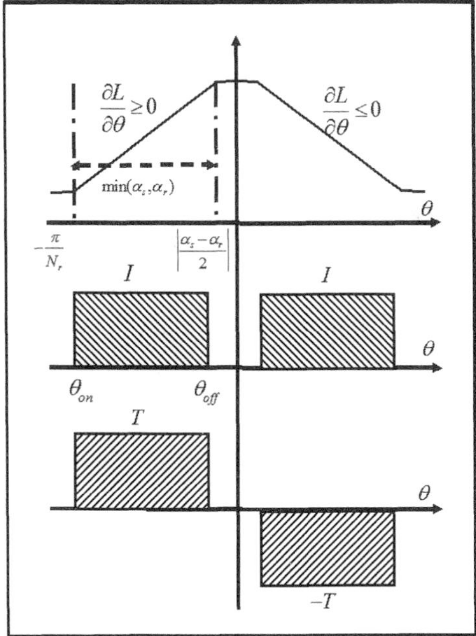

Figure 14.5 Commutation in SRM drives.

14.4 Open Loop Control Strategy for Torque

By the proper selection of the control variables, commutation instants, and reference current, an open loop control strategy for SRM drive can be designed. The open loop control strategy is comprised of the following steps:

- Detection of the initial rotor position.
- Computation of the commutation thresholds in accordance with the sign of torque, current level, and speed.

- Monitoring of the rotor position and selection of the active phases.
- A control strategy for regulation of the phase current at low speeds.

Each step is explained in detail in the subsequent sections.

14.4.1 Detection of the Initial Rotor Position

The main task at rotor standstill is to detect the most proper phase for initial excitation. Once this is established, according to the direction of rotation, a sequence of stator phase excitation will be put in place. The major difficulty in using commercially available encoders is that they do not provide a position reference. Therefore, the easiest way to find rotor position for motor startup is to align one of the stator phases with the rotor. This can be achieved by exciting an arbitrary stator phase with an adequate current for a short period of time. Once the rotor is in an aligned position, a reference initial position can then be established. This method requires an initial movement by the rotor, which may not be acceptable in some applications. In these cases, the incorporation of a sensorless scheme at standstill is sought out. Although the explanation of sensorless control strategies for rotor position detection is beyond the extent of this chapter, due to its critical role, the detection of rotor position at standstill is explained here.

To detect rotor position at standstill, a series of voltage pulses with fixed and sufficiently short duration are applied to all phases. By consequent comparison between the magnitudes of the resulting peak currents, the most appropriate phase for conduction is selected. Figure 14.6 shows a set of normalized inductance profiles for a 12/8 SRM drive.

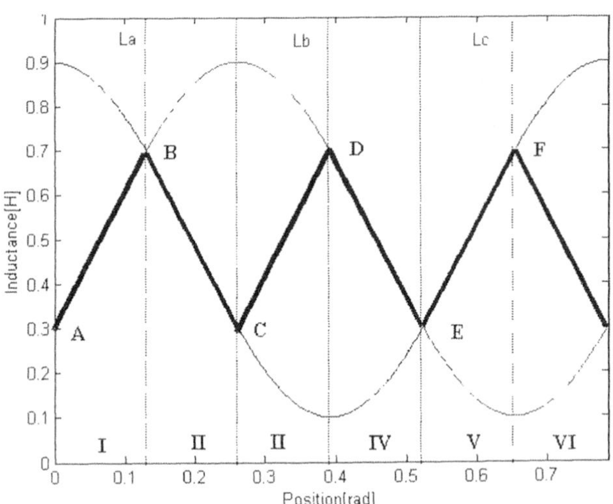

Figure 14.6 Assignment of various regions according to inductances in a 12/8 SRM drive.

A full electrical period is divided into six separate regions according to the magnitudes of the inductances. Due to the absence of the induced voltage and small amplitude of currents, one can prove that the following relationship will hold for the magnitudes of measured currents:

$$I_{ABC} = \frac{V_{Bus}\Delta T}{L_{ABC}} \tag{14.2}$$

where ΔT, V_{Bus}, and L_{ABC} stand for duration of pulses, dc link voltage, and phase inductances, respectively. Table I summarizes the detection process for a 12/8 SRM drive. Once the range of position is detected, the proper phase for starting can be easily determined. Furthermore, in each region there exists a phase that offers a linear inductance characteristic. This phase can be used for the computation of rotor position using (14.2).

Table I Detection of best phase to excite at standstill

Region	Condition	Rotor angle [mech]
I	$I_A < I_B < I_C$	$0 < \theta^* < 7.5^0$
II	$I_B < I_A < I_C$	$7.5^0 < \theta^* < 15^0$
III	$I_B < I_C < I_A$	$15^0 < \theta^* < 22.5^0$
IV	$I_C < I_B < I_A$	$22.5^0 < \theta^* < 30^0$
V	$I_C < I_A < I_B$	$30^0 < \theta^* < 37.5^0$
VI	$I_A < I_C < I_B$	$37.5^0 < \theta^* < 45^0$

The flowchart shown in Fig. 14.7 summarizes the detection process at standstill.

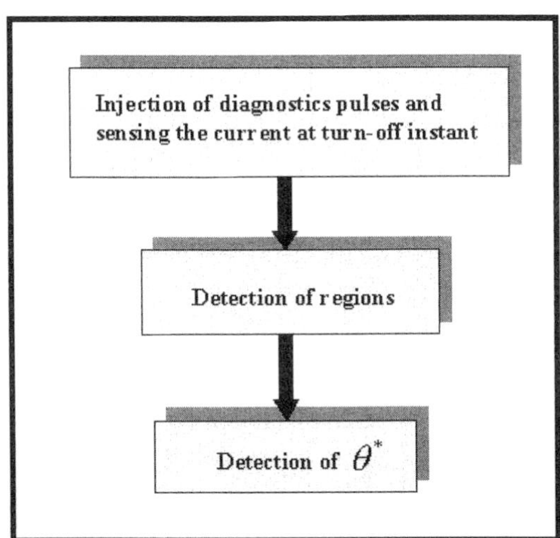

Figure 14.7 Detection of rotor position at standstill.

14.4.2 Computation of the Commutation Thresholds

In the next step, the commutation angles for each phase should be computed and stored in memory. If the commutation angles are fixed, computing the thresholds is relatively straightforward. It must be noted that within each electrical cycle, every phase should be excited only once. In addition, in a symmetric machine, phases are shifted by

$$\Delta\theta = \frac{(N_s - N_r)360^0}{N_s} \tag{14.3}$$

where N_s and N_r stand for the number of rotor and stator poles. Given a reference for rotor position such as the aligned rotor position with phase-A, one can compute and store the commutation instants for each phase. The commutation thresholds are usually converted into a proper scale so they can be compared with the value of a counter which tracks the number of incoming pulses from the position sensor. If a particular encoder can generate N pulses per mechanical revolution, then every mechanical degree corresponds to $4.N/360$ pulses received by the QEP counter of the LF2407. The quadrature input of the LF2407 is designed for commercially available encoders with quadrature output. Notably, this component can increase the resolution of the sensed position by a factor of four.

If optimal performance of the machine is targeted, the effects of rotational speed and current must be taken into account. Figure 14.8 shows a typical current pulse for SRM drive. To achieve optimal control, the delay angles during the turn-on and turn-off process need to be taken into account. By neglecting the effects of motional back-EMF in the neighborhood of commutation, which is a valid assumption as turn-on and turn-off instants occur close to unaligned and aligned position, respectively, one can calculate the delay angles as

$$\begin{aligned} \theta_{on-delay} &= \frac{\omega L_u}{r} \ln\left(\frac{V}{V - rI_{\max}}\right) \\ \theta_{off-delay} &\approx \theta_{on-delay}\left(\frac{L_a I_{\max}}{L_u}\right) \end{aligned} \tag{14.4}$$

where L_u, L_a, ω, V, and r denote unaligned inductance, aligned inductance, angular speed, bus voltage, and stator phase resistance, respectively. The dependency of the aligned position inductance upon maximum phase current is an indication of the nonlinear effects of saturation that need to be taken into account. As the speed and level of current increases, one needs to adopt the commutation angles using (14.4). As can be seen, the dependency of commutation angle upon the angular speed is linear while its dependency upon the maximum phase current has a very nonlinear relationship.

14.4.3 Monitoring of the Rotor Position and Selection of the Active Phases

Once the previous steps are done, one can start with the main control tasks, namely, enforcing the conduction band and regulating the current. The block diagram depicted in Fig. 14.9 shows the structure used in a typical algorithm, which forms the basic control strategy of the SRM drive. Monitoring the rotor position is a relatively easy task with a LF2407. By properly initializing the second timer in the event manager, it can be programmed to act as a counter for QEP encoder input. The content of this counter can be accessed at any stage of the program. This makes the monitoring of the rotor position an easy task. For the first task in the interrupt service routine, the current value of the rotor position will be compared against the commutation thresholds, and phases that should be on will be identified. In the next step, the current in active phases where an active phase is referred to as a phase that is turned on will be regulated.

14.4.4 A Control Strategy for Regulation of the Phase Current at Low Speeds

At low speeds where the induced EMF is small, a method for control of the phase current is necessary. In the absence of such routines, the phase current will increase exponentially, possibly damaging the semiconductor devices or motor windings. Hysteresis and PWM control strategies are commonly used for regulating the phase current at low speeds. At higher speeds, the presence of a significantly larger back-EMF limits the growth of the phase current and there is no need for such regulation schemes. The profile of the regulated current depends on the control objective. In most applications, a flat-topped or square shaped current pulse will be used. Figure 14.10 shows a regulated current waveform along with the gate pulse that is recorded at low speed region.

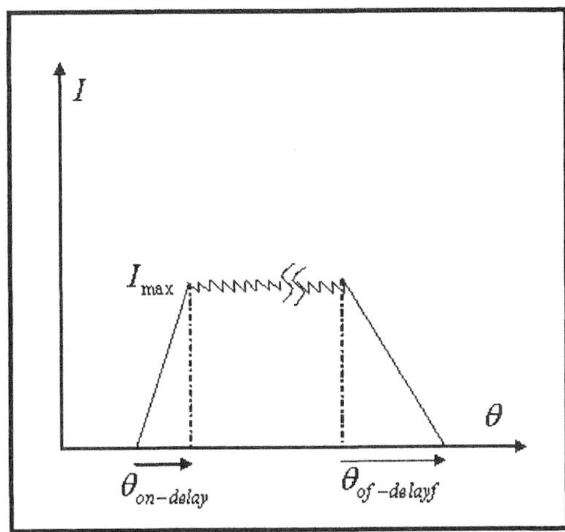

Figure 14.8 A typical current pulse at low speeds.

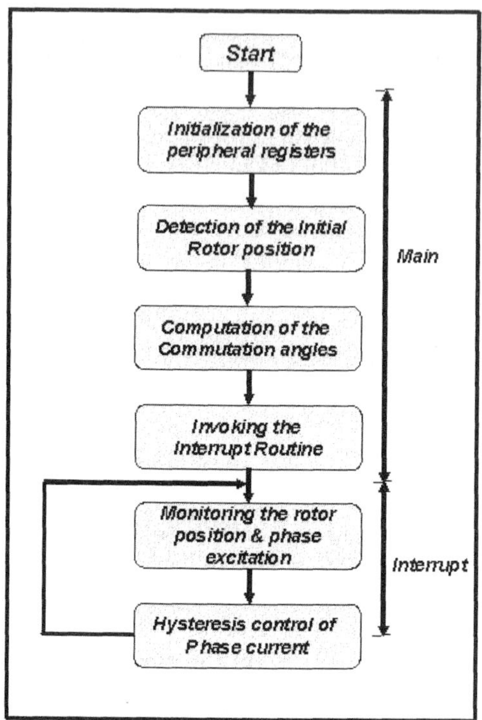

Figure 14.9 Block diagram of the basic control in SRM drives

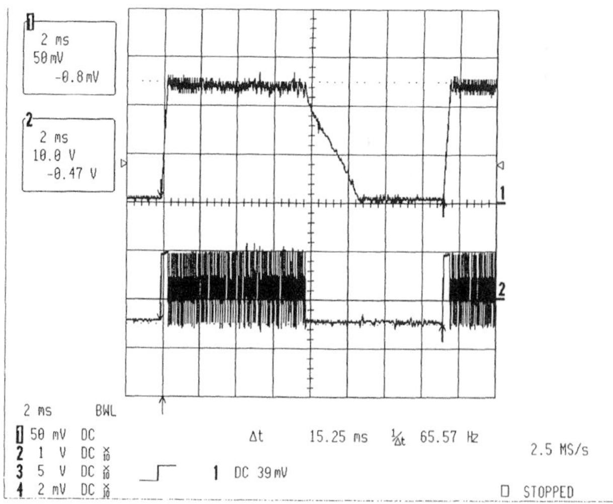

Figure 14.10 Phase current waveform and the gating signal without optimization
Reference current = 5.5 A; conduction angle = 180°(electrical)
(operating speed = 980 RPM; output power = 120 W)

In order to conduct hysteresis control, the currents in active phases need to be sensed. Once the phase current is sampled, it needs to be converted into digital form. This can be done using the on-chip analog-to-digital converter of the LF2407. The control rules for a classic two switch per phase inverter shown in Fig. 14.4 are given by:

- If $I_{\min} \geq I$, then both switches are on. This results in applying the bus voltage across the coil terminals.
- If $I_{\max} \leq I$, then both switches are turned off. This results in applying the negative bus voltage across the coil terminals.
- If $I_{\min} \leq I \leq I_{\max}$, there is no need to make any changes in the status of the switches (i.e., if the switches are on they remain on and if they are off they remain off).

By simple comparison between the sampled current and current limits, one can develop a hysteresis control strategy. Since the current is sampled during each interrupt service routine, the time period of the interrupt should be sufficiently small to allow for a tight regulation. Because in most practical cases only two phases conduct simultaneously, and given the speed of 30 MIPS (millions of instructions per seconds) in LF2407 (or 40MIPS for LF2407A) the interrupt service time should be very small.

PWM technique can also be used in control of the phase current in SRM drive. The LF2407 offers a fully controllable set of PWM signals via the Compare Units. The frequency and duty cycle of these PWM pulses can be adjusted at any stage of the program by the setting of two peripheral registers. This valuable feature can easily accommodate the control needs of a three-phase SRM drive system. In the case of a four-phase SRM drive, the fourth PWM signal can be generated by using one of the timer compare outputs. The block diagram depicted in Fig. 14.11 shows a typical PWM regulator as used for the control of the phase current in a SRM drive. It must be noted that due to the variable time constant of the stator windings, a fixed set of gains in the controller will not be sufficient, and a gain scheduling technique should be added. One practical way to achieve this is to obtain the gains corresponding to the aligned and unaligned rotor time constants, and then use a linear interpolation method to find the controller gains at the intermediate positions.

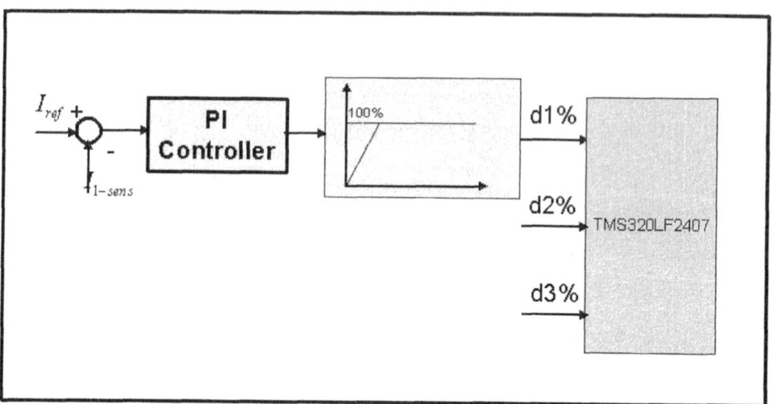

Figure 14.11 Block diagram of the PWM current control in SRM drive.

The block diagram in Fig. 14.12 summarizes the various steps along with the peripherals used in the LF2407. The main inputs to the program consist of commutation angles and the current profile. The quadrature outputs of an encoder have been used to determine the rotor position and angular speed of the drive. The phase currents have been sampled and converted into digital form to be used in current control. The output gates have been chosen from the general purpose input/output (GPIO) pins. The interface, conditioning circuit, and buffers are not shown in this picture. The control routine, used for the detection of active phases and hysteresis/PWM control of the phase currents, is combined in the software to form the final gating signal.

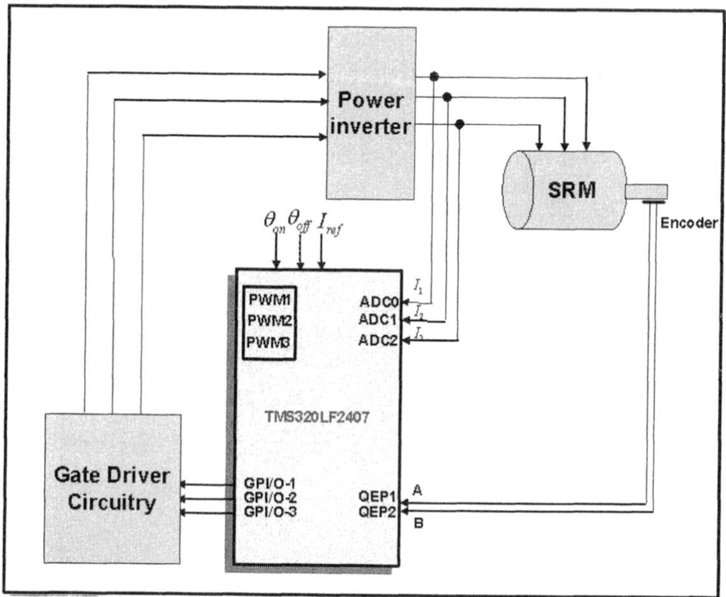

Figure 14.12 Block diagram of the basic control in SRM drive system.

Once the basic operation of the SRM drive is established, one can design and develop closed loop forms of the control. In the following sections, closed loop torque and speed control routines in the SRM drive are discussed, including four-quadrant operation of the drive.

14.5 Closed Loop Torque Control of the SRM Drive

As SRM technology begins to emerge in the form of a viable candidate for industrial applications, the significance of reliable operation under closed-loop torque, speed, and position control increases. Figure 14.13 depicts a typical cascaded control configuration for SRM drives. The main control block is responsible for generating the gate signals for the power switches. It also performs current regulation and phase commutation functions. In order to perform these tasks, it requires reference current, commutation instants, and a sequence of excitation. The torque controller provides the reference current, while the information regarding the commutation is obtained from a separate block that coordinates motoring, generating, and direction of rotation, as demanded by the various types of control. The various feedback information is generated using either estimators or transducers.

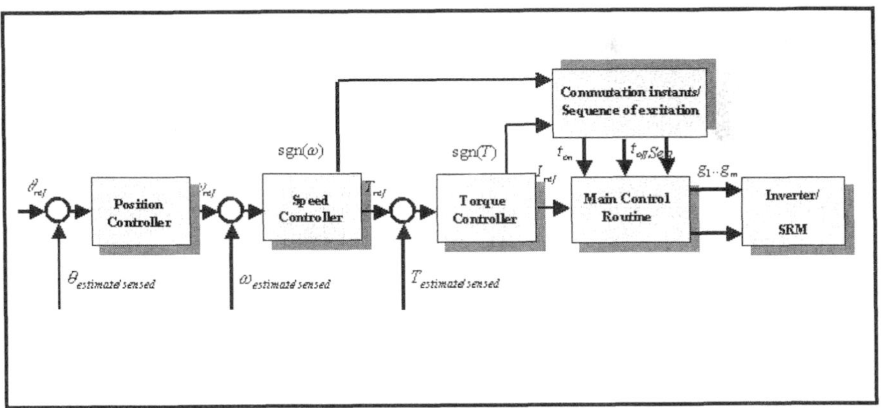

Figure 14.13 Cascaded control configuration for a SRM drive system.

Depending on the application, an adjustable speed motor drive may operate in various quadrants of the torque/speed plane. For instance, in a water pump application, where control of the output pressure is targeted, torque control in one quadrant is sufficient, whereas in an integrated starter/alternator, four-quadrant operation is necessary. Figure 14.14 shows the minimum requirement of an adjustable speed motor drive for performing torque, speed, and position control tasks. A speed controller may issue positive (motoring) or negative (generating) torque commands to regulate the speed. In a similar way, a position controller will ask for positive (clockwise) and negative (counter clockwise) speed commands. The accommodation of such commands will span all four quadrants of operation in the torque/speed plane. As a result, four-quadrant operation is a necessity for many applications in which positioning the rotor is an objective. In order to achieve four-quadrant operation in SRM drives, the direction of rotation in the air gap field needs to be altered. In addition, to generate negative torque during generation mode, the conduction band of the phase should be located in a region with negative inductance slope.

Figure 14.15 depicts a general block diagram of the closed loop torque control system. The main modules in this figure are:

- An estimator for the average/instantaneous electromagnetic torque.
- A feed-forward function for fast and convergent tracking of the commanded torque.
- A computational block to determine commutation instants according to the sign of demanded torque and magnitude of the phase current.

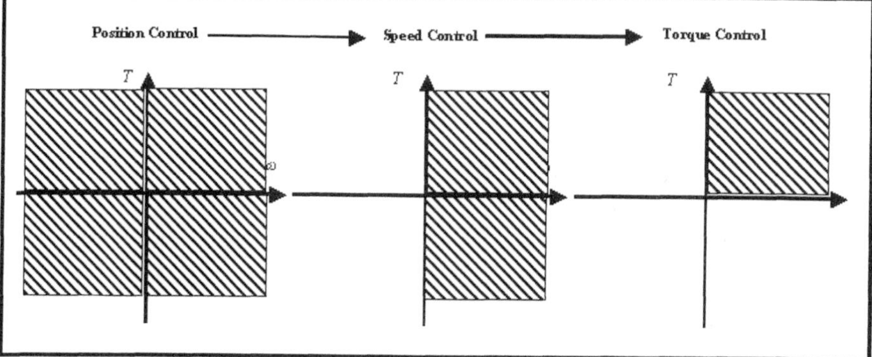

Figure 14.14 Minimum requirement of an adjustable speed drive for performing torque, speed, and position control.

The estimator for average/instantaneous electromagnetic torque is designed based on (14.1). The design also incorporates an analytical model of the phase inductance/flux linkage as shown below:

$$L(i,\theta) = L_0(i) + L_1(i)\cos(N_r\theta) + L_2(i)\cos(2N_r\theta) \tag{14.5}$$

where L_0, L_1, and L_2 represent polynomials, that reflect the nonlinear effects of saturation. Moreover, the inverse mapping of the torque estimator is used to form a feed-forward function. In the absence of the torque sensor/estimator, this feed-forward function can be used effectively to perform open loop control of the torque. The use of a feed-forward controller accelerates the convergence of the overall torque tracking. The partial mismatch between reference and estimated torque is then compensated via a *PI* controller. It must be noted that the introduction of the measured torque into the control system requires an additional analog-to-digital conversion. Figure 14.16 shows a comparison between the estimated and measured torque in a 12/8 SRM drive at steady state when responding to a periodic ramp function in closed loop control. The average torque estimator shows good accuracy. The existence of a 0.4 Nm averaging error is due to the fact that iron and stray losses are not included in the torque estimator. In order to perform this test, a permanent magnet drive acting as an active load was set in a speed control loop running in the same direction at 800 rpm.

As mentioned earlier, operation in all four quadrants of the torque vs. speed plane is a requirement for many applications. Given the symmetric shape of the inductance profile with respect to the aligned rotor position, one can expect that for a given conduction band at a constant speed, current waveforms during motoring and generating should be a mirror image of each other. However, one should note that the back-EMF during generation acts as a voltage source resulting in an increase of phase current even after a phase is shut down. This may cause some complications in terms of stability at high speeds. In order to alter the direction of rotation, the only necessary step is to change the sequence of excitation. Notably the sequence of excitation among stator phases is opposite of the direction of rotation. The transition between two modes needs to be quick and smooth. Upon the receipt of a command requesting a change in direction, the excited phase needs to be turned off to avoid generating additional torque. Regenerative braking should be performed simultaneously. This requires the detection of a phase in which the inductance profile has a negative slope. The operation in generation mode continues until the speed decays to zero or a tolerable near zero speed. At this time, all the phases will be cleared and a new sequence of excitation can be implemented. Speed reversal during generating is not a usual case because the direction of rotation is dictated by the prime mover. In the case that the speed reversal is initiated by the prime mover, the SRM controller needs to be notified. Otherwise, a mechanism for the detection of rotation direction should be in place. Such a mechanism would detect any unexpected change of mode, i.e., motoring to generating.

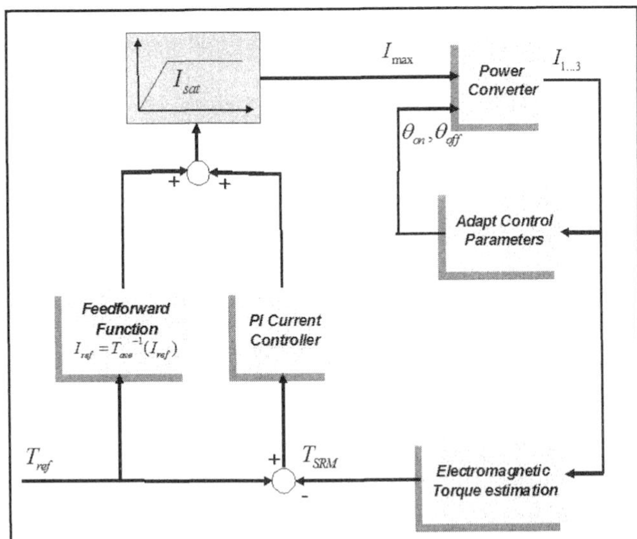

Figure 14.15 General block diagram of the torque control system.

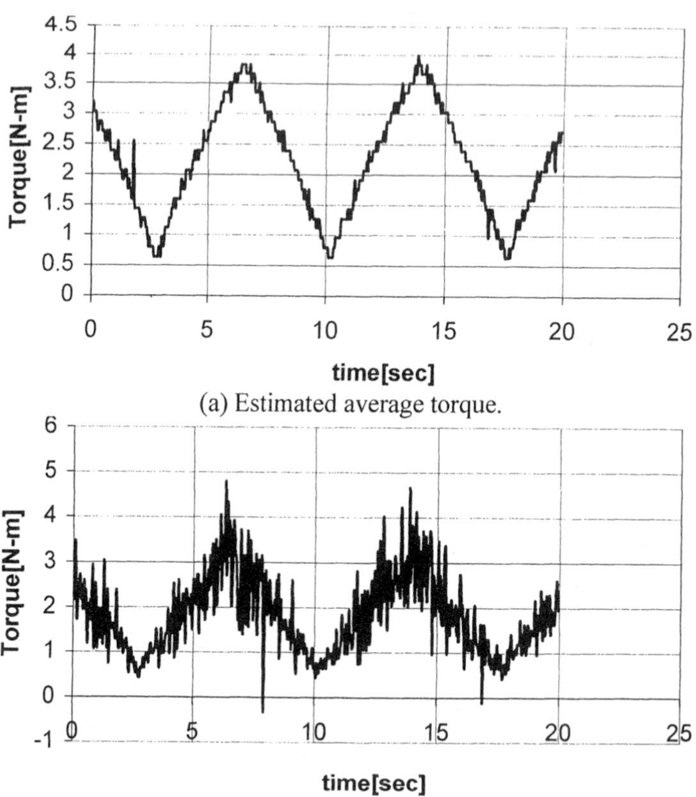

Figure 14.16 Comparison between measured and estimated torque.

14.6 Closed Loop Speed Control of the SRM Drive

As it is the next step in developing a high performance SRM drive, speed control is explained. As shown in Fig. 14.17, a cascaded type of control can be used to perform closed loop speed control. The speed can be sensed using the position information that is already provided by the encoder. Because the SRM is a synchronous machine, one may choose the electrical frequency of excitation for control purposes. The relationship between mechanical and electrical speeds is given by

$$\omega_e = N_r \omega_m \tag{14.6}$$

where N_r is the number of rotor poles. Ultimately, success in performing tightly regulated speed control depends upon the performance of the inner torque control system as depicted in Fig. 14.17. It is recommended that a feed-forward function be used to mitigate the initial transients in issuing commands to the torque control system.

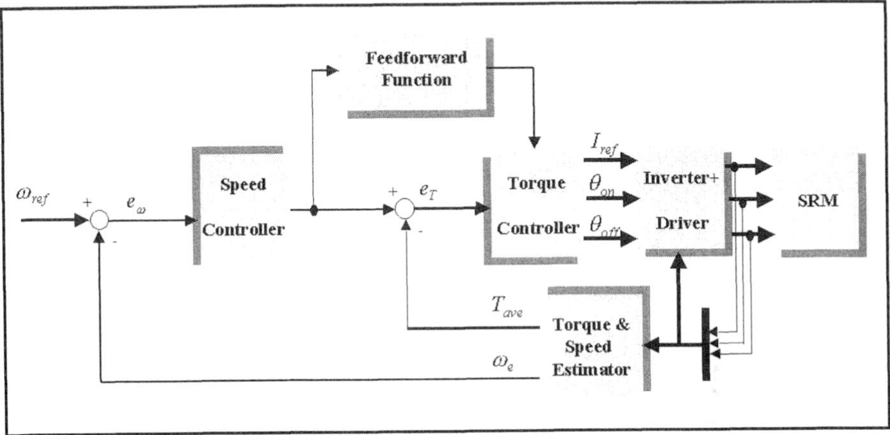

Figure 14.17 Closed loop speed control of SRM drive.

14.7 Summary

SRM drives are making their entry into the adjustable speed motor drive market. To take full advantage of their capacities, the development of high-performance control strategies has turned into a necessity. The advent of cost-effective DSP-based controllers provides an opportunity to engineer for this need in an effective way. A successful implementation of these methodologies demands a good understanding of the torque generation process. Basic control methods for the SRM drive have been discussed. These include the principles of design for closed loop control strategies. More advanced technologies, such as position sensorless and adaptive control, are also being investigated by many researchers across the globe, and there have been great advances in these areas as well. It is expected that developments in better efficiency, fault tolerance, and compactness will come about as a result of these efforts in years to come.

14.8 Algorithm for Running the SRM Drive using an Optical Encoder

A simple algorithm for running a four-phase SRM drive using a LF2407 DSP controller is shown in this section. One may change t-on and t-off to adjust the commutation angles. In order to adjust the current magnitude, HLIMIT and LLIMIT should be tuned. A level-three interrupt has been used to perform the control of the conduction band and phase current in an encoder-based architecture. By aligning one phase, a reference for rotor position is provided. Next, the commutation instants for each phase have been computed and the interrupt routine has been invoked. Because there is no control on the current during the aligning process, a current limiting algorithm is required. To obtain a better understanding of the algorithm, the following flow chart is prepared

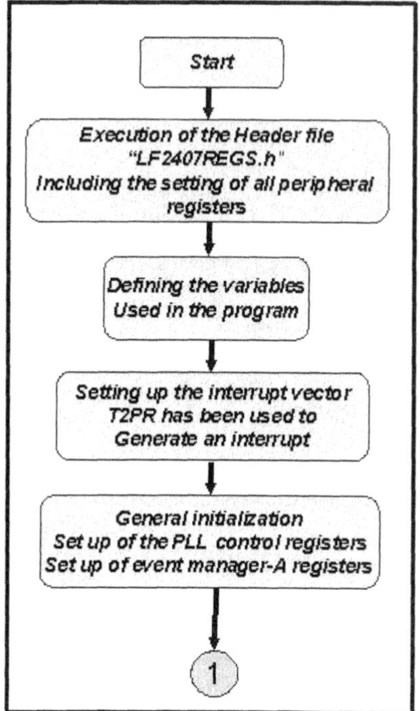

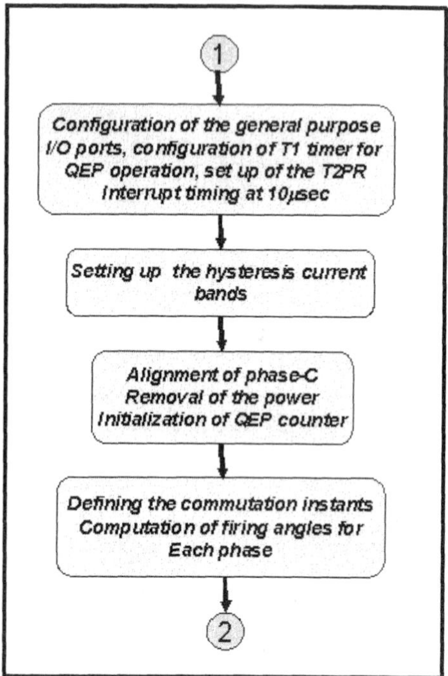

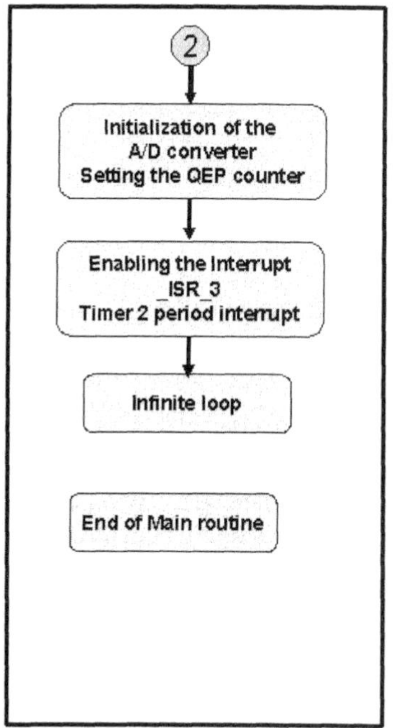

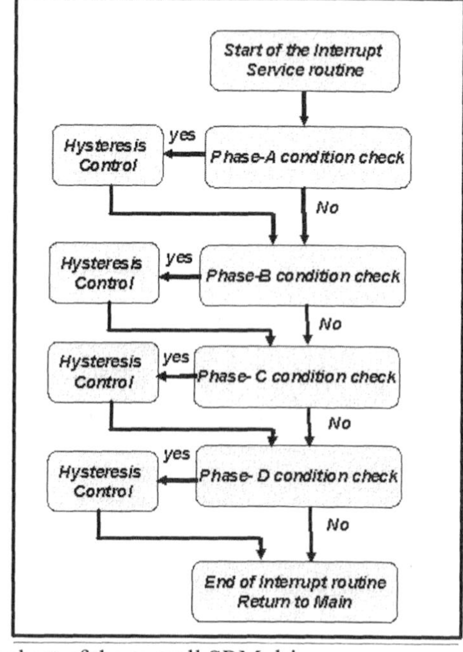

Figure 14.18 Flow chart of the overall SRM drive.

Chapter 15

DSP-BASED CONTROL OF MATRIX CONVERTERS

15.1 Introduction

Traditionally, ac voltages and currents having variable amplitude and frequency are generated in PWM modulated voltage-source inverters (PWM-VSI). These inverters need dc power, which is usually supplied by a diode bridge rectifier or a PWM active rectifier. This can be considered indirect power conversion because the topology is based on two types of power conversions via a dc link (capacitor). Indirect power converters perform the ac/ac power conversion by converting ac to dc through a rectifier, and then converting dc back to ac via an inverter. This operation is illustrated in Fig. 15.1.

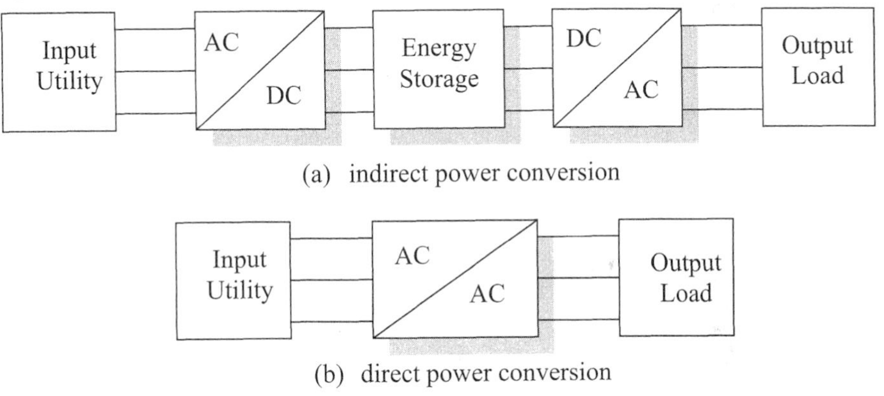

Figure 15.1 Two power conversion schemes.

An alternative to indirect power converters is the direct power converter such as a cycloconverter or a forced-commutated cycloconverter (FCC). Since both of these converters perform true direct ac/ac power conversion, they need no intermediate power converting process. Cycloconverters, using thyristors, are typically utilized for power applications in the megawatt-range. However, the inability of the thyristors to control turn off limits the output frequency to half the input frequency.

The FCC, otherwise known as a matrix converter, has no limit on output frequency due to the fact that it uses semiconductor switches with controlled turn-off capability. Examples of these semiconductor switches include the IGBT, MOSFET, and MCT. Since being introduced in 1976, the matrix converter has received an increased amount of interest because it lacks the dc link and posses the capability of bi-directional power flow control. The matrix converter has been studied intensely as an alternative to indirect power converter systems.

15.2 Topology and Characteristics

The matrix converter is a single-stage converter, and has an array of m by n bi-directional switches which directly connect an m-phase voltage source to an n-phase load. The number of input phases must be at least three, and the number of output phases can be chosen from one to theoretically infinity. Figure 15.2 shows a schematic representation of the most plausible application in which a matrix converter connects a three-phase voltage source to a three-phase load. The matrix converter is characterized by its ability to connect any input phase, a, b, or c to any output phase A, B, or C at any instant. This allows bi-directional power flow and sinusoidal input currents by directly interconnecting the input and output voltage system via bi-directional switches. The advantages and disadvantages of the matrix converter, when compared with conventional indirect power conversion, can be listed as follows:

A. Advantages

- No dc link capacitor or inductor
- Sinusoidal input and output currents
- Possible power factor control
- Four-quadrant operation
- Compact and simple design
- Regeneration capability

B. Disadvantages

- Reduced maximum voltage transfer ratio (0.866)
- Many bi-directional switches needed
- Increased complexity of control
- Sensitivity to input voltage disturbances
- Complex commutation method

Although the matrix converter has several disadvantages, these drawbacks have been mostly overcome. For example, the reduced voltage transfer ratio problem has been worked around by designing motors that reach maximum flux at the reduced matrix converter output voltage. The problem of physical space needed by the increased number of switches has been overcome by the development of power electronic building blocks such as the Eupec ECONOMAC™ matrix module, or by modifying the matrix converter topologies to use a fewer number of switches. The complexity of control and sensitivity to input voltage disturbances has also been solved by the development of different control algorithms. In all, the progress of the matrix converter has significantly improved its performance, rendering it an acceptable choice for compact and integrated converter-motor drives.

Because the matrix converter is fed by an input voltage source and connected to an inductive load, the following two basic rules must be always followed.

- Any two input phase voltages *must not* be connected to the same output line to avoid a short-circuit condition.
- Any output phase cannot be opened to avoid the interruption of inductive loads.

By defining the switching functions of each bidirectional switch as

$$s_{ij}(t) = \begin{cases} 1, & S_{ij} \quad closed \\ 0, & S_{ij} \quad open \end{cases} \quad \text{where,} \quad i \in \{a,b,c\}, \quad j \in \{A,B,C\} \tag{15.1}$$

the above two constraints can be expressed by

$$s_{aj} + s_{bj} + s_{cj} = 1, \qquad j \in \{A,B,C\} \tag{15.2}$$

With these constraints, the 3 by 3 matrix converter can allow only 27 possible switching states among the possible 512 switching combinations.

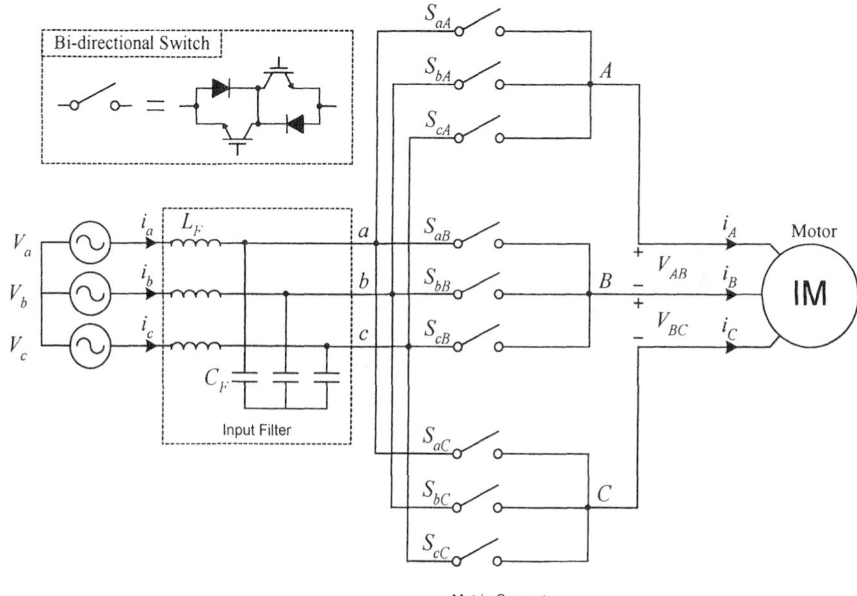

Figure 15.2 The three-phase matrix converter topology.

15.3 Control Algorithms

Two control schemes, the Venturini method and the space vector modulation method (SVM), provide independent control of the magnitude and the frequency of the output voltages.

15.3.1 Venturini Method

In 1980, Venturini and Alesina proposed a PWM modulation method for the matrix converter, known to as the "Venturini method". The Venturini method can be summarized as follows: for a given set of three-phase input voltages, a desired set of three-phase output voltages can be synthesized by sequential piecewise sampling of the input waveforms. The duration of each sample is derived mathematically to ensure that the average value of the actual output waveform within each sampling cycle tracks the required output waveforms. This method is also used to control the three-phase input currents.

If the input and output voltage vectors are defined by

$$\vec{V}_{in} = \begin{bmatrix} V_a(t) \\ V_b(t) \\ V_c(t) \end{bmatrix}, \qquad \vec{V}_{out} = \begin{bmatrix} V_A(t) \\ V_B(t) \\ V_C(t) \end{bmatrix} \tag{15.3}$$

the relationship between them can be written as

$$\begin{gathered} \vec{V}_{out} = \vec{M} \cdot \vec{V}_{in} \\ \begin{bmatrix} V_A(t) \\ V_B(t) \\ V_C(t) \end{bmatrix} = \underbrace{\begin{bmatrix} m_{11}(t) & m_{12}(t) & m_{13}(t) \\ m_{21}(t) & m_{22}(t) & m_{23}(t) \\ m_{31}(t) & m_{32}(t) & m_{33}(t) \end{bmatrix}}_{\vec{M}} \begin{bmatrix} V_a(t) \\ V_b(t) \\ V_c(t) \end{bmatrix} \end{gathered} \tag{15.4}$$

where $\vec{M}$ is the instantaneous transfer function matrix. From the input-output power balance ($P_{in}=P_{out}$), the following relationships are valid for the input and output currents:

$$\begin{gathered} \dot{i}_{in} = \vec{M}^T \cdot \dot{i}_{out} \\ \underbrace{\begin{bmatrix} i_a(t) \\ i_b(t) \\ i_c(t) \end{bmatrix}}_{\dot{i}_{in}} = \underbrace{\begin{bmatrix} m_{11}(t) & m_{21}(t) & m_{31}(t) \\ m_{12}(t) & m_{22}(t) & m_{32}(t) \\ m_{13}(t) & m_{23}(t) & m_{33}(t) \end{bmatrix}}_{\vec{M}^T} \underbrace{\begin{bmatrix} i_A(t) \\ i_B(t) \\ i_C(t) \end{bmatrix}}_{\dot{i}_{out}} \end{gathered} \tag{15.5}$$

where $\vec{M}^T$ is the transpose of matrix $\vec{M}$.

Each entity of the instantaneous transfer function matrix, $m_{ij}(t)$ $(i,j=1,2,3)$, represents the duty cycle function of a bi-directional switch within one switching period. The duty cycle functions are limited by the existence constraint

$$0 \le m_{ij}(t) \le 1 \qquad (i,j=1,2,3) \tag{15.6}$$

and the restriction imposed on the matrix converter switches by (15.2)

$$\sum_{j=1}^{3} m_{ij}(t) = 1 \qquad (i=1,2,3) \tag{15.7}$$

The maximum voltage transfer ratio of input to output voltage was limited by 0.5 in the initial approach of the Venturini method. Later, this method was further modified in order to increase the maximum voltage transfer ratio to 0.866.

15.3.2 The Initial Approach

We first find the transfer function matrix $\overrightarrow{M}$, which consists of the duty cycle functions of the nine bi-directional switches. For a set of three-phase input supply voltages

$$[V_i(t)] = \begin{bmatrix} V_a(t) \\ V_b(t) \\ V_c(t) \end{bmatrix} = \begin{bmatrix} V_{im}\cos(\omega_i t) \\ V_{im}\cos(\omega_i t - 2\pi/3) \\ V_{im}\cos(\omega_i t + 2\pi/3) \end{bmatrix} \tag{15.8}$$

where V_{im} is the amplitude of the input voltage and ω_i is the input angular frequency. The output voltages are generated as average values of the piecewise sampling of the input supply waveforms by three switching sequences within sampling time T_s.

The three output phase voltages are related to the input supply voltages with the transfer function matrix given by

$$[V_o(t)] = \begin{bmatrix} V_A(t) \\ V_B(t) \\ V_C(t) \end{bmatrix} = \begin{bmatrix} V_{om}\cos(\omega_o t + \theta_o) \\ V_{om}\cos(\omega_o t + \theta_o - 2\pi/3) \\ V_{om}\cos(\omega_o t + \theta_o + 2\pi/3) \end{bmatrix} \tag{15.9}$$

where

$$\begin{bmatrix} V_A(t) \\ V_B(t) \\ V_C(t) \end{bmatrix} = \begin{bmatrix} m_1(t) & m_2(t) & m_3(t) \\ m_2(t) & m_3(t) & m_1(t) \\ m_3(t) & m_1(t) & m_2(t) \end{bmatrix} \begin{bmatrix} V_a(t) \\ V_b(t) \\ V_c(t) \end{bmatrix} \tag{15.10}$$

By solving (15.10), mathematical expressions of the switch duty cycles can be obtained.

$$\begin{aligned} m_1(t) &= \frac{T_1}{T_s} = \frac{1}{3} + \frac{2}{3} q\cos(\omega_o - \omega_i)t \\ m_2(t) &= \frac{T_2}{T_s} = \frac{1}{3} + \frac{2}{3} q\cos((\omega_o - \omega_i)t - \frac{2\pi}{3}) \\ m_3(t) &= \frac{T_3}{T_s} = \frac{1}{3} + \frac{2}{3} q\cos((\omega_o - \omega_i)t + \frac{2\pi}{3}) \end{aligned} \tag{15.11}$$

where q is the voltage transfer ratio ($q = V_{om}/V_{im}$). These modulation functions are used to control the matrix converter switches in order to obtain sinusoidal input as well as sinusoidal output currents.

For complete generation of the output voltage waveforms with any output frequency, the desired output voltages must be entirely contained within the continuous envelope formed by the input voltages. It is clear that this constraint limits the available maximum voltage transfer ratio to 0.5 as shown in Fig. 15.3.

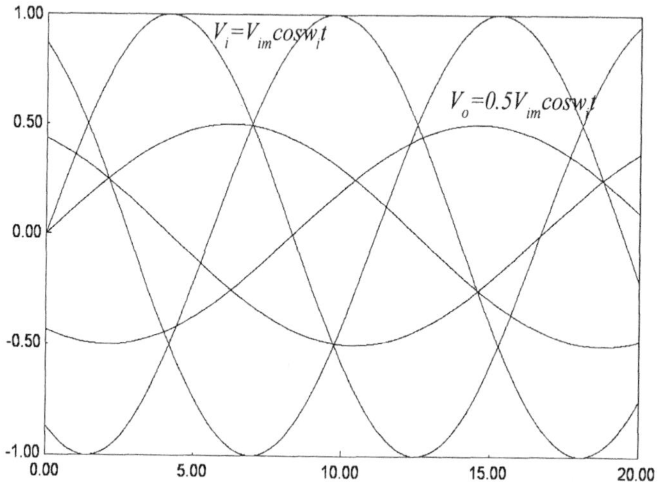

Figure 15.3 Envelope of the three-phase input voltages and output voltage of maximum amplitude $0.5V_{im}$.

15.3.3 The Enhanced Approach

The output voltage is limited to half of the input voltage with the previously discussed control scheme. In enhanced control scheme, the voltage transfer ratio is raised to 0.866. In this new scheme, the common-mode third-harmonic of the input frequency is subtracted from the input-phase voltages in order to extend the output voltage limitation. The optimum amplitude of the third-harmonic of the input frequency was found to be $V_{im}/4$, which allows for a 0.75 maximum voltage transfer ratio by enlarging the area within the input voltage envelope. Note that this subtraction is equivalent to adding the third-harmonic of the input frequency to the output phase voltage, which is shown graphically in Fig. 15.4.

Further increasing the voltage transfer ratio can be obtained by subtracting the third harmonic of the output frequency from the desired output phase voltage. By decreasing the peak-to-peak value of the output phase voltage, a voltage transfer ratio of 0.866 can be obtained, which is the maximum voltage gain a matrix converter can generate. The optimum amplitude of the third-harmonic of the input frequency was found to be $V_{om}/6$. Fig. 15.5 shows how third-harmonic injection can enlarge the maximum value of the output voltage. Third-harmonic injection of the input and output frequencies into the target output voltages has no effect on isolated-neutral three-phase loads due to third harmonic cancellation in these systems.

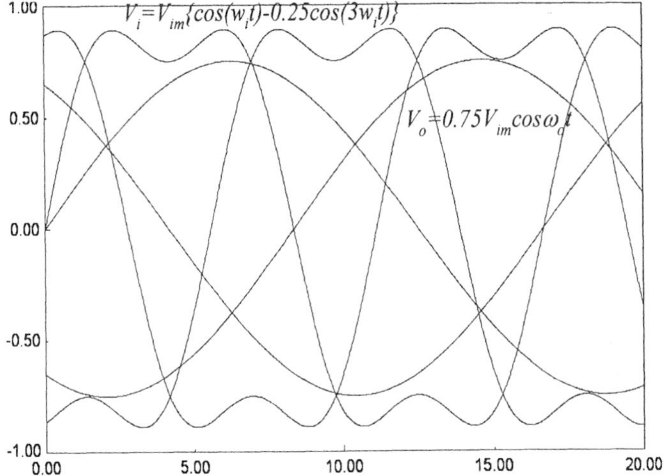

Figure 15.4 Envelope of the third-harmonic injected input voltages and output voltages with $0.75V_{im}$ maximum value.

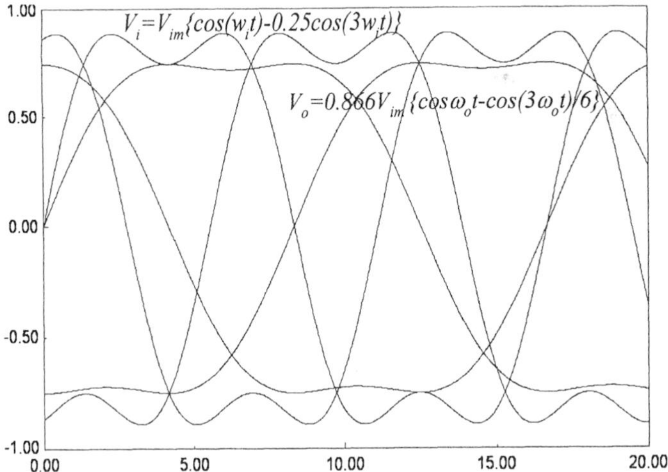

Figure 15.5 Envelope of the third-harmonic injected input voltages and third-harmonic injected output voltages with $0.866V_{im}$ maximum value.

The two-step third-harmonic injection described above results in the desired output voltage expression.

$$\begin{bmatrix} V_A(t) \\ V_B(t) \\ V_C(t) \end{bmatrix} = \begin{bmatrix} V_{om}\cos(\omega_o t) \\ V_{om}\cos(\omega_o t - 2\pi/3) \\ V_{om}\cos(\omega_o t + 2\pi/3) \end{bmatrix} - \frac{V_{om}}{6}\begin{bmatrix} \cos(3\omega_o t) \\ \cos(3\omega_o t) \\ \cos(3\omega_o t) \end{bmatrix} + \frac{V_{im}}{4}\begin{bmatrix} \cos(3\omega_i t) \\ \cos(3\omega_i t) \\ \cos(3\omega_i t) \end{bmatrix} \tag{15.12}$$

$$= \begin{bmatrix} m_{11}(t) & m_{12}(t) & m_{13}(t) \\ m_{21}(t) & m_{22}(t) & m_{23}(t) \\ m_{31}(t) & m_{32}(t) & m_{33}(t) \end{bmatrix} \begin{bmatrix} V_a(t) \\ V_b(t) \\ V_c(t) \end{bmatrix}$$

By a mathematical approach, the switching functions required to obtain the target output voltages expressed in (15.12) can be found by

$$\begin{aligned} m_{ij}(t) &= \frac{1}{3} + \frac{2}{3} q \cos(\omega_i t - 2(j-1)\frac{\pi}{3}) \\ &\cdot \{\cos(\omega_o t - 2(i-1)\frac{\pi}{3}) - \frac{1}{6}\cos(3\omega_o t) + \frac{1}{2\sqrt{3}}\cos(3\omega_i t)\} \\ &- \frac{2}{9\sqrt{3}} q \{\cos(4\omega_i t - 2(j-1)\frac{\pi}{3}) - \cos(2\omega_i t + 2(j-1)\frac{\pi}{3})\} \end{aligned} \quad (15.13)$$

15.4 Space Vector Modulation

In 1989, Huber presented a SVM algorithm for the matrix converter. The first version of the SVM technique focused on only output sinusoidal voltage generation by applying the SVM technique to output stage. Later Huber presented the full version of SVM using the space vector techniques in the input stage as well as the output stage, which can simultaneously achieve sinusoidal input and output currents. This algorithm has given a significant impact by enabling the SVM used in the conventional PWM-VSI to be applied to the matrix converter. While the Venturini method is inherently less flexible because of its mathematical approach, the SVM algorithm allows the matrix converter to use a variety of control methods.

The SVM technique is derived from the fact that the matrix converter has exactly the same operation of its equivalent indirect counterpart. The matrix converter is essentially a rectifier stage and inverter stage with a fictitious dc link. Figure 15.6 shows the equivalent circuit of the matrix converter consisting of bi-directional switches. It is clear that the matrix converter has the same input and output voltages and currents as its equivalent circuit shown in Fig. 15.7. The ac/ac power conversion is now independently performed in the rectifier stage and the inverter stage via the fictitious dc link. The fictitious dc link voltage is built by chops of the input voltages through the rectifier stage. Applying the inversion algorithm to the fictitious dc link voltage as in a conventional PWM-VSI generates the output voltages.

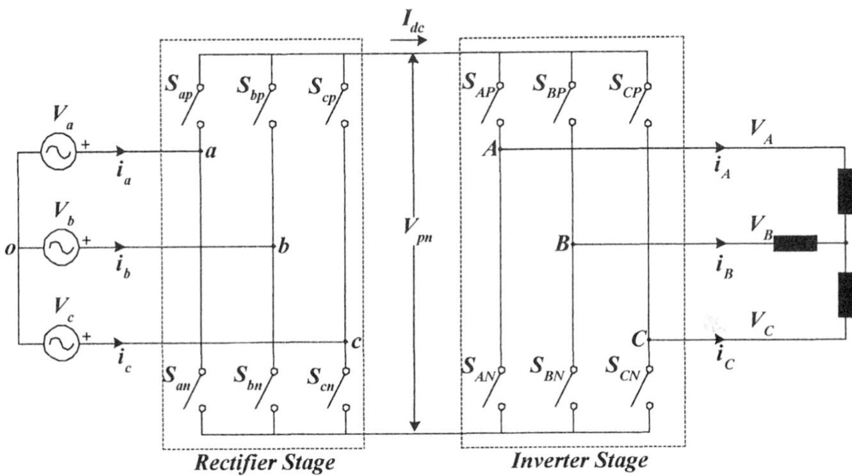

Figure 15.6 The equivalent circuit of the matrix converter used in a derivation of the SVM.

In the inverter stage, the output voltage space vector is defined in terms of the line-to-line voltages by

$$\vec{V}_o(t) = \frac{2}{3}(V_{AB} + V_{BC} \cdot e^{j120°} + V_{CA} \cdot e^{-j120°}) \tag{15.14}$$

In the complex plane, $\vec{V}_o(t)$ is a vector rotating at an angular output frequency ω_o with a constant length ($\sqrt{3}V_{om}$). This output voltage vector is synthesized by time averaging using the six active vectors and two zero vectors. Figure 15.8 shows an example of how the output voltage vector could be synthesized when it lies in Sector I.

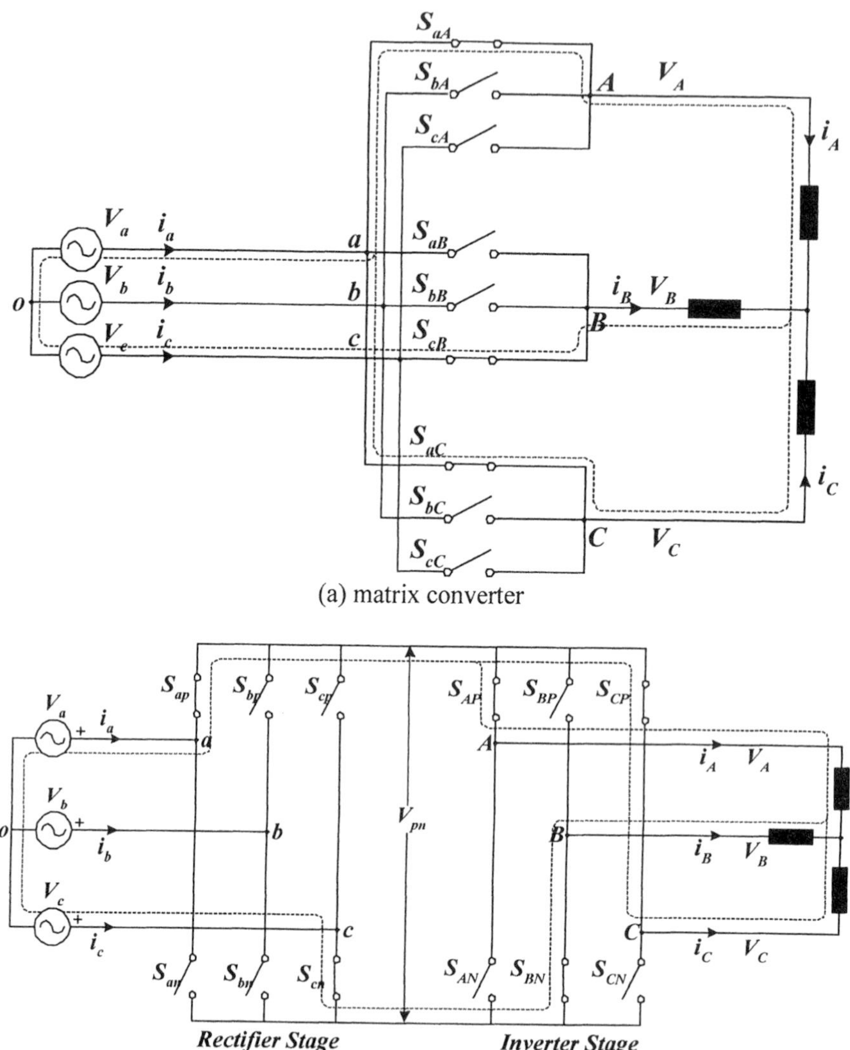

(a) matrix converter

(b) equivalent circuit

Figure 15.7 Equivalent operation of the matrix converter and its equivalent circuit.

The duty cycles of the active and zero vectors are found by

$$d\alpha = \frac{T_\alpha}{T_s} = mv \cdot \sin(60° - \theta sv)$$

$$d\beta = \frac{T_\beta}{T_s} = mv \cdot \sin(\theta sv) \tag{15.15}$$

$$d_{ov} = \frac{T_{ov}}{T_s} = 1 - d_\alpha - d_\beta$$

The modulation index, m_v in the inverter stage is defined as

$$m_v = \frac{\sqrt{3}V_{om}}{V_{pn}}, \qquad 0 \le m_v \le 1 \tag{15.16}$$

The main objective of the rectifier stage is to draw sinusoidal input currents with a controllable displacement angle as well as maintain a dc voltage in the fictitious dc link. Since the input currents are sinusoidal, the output of the rectifier stage can be considered as a dc current source, I_{dc}. The SVM of the input current vector is completely analogous to the SVM of the inverter stage. Figure 15.9 shows the input current vector and the current switching hexagon.

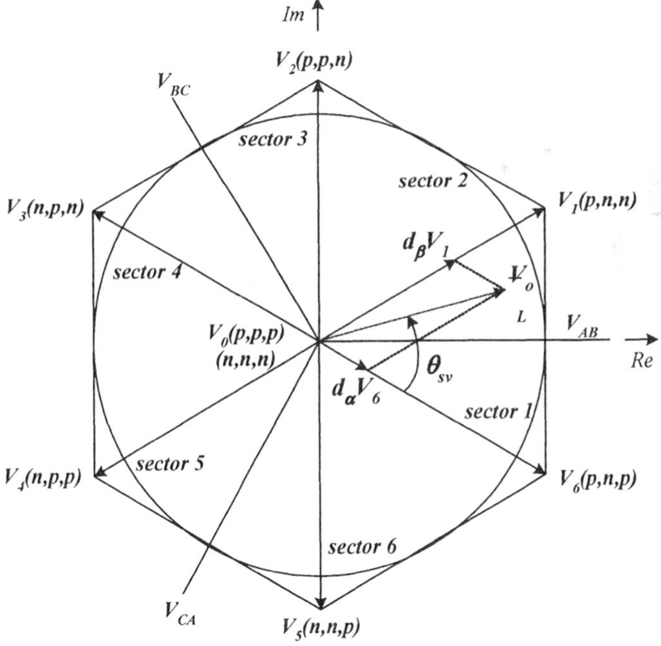

Figure 15.8 Output voltage space vectors and example of output voltage vector synthesis.

The duty ratios of the rectifier stage are

$$d\mu = \frac{T_\mu}{T_s} = mc \cdot \sin(60° - \theta sc)$$

$$d\upsilon = \frac{T_\upsilon}{T_s} = mc \cdot \sin(\theta sc) \quad (15.17)$$

$$d_{0c} = \frac{T_{0c}}{T_s} = 1 - d_\mu - d_\upsilon$$

The modulation index, m_c in the rectifier stage is defined as

$$m_c = \frac{I_{im}}{I_{dc}}, \qquad 0 \le m_c \le 1 \quad (15.18)$$

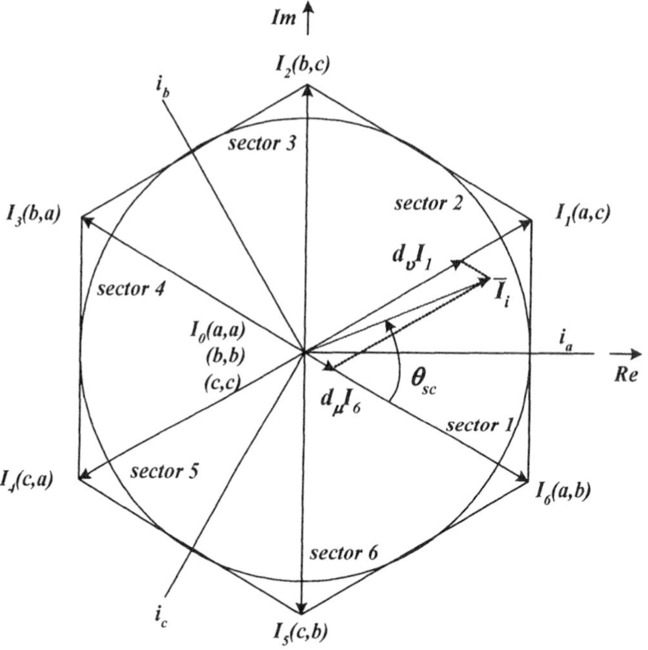

Figure 15.9 Input current space vectors and example of input current vector synthesis

Two space vectors, input current and output voltage, are independently applied to the rectifier and inverter stages to obtain sinusoidal input currents and sinusoidal output voltages. The two modulations are finally combined to control the matrix converter.

$$d\alpha\mu = T\alpha\mu / Ts = d\alpha \cdot d\mu = m \cdot \sin(60° - \theta sv) \cdot \sin(60° - \theta sc)$$

$$d\beta\mu = T\beta\mu / Ts = d\beta \cdot d\mu = m \cdot \sin(\theta sv) \cdot \sin(60° - \theta sc)$$

$$d\alpha v = T\alpha v / Ts = d\alpha \cdot dv = m \cdot \sin(60° - \theta sv) \cdot \sin(\theta sc)$$
$$d\beta v = T\beta v / Ts = d\beta \cdot dv = m \cdot \sin(\theta sv) \cdot \sin(\theta sc) \quad (15.19)$$

The voltage transfer ratio, *m*, of the matrix converter is proportional to the product of the two modulation indexes, m_c and m_v.

$$0 \le m = mc \cdot mv \le 1 \quad (15.20)$$

15.5 Bidirectional Switch

The matrix converter requires bidirectional switches capable of conducting current and blocking voltage in both directions. Since there is no such device currently available, it is constructed by a combination of diodes and unidirectional switches such as the IGBT.

Figure 15.10(a) shows the diode bridge bidirectional switch cell configuration used in the earlier versions of the matrix converter. The main advantage is that the current conduction is carried by only one IGBT, thus, one gate driver is needed per switch cell. However, device losses are relatively high because there are three devices in each conduction path. The current direction through the switch cell cannot be controlled. This is a disadvantage because most commutation technologies require the independent control of each IGBT in a cell.

The common emitter (or collector) bidirectional switch arrangement consists of two diodes and two IGBTs connected in parallel as shown in Fig. 15.10(b) and (c). Each IGBT can independently control the direction of the current from source to load, or load to source. Conduction losses are also reduced because only two devices carry the current. The common collector bidirectional switch cell arrangement also requires less isolated gate drive power than the common emitter bidirectional switch cell.

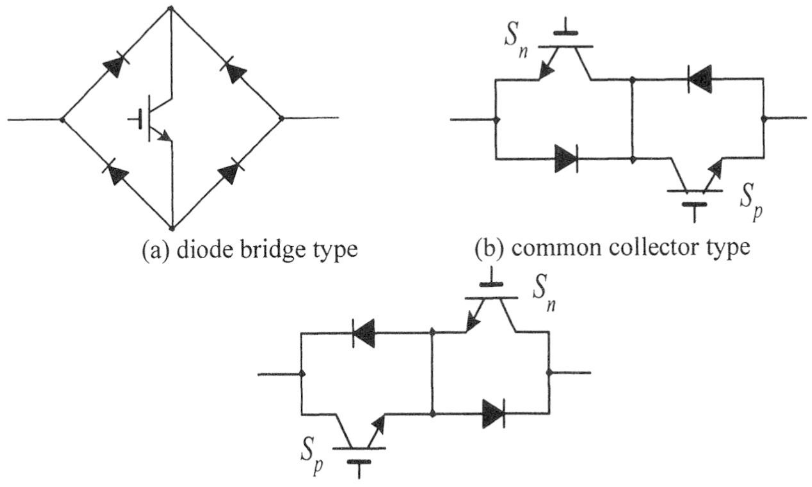

(a) diode bridge type (b) common collector type

(c) common emitter type

Figure 15.10 Bidirectional switch cell.

15.6 Current Commutation

Reliable current commutation between switches in the matrix converter is more difficult to achieve than conventional PWM inverters because there are no natural freewheeling paths provided by the diodes in inverters. Therefore, commutation has to be actively controlled at all times. No two bi-directional switches are switched on at any instant, as it would result in line-to-line short circuits and the destruction of the converter. Also, the bi-directional switches for one output phase should not all be turned off at any instant, as it would result in the absence of a path for the inductive load current, causing large over-voltages. These two considerations cause a conflict because semiconductor devices cannot be switched instantaneously due to propagation delays and finite switching times.

A common method of current commutation uses a four-step commutation strategy in which it is necessary to know the direction of the output current through the switches. Figure 15.11 shows a schematic condition to commutate an output phase current from bi-directional switch S_1 from S_2. Before the commutation, both IGBTs (S_{1c} and S_{1nc}) in the switch cell S_1 are gated to allow both directions of current flow. Assume that the load current is in the direction shown in Fig. 15.11. When a commutation from S_1 to S_2 is required, the current direction information is used to determine which unidirectional switch in S_1 is not conducting. This unidirectional switch is turned off first. In this case, the unidirectional switch S_{1nc} is turned off. Then, the next conducting switch in the incoming switch cell S_2 is gated (S_{2c} in this example). The load current transfers to the incoming switch either at this point or when the outgoing switch (S_{1c}) is turned off according to amplitude of input voltages. The remaining unidirectional switch in the incoming device (S_{2nc}) is turned on to allow for current reversal. This process is shown in a timing diagram in Fig. 15.12. The delay between each switching event is determined by the device characteristics in order to allow enough switching time for the device.

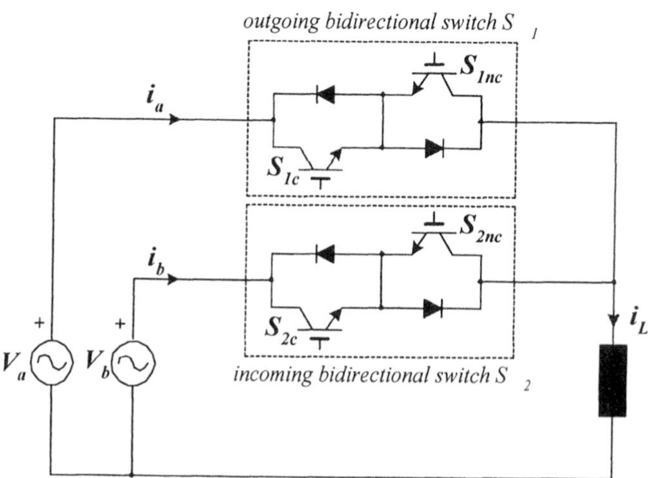

Figure 15.11 Schematic condition to commutate an output phase current from a switch S_1 from S_2.

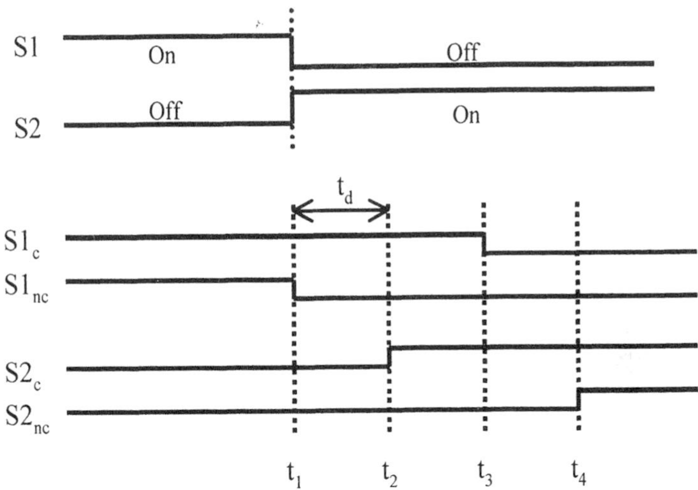

Figure 15.12 Switching diagram of four-step commutation.

15.7 Overall Structure of Three-Phase Matrix Converter

15.7.1 Controller

A powerful DSP such as the LF2407 is a necessity for controlling matrix converters due to their complex switching operation. Either the Venturini or SVM algorithm explained previously can be implemented on the DSP. In general, the DSP output ports cannot provide the required 18 PWM signals. Therefore, nine switching signals are generated by the DSP and fed to a Programmable Logic Device (PLD) such as an FPGA or PAL.

15.7.2 Interface Parts

Four-step commutation as explained above, is implemented via an interface board using a PLD. Input signals of the PLD are nine switching signals for nine bi-directional switches, which are provided from the DSP, and information of three output line-current directions, which is provided through current sensors and simple voltage comparators. Output signals of the PLD are sent to 18 gate drives that turn the individual unidirectional switches on or off.

15.7.3 Power Circuits

A matrix converter can be implemented by a matrix module (Eupec ECONOMAC) which consists of 18 IGBTs. Several additional power circuits are required to guarantee and protect the matrix converter and controller operation. Input filters should be used at the input of the matrix converter to reduce the switching harmonics present in the input current. The requirements for the input filter includes a cutoff frequency lower than the switching frequency of the matrix converter, minimal reactive power at the grid frequency, minimal volume and weight, and minimal input filter inductance voltage drop at rated current in order to avoid a reduction in the voltage transfer ratio. Careful design of the input filter is required because a bad filter design can affect the output currents as well as the input currents of the converter.

A clamp circuit is also required to protect the matrix converter from over-voltages in the input or output side. Over-voltages could appear on the input side due to input line perturbations. Over-voltages can also appear from the output side if an over-current fault exists. In these situations, a clamp circuit can protect the matrix converter by the charging of a DC capacitor through diodes. The clamping circuit uses 12 fast-recovery diodes to connect the capacitor to the input and output terminals. The complete matrix converter as discussed above is shown in Fig. 15.13.

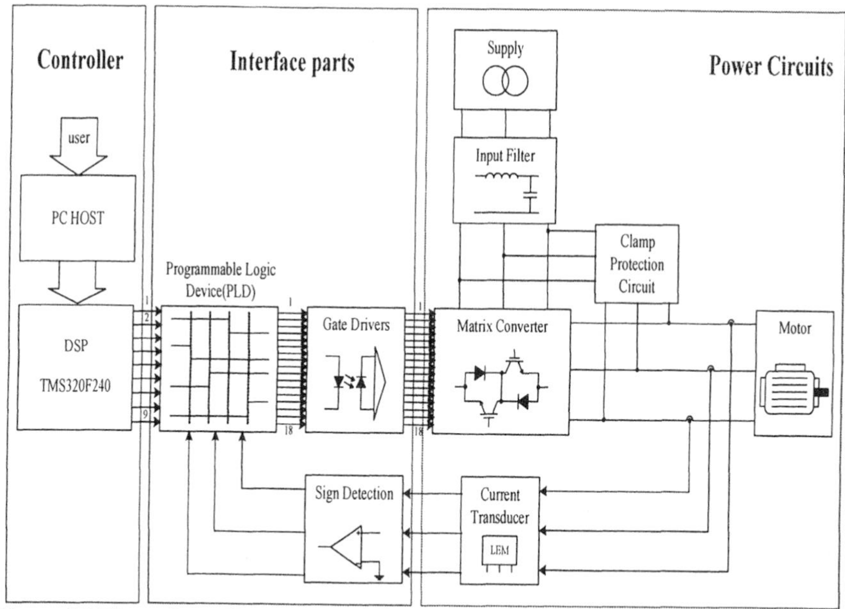

Figure 15.13 Overall structure of three-phase matrix converter implementation.

15.8 Implementation of the Venturini Algorithm using the LF2407

Since matrix converters require the control of 18 unidirectional switches, DSP-based implementation is essential. The matrix converter algorithms are computationally intensive and require real-time calculation of the switch duty cycles. The requirements for generating PWM control signals for matrix converters in real-time include the fast computation of the switch duty cycles (within only one switching period) and the generation of the control pulses. The DSP-based implementation of PWM techniques is important in making the matrix converters practicable. In this section, the implementation of the first version of the Venturini algorithm using a LF2407 Evaluation Module (EVM) is introduced.

The Venturini algorithm is totally different from conventional PWM algorithms such as the SPWM (sinusoidal PWM) or SVPWM algorithm for inverters. It cannot be implemented using the standard PWM modules on the LF2407 without external logic circuitry. Therefore, an implementation method will be introduced that does not require the use of external logic circuits.

DSP implementation of the Venturini algorithm discussed above includes two main procedures:

- Calculation of the switch duty cycle (T_1, T_2, and T_3) based on the input and output voltage waveforms using (15.5)-(15.7) within one switching period.
- Generation of the switch control pulses with the predefined pattern according to the duty cycles.

The calculations should be performed for every switching period. The sample period and duty cycles calculated during one period are used as control variables for the next set of calculations. To do this, calculation results can be stored first to memory and subsequently loaded into the timer period register. Once the timer period is loaded, the switching pulses are then generated for nine switches. These procedures are performed using the timer operation as shown Fig. 15.14.

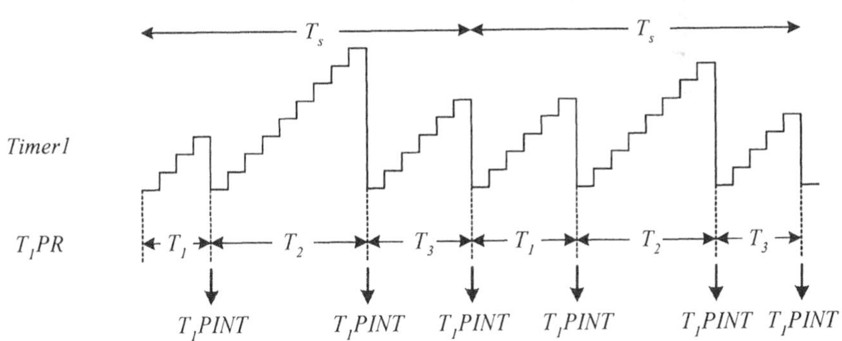

Figure 15.14 Operation of the timer counter and the period interrupt.

At the first instant of a switching period, a timer period register is loaded with the duty cycle T_1 calculated in the previous period. The predefined control pulses corresponding to T_1 are sent to switches S_1, S_6, and S_8. The timer then starts the continuous up-counting operation. As soon as the value in its counter is equal to that of the period register, the period interrupt T_1PINT occurs. The timer operation is disabled in the interrupt service routine and the timer counter resets to 0. At the same time, the duty cycle for T_2 is loaded into the period register, and the corresponding gating pulses are generated for switches S_2, S_4, and S_9. Therefore, commutation can take place from S_1 to S_2, S_6 to S_4, and S_8 to S_9. The timer restarts the continuous up-counting operation again and the same procedure is repeated for the duty cycle T_3.

The calculation of the duty cycles is performed while the gating signals are being generated, and is therefore completed within T_s. The three resulting duty cycles are used for the '*ON*' durations of nine switches in the next sampling period. Real-time calculation of the duty cycles is achieved without affecting the generation of the pulse signals. The gating signals produced are shown in Fig. 15.15.

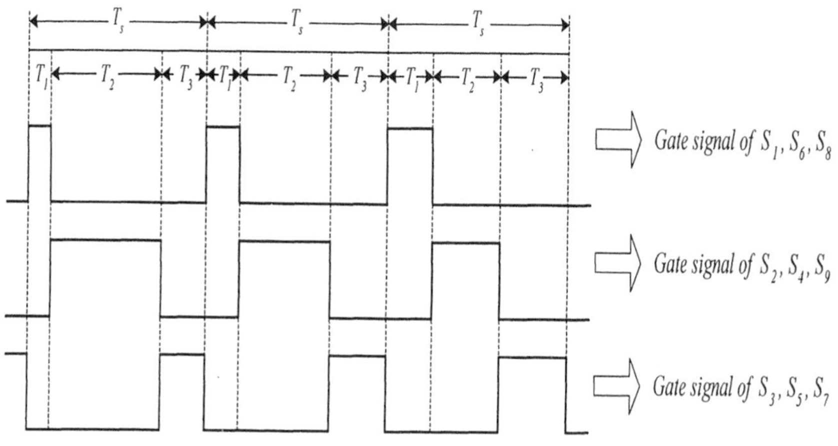

Figure 15.15 DSP output signals.

Note that only one timer is needed in generating the gating pulses because the same commutation sequence is used for three bidirectional switches in three output legs. If different commutation sequences are desired, then each sequence will need its own timer. Using different commutating sequences will result in the same performance of the converter except that the harmonic components of the output currents will vary.

Figure 15.16 shows flow charts of the main algorithm and the period interrupt service routine used with DSP implementation.

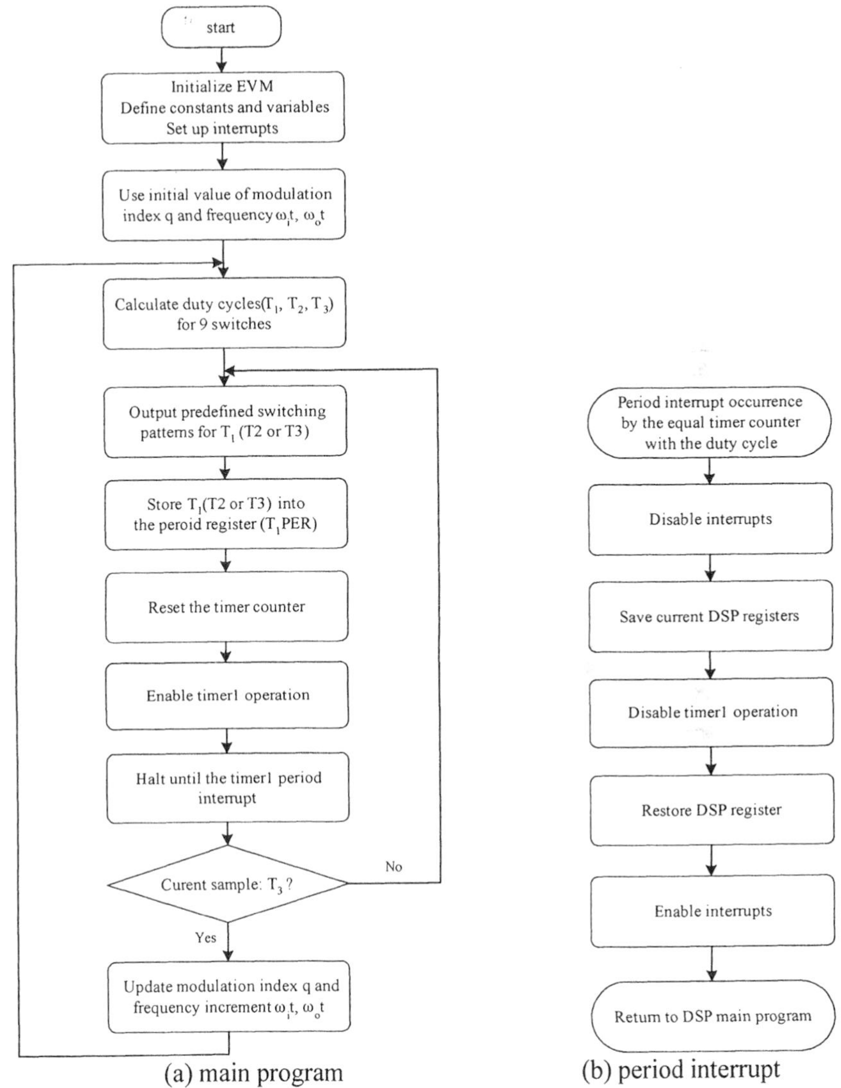

Figure 15.16 Flow chart of the Venturini-based DSP program.

References

1. L. Gyugyi and B. Pelly, *Static Frequency Changers*. New York: Wiley 1976.
2. M. Venturini, "A new sine wave in, sine wave out conversion technique eliminates reactive elements," *Proceedings of Powercon* 7, vol. E, pp. E3-1-E3-15, 1980.

3. A. Alesina and M. Venturini, "Analysis and design of optimum-amplitude nine-switch direct AC-AC converters," *IEEE Transactions on Power Electronics*, vol. 4, no. 1, pp. 101-112, Jan. 1989.
4. L. Huber and D. Borojevic, "Space vector modulator for forced commutated cycloconverter," IEEE Industry Applications Society Annual Meeting, vol. 1 pp. 871-876, 1989.
5. L. Huber, and D. Borojevic, "Space vector modulated three-phase to three-phase matrix converter with input power factor correction," *IEEE Transactions on Industry Applications*, vol. 31, no. 6, pp. 1234-1246, 1995.
6. L. Zhang, C. Watthanasarn and W. Spepherd, "Control of ac-ac matrix converters for unbalanced and/or distorted supply voltage," in Proc. PESC'01, vol. 2, pp. 1108-1113.
7. P. W. Wheeler, J. Rodriguez, J.C. Clare, L. Empringham and A. Weinstein, "Matrix converters: A technology review," *IEEE Transactions on Industry Electronics*, vol. 49, no. 2, pp. 276-288, 2002.
8. C. Watthanasarn, Optimal control and application of ac-ac matrix converters, Ph.D. thesis, University of Bradford, UK, 1997.
9. TMS320F/C240 DSP Controllers, Peripheral Library and Specific Devices, Texas Instruments, 1999.
10. TMS320F/C240 DSP Controllers, CPU and Instruction Set, Texas Instruments, 1999.
11. P. Nielsen, F. Blaabjerg, and J. Pedersen, "Novel solutions for protection of the matrix converter to three-phase induction machine" in *Conf. Rec. IEEE-IAS Annual Meeting*, pp. 1447-1454, 1997.

Appendix A

DEVELOPMENT OF FIELD-ORIENTED CONTROL INDUCTION MOTOR USING VISSIM™

A.1 Introduction

VisSim™ software by Visual Solutions is a graphical user interface (GUI) that allows the user to develop and simulate control algorithms without having to write lines of code. Developing control algorithms is done by interconnecting discrete graphical or functional blocks to make a block diagram. Several individual blocks may be combined to form a compound block as seen in the "VHz + SVGEN / MF..." block of Fig. A.1. Once an algorithm is developed, it can be simulated on the personal computer (PC), jointly run by the PC and LF2407, or run totally on the LF2407 DSP.

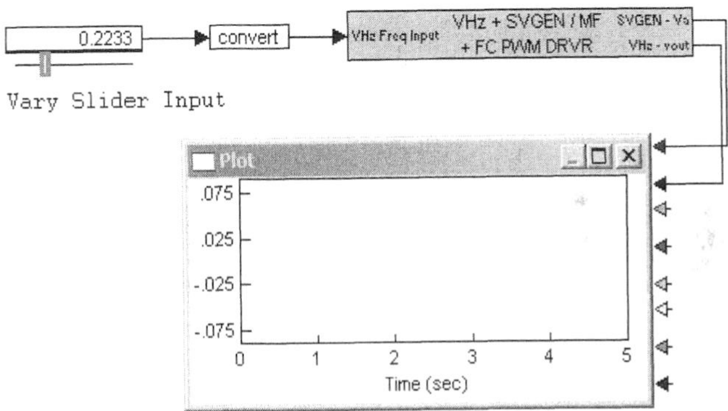

Figure A.1 A VisSim™ block diagram.

A.2 Overview of VisSim™ Placing and Wiring Blocks

Functional blocks can be placed and wired to implement the desired algorithm. All previously made blocks within a schematic can be found and are organized with respect to their associated groupings by going to the *Blocks* menu shown in Fig. A.2. VisSim contains a number of standard discrete functional blocks for linear, nonlinear, continuous, discrete-time, time varying, and hybrid system design such as:

- Animation
- Annotation
- Arithmetic
- Boolean
- Matrix Operations
- Nonlinear
- Optimization
- Random Generator

- DDE
- Integration
- Linear Systems
- MatLab Interface
- Signal Consumer
- Signal Producer
- Time Delay
- Transcendental

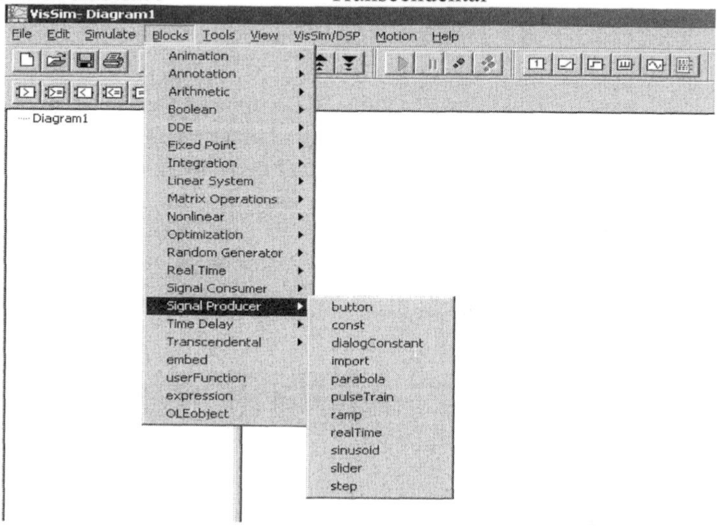

Figure A.2 View of the many block categories under the *Blocks* menu.

In addition to the many general purpose functional blocks, there are blocks written exclusively for the LF2407 DSP. These DSP blocks are hand-written assembly routines developed by Texas Instruments. Since these blocks are hand-coded and already optimized, they execute quicker when running on the LF2407 than a user-created compound block of equivalent functionality. The special DSP functional blocks are useful when fast execution is critical on the LF2407.

A.2.1 Developing a Control Algorithm in VisSim™

To become successful in developing DSP-based control algorithms in VisSim™, one should follow a specific design flow. First, a block diagram should be developed and simulated on the computer. At this stage the design may be verified and tested mathematically. All efforts should be made to use integer numbers whenever possible because the LF2407 is a fixed-point DSP. Doing so will aid in transitioning the block design from the computer to the DSP if desired.

The next step of the development process is to run selected portions of the design on the LF2407 and the rest of the design on the computer. For example, if a space vector block is being used, the appropriate LF2407 DSP functional block can replace the simulation only space vector block. This will allow the space vector block to run on the LF2407 with the rest of the algorithm running on the computer.

The third and final step in the design process is to convert the whole design to run solely on the DSP. This ensures the whole design is DSP compatible. To accomplish this, the following inversions must be done. First, all floating point values must be converted to fixed point. Second, all non-DSP compatible blocks must be substituted. Third, the algorithm should incorporate DSP-specific blocks wherever possible. Finally, the block design should be entirely encapsulated to form a compound block. These steps will help the design to run as smoothly as possible on the DSP.

Once a design has been converted for DSP execution, VisSim™ works with Code Composer Studio to compile and build a DSP executable program. VisSim™ contains many DSP options which must be configured according to the desired operation before compiling the design. According to the configuration options specified by the user, VisSim™ first converts the various functional blocks into C code. Next, Code Composer compiles the C code into native C2xx assembly. Once the assembly file has been created, it automatically creates the **.out* file which will be executed on the DSP. These conversions, from the functional block to C to assembly code are performed automatically by VisSim™ and Code Composer. All the user needs to do is configure the design and VisSim™ for the desired operation.

A.3 Computer Simulation of Vector Control of Three-Phase Induction Motor Using VisSim™

The VisSim™ design featured in this section simulates the Field Oriented Control (FOC) method of controlling an induction motor shown in Fig. A.3. All parameters, processes, and variables are modeled mathematically. The VisSim™ block connections shown below represent their respective mathematical formulas.

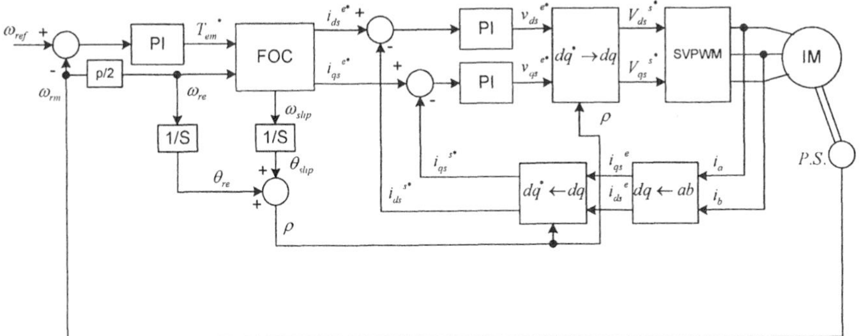

Figure A.3 Full block diagram of the FOC of an induction motor.

In this design, the speed of the rotor is used to determine the desired rotor flux which will be induced in the air-gap of the induction motor. To accomplish this, `lamda_r*` is set to the constant value of 0.4 for this design. For rotor speeds less than the base (rated) speed, the rotor flux command is set equal to 0.4. For speeds above the base speed, the rotor flux needs to be decreased in an inverse, non-linear fashion. We need to develop a control strategy that varies the frequency and voltage of the signal applied to the windings of the induction motor, while also controlling the flux created in the air-gap.

This design contains many compound blocks which are groupings of discrete functional blocks. The first compound block we will discuss is the mathematical model of the induction motor (IM).

A.3.1 Induction Motor (IM) Dynamic Model

Our objective is to simulate a three-phase, four-pole IM in the stationary reference frame. Properly modeling an IM is a difficult task. This is due to unpredictable parameter variations such as temperature and magnetic flux saturation. In spite of all this, we can obtain decent results and a good general feel for how the motor operates.

For small dynamic stability analysis, a rotating reference frame yields steady-state values of steady-state voltages under balanced conditions. A model that includes the stator and rotor flux reference along the *d-q* axes is needed. This is because these values are used to calculate the voltages induced in the rotor.

The IM will be used by the algorithm to derive the FOC method. Once again, calculations are greatly simplified when all quantities are referenced along the *d-q* reference frame; however, in this case, the *d-q* reference frame is stationary. An additional advantage is that transforming to the *d-q-0* coordinate system in any reference frame removes the time-varying inductances associated with the IM. The IM used in this simulation has the following electrical parameters:

P = Pole pairs, 4
r_s = stator resistance, 0.531 Ω
r_r = rotor resistance, 0.408 Ω
L_s = stator inductance, 10.99 mH
L_r = rotor inductance, 10.99 mH
L_m = magnetizing inductance, 8.47 mH
J = rotor of inertia, 0.1×10^{-3} kg.m^2

We will first explore the equations used to create the mathematical model of the IM. After the relevant equations are listed, the corresponding block diagrams will be displayed when combined together. These combined together compose the IM block in Fig. A.3. Equations (A.1) and (A.2) specify the stator *d-q* voltages and flux linkages.

$$
\begin{aligned}
v_{ds} &= r_s i_{ds} + \frac{d\lambda_{ds}}{dt} - \omega\lambda_{qs} \\
\frac{d\lambda_{ds}}{dt} &= v_{ds} - r_s i_{ds} + \omega\lambda_{qs} \\
\lambda_{ds} &= \int \left(v_{ds} - r_s i_{ds} + \omega\lambda_{qs} \right)
\end{aligned} \tag{A.1}
$$

$$
\begin{aligned}
v_{qs} &= r_s i_{qs} + \frac{d\lambda_{qs}}{dt} + \omega\lambda_{ds} \\
\frac{d\lambda_{qs}}{dt} &= v_{qs} - r_s i_{qs} - \omega\lambda_{ds} \\
\lambda_{qs} &= \int \left(v_{qs} - r_s i_{qs} - \omega\lambda_{ds} \right)
\end{aligned} \tag{A.2}
$$

Equation (A.3) specifies the rotor *d-q* voltages and flux linkages.

$$
\begin{aligned}
v_{dr} &= 0 = r_r i_{dr} + p\lambda_{dr} - (\omega - \omega_r)\,\lambda_{qr} \\
\lambda_{dr} &= \int \left(-r_r i_{dr} + (\omega - \omega_r)\,\lambda_{qr} \right)
\end{aligned} \tag{A.3}
$$

$$
\begin{aligned}
v_{qr} &= 0 = r_r i_{qr} + p\lambda_{qr} + (\omega - \omega_r)\,\lambda_{dr} \\
\lambda_{qr} &= -\int \left(r_r i_{qr} + (\omega - \omega_r)\,\lambda_{dr} \right)
\end{aligned} \tag{A.4}
$$

The above equations in matrix form are given by

$$
\lambda = \begin{bmatrix} \lambda_{ds} \\ \lambda_{qs} \\ \lambda_{dr} \\ \lambda_{qr} \end{bmatrix} \tag{A.5}
$$

The block diagram in Fig. A.4 is used to model the stator and rotor flux linkages for the induction motor.

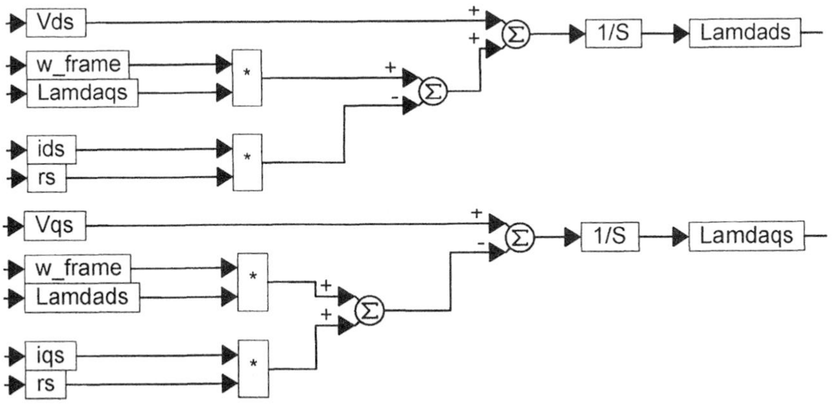

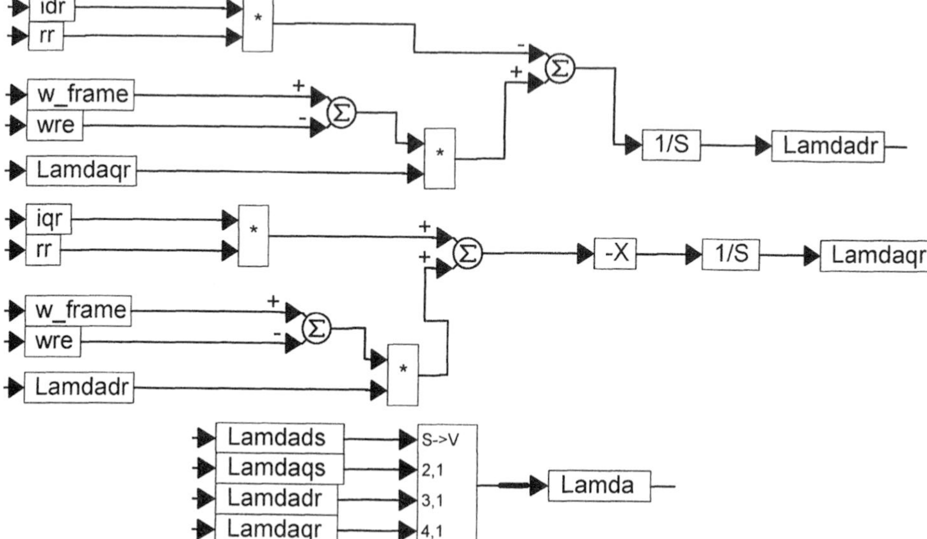

Figure A.4 VisSim™ block diagram of *d-q* axis stator / rotor flux linkages and matrix.

Now that the flux linkages have been modeled, we can specify our inductance values for our mathematical IM. We can put these values in matrix form and specify the inductance as

$$L = \begin{bmatrix} L_s & 0 & L_m & 0 \\ 0 & L_s & 0 & L_m \\ L_m & 0 & L_r & 0 \\ 0 & L_m & 0 & L_r \end{bmatrix} \tag{A.6}$$

To create the VisSim™ block diagram of this matrix, see Fig. A.5.

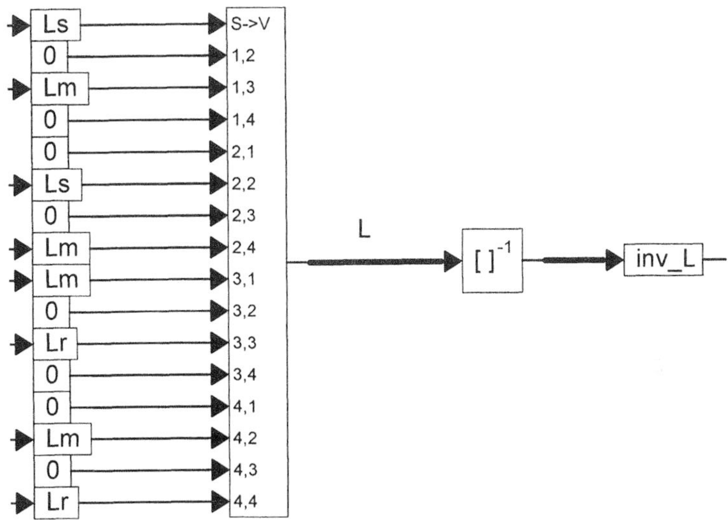

Figure A.5 VisSim™ block diagram of stator, rotor, and mutual inductances in matrix form.

Now that we have defined our flux linkages and inductances, we can define the electrical torque and rotor speed equations as follows:

$$\begin{bmatrix} \lambda_{ds} \\ \lambda_{qs} \\ \lambda_{dr} \\ \lambda_{qs} \end{bmatrix} = \begin{bmatrix} L_s & 0 & L_m & 0 \\ 0 & Ls & 0 & L_m \\ L_m & 0 & L_r & 0 \\ 0 & L_m & 0 & L_r \end{bmatrix} * \begin{bmatrix} i_{ds} \\ i_{qs} \\ i_{dr} \\ i_{qs} \end{bmatrix} \tag{A.7}$$

$$T_e = \frac{3P}{2}\left(i_{qs}\lambda_{ds} - i_{ds}\lambda_{qs}\right) \tag{A.8}$$

$$d\omega_{re} = \frac{P}{2J}\left(T_e - T_L\right) \tag{A.9}$$

$$\omega_{rm} = \frac{2}{P}\omega_{re} \quad rad/s$$
$$\omega_{rm} = \frac{60\omega_{re}}{P\pi} \quad rev/\min \tag{A.10}$$

We also need to specify the motor parameters listed in the beginning of this section. The block diagram in Fig. A.6 shows the motor parameters and other constants used with the simulation of the IM.

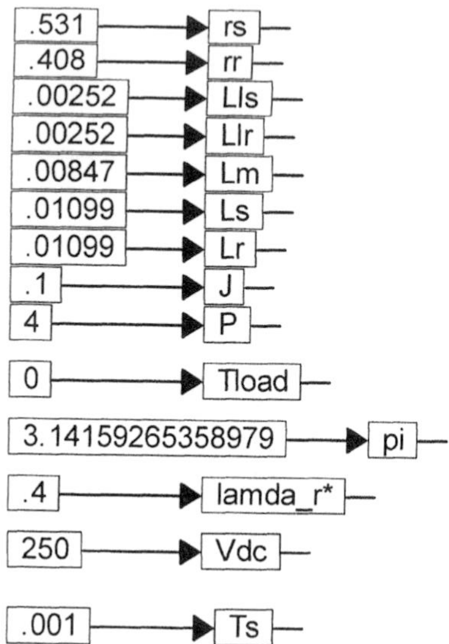

Figure A.6 Motor parameters and other constants used in simulation.

We can now model the electromechanical torque and speed of the motor in our simulation. The block diagram in Fig. A.7 models the torque and speed of the IM with respect to the other parameters modeled.

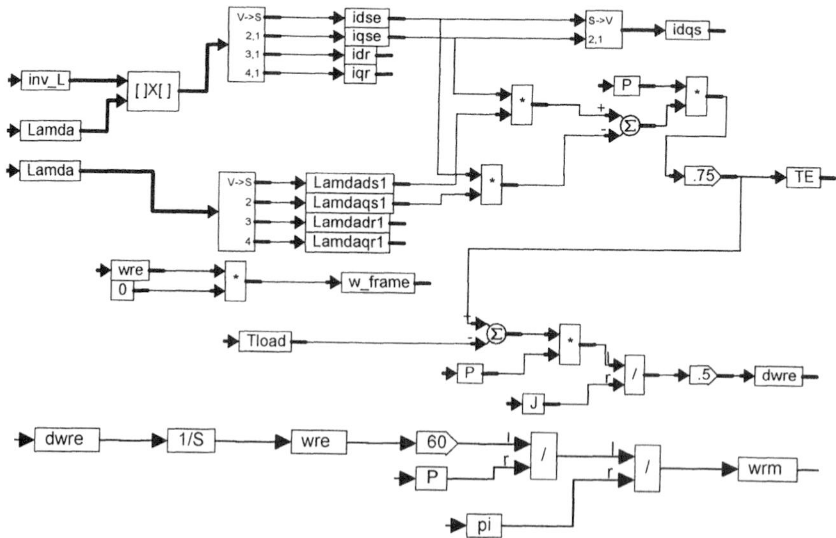

Figure A.7 Block diagram of the stator and rotor flux linkages, electromechanical torque, and speed.

A.3.2 Field Oriented Control (FOC) Block

We will now discuss the details of the FOC block sub-system region of Fig. A.3. Like the IM block discussed previously the FOC is composed of many discrete functional blocks made into a compound block. This compound block is then connected with other discrete blocks to form the control sub-system shown in Fig. A.8.

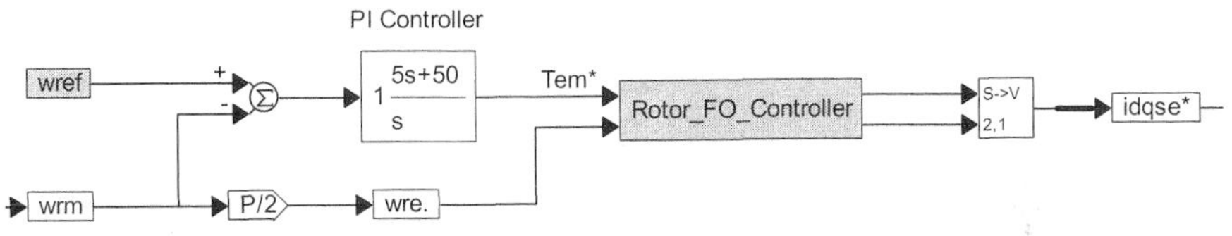

Figure A.8 The rotor field oriented control (FOC) sub-system.

There are two inputs to the FOC sub-system: wre and Tem*. The respective outputs of the FOC sub-system are the command values, iqse* and idse*. These are in the field-oriented rotor reference frame. The fundamental intention of this design is that the FOC will cause the rotor to generate a speed profile that follows the commanded speed input, wref. To do this, the commanded speed signal is first fed into the system where it is subtracted from the measured speed of the rotor. The error generated is then fed into a torque controller block. The torque control is done by the PI controller that generates the torque command Tem*. This torque command is used to set the electromagnetic torque induced within the induction motor by calculating an appropriate iqse* command based on the generated Tem*, along with the user defined parameters lamda_r*, and theta_r (the integral of the speed of the rotor). These values are then fed into the Rotor_FO_controller compound block. The insides of the Rotor_FO_Controller block are shown in Fig. A.9.

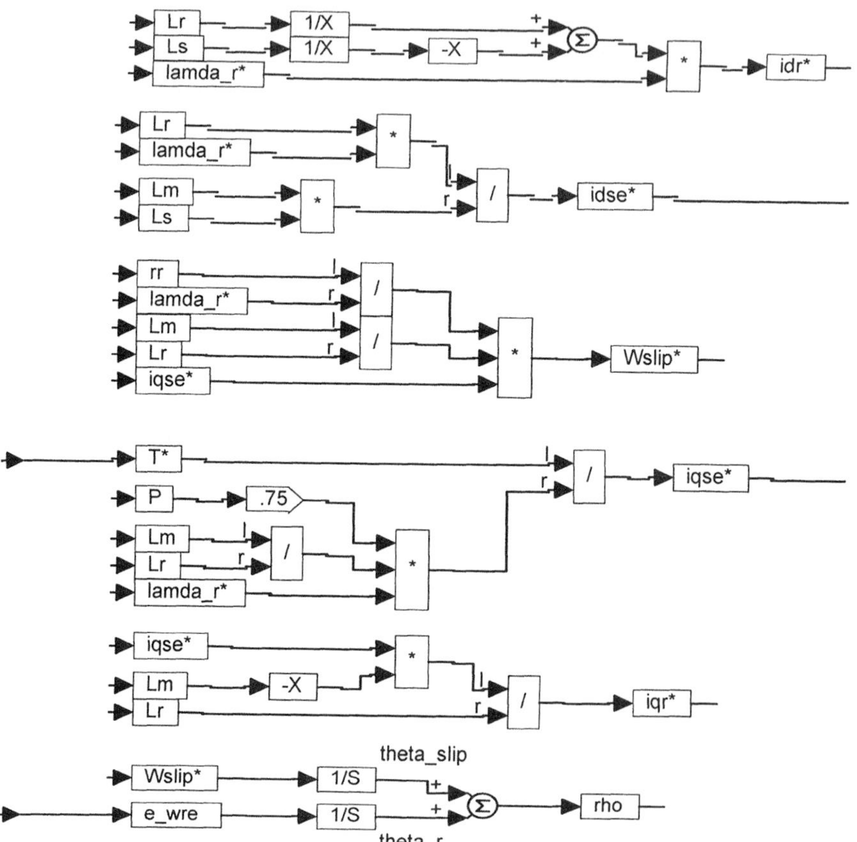

Figure A.9 Inside the Rotor_FO_Controller compound block.

The Rotor_FO_Controller block mathematically calculates the values of iqse*, idse*, the angle rho which is sum of the slip angle, theta_slip, and the rotor angle theta_r. These are the outputs of this block and are the inputs for the next blocks. The Rotor_FO_controller and the other blocks shown above regulate the sinusoidal currents applied to the current controllers according to the commanded electromagnetic torque, Tem*, and the constant rotor flux, lamda_r*. The next block utilizes the inverse Clarke and Park transformations. These methods are used to transform values referenced along the rotating reference frame into values referenced in the stationary frame. Isd and isq are then used to set the magnitude of the current sinusoidal values.

A.3.3 The dq→dq Sub-System*

The dq*→dq sub-system in Fig. A.3 consists of the dq*→dq block and the two PI controllers which feed the block. The two PI controllers shown in Fig. A.10 output the desired command values for the *dq*-axis stator voltages (Vds^{e*} and Vqs^{e*}).

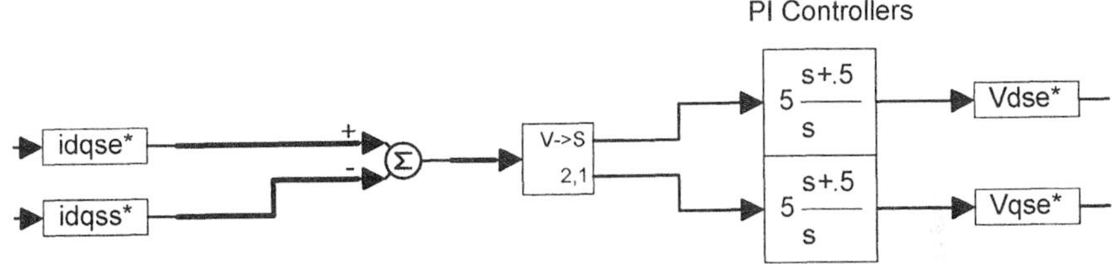

Figure A.10 PI blocks after FOC compound block output.

The command values for the *dq*-axis stator voltages (Vds^{e*} and Vqs^{e*}) are then used in (A.11) to determine Vqs^{s*} and Vds^{s*}. Equation (A.12) is implemented via the dq*→dq block. The insides of the dq*→dq block are also shown in Fig. A.11.

$$\begin{aligned} v_{qs}^{\ s*} &= v_{qs}^{\ e*} \cos\rho + v_{ds}^{\ e*} \sin\rho \\ v_{ds}^{\ s*} &= -v_{qs}^{\ e*} \sin\rho + v_{ds}^{\ e*} \cos\rho \end{aligned} \tag{A.11}$$

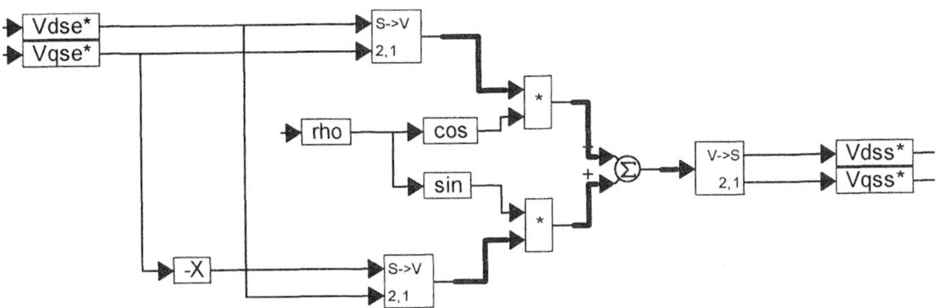

Figure A.11 Command values for *dq*-axis stator voltages (dq*→dq block).

A.3.4 The Space Vector PWM (SVPWM) Sub-System

The SVPWM subsystem is the SVPWM block shown in Fig. A.3. This sub-system generates the a-b-c stator voltages and then converts them into the *d-q* reference frame voltages for use by the induction motor model. The block diagram in Fig. A.12 contains the SVPWM block.

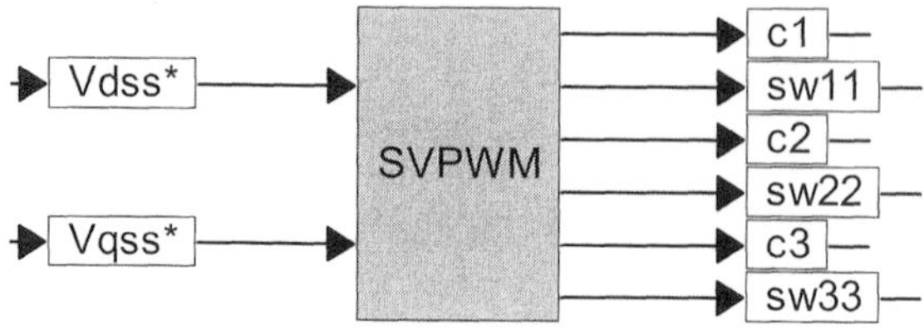

Figure A.12 Voltage-fed six-step space-vector PWM inverter.

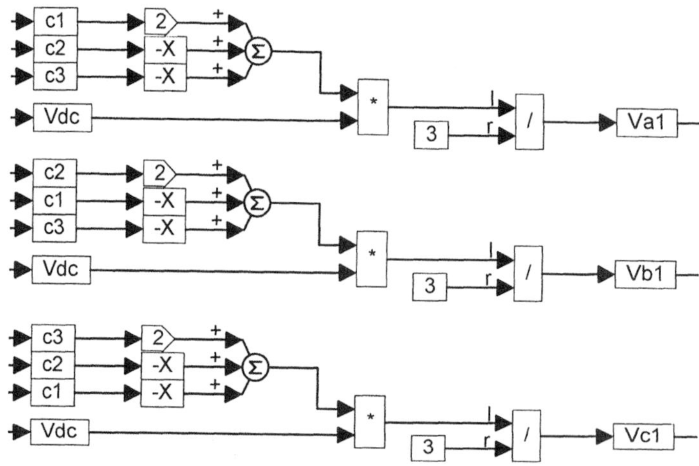

Figure A.13 Generation of *a-b-c* motor stator voltages from SVPWM.

Because the induction motor model equations use the d-q voltages, it is necessary to translate the a-b-c voltages (which would normally be fed into a real motor) into the d-q frame. This task is done by the block diagram in Fig. A.14.

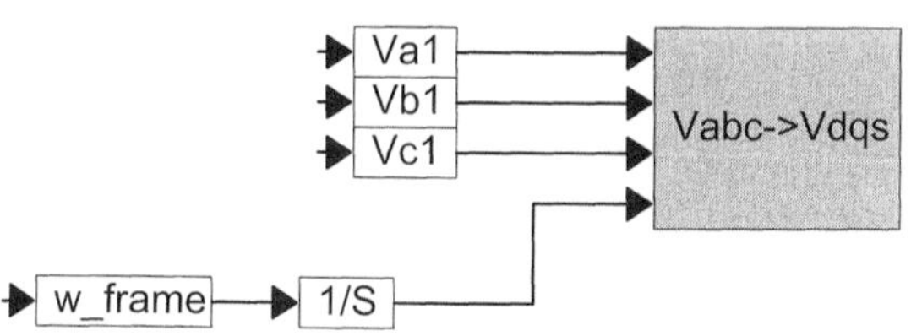

Figure A.14 Transformation block of *a-b-c* voltages to *d-q* reference frame voltages for induction motor.

The inside of the Vabc→Vdqs compound block is shown in Fig. A.15.

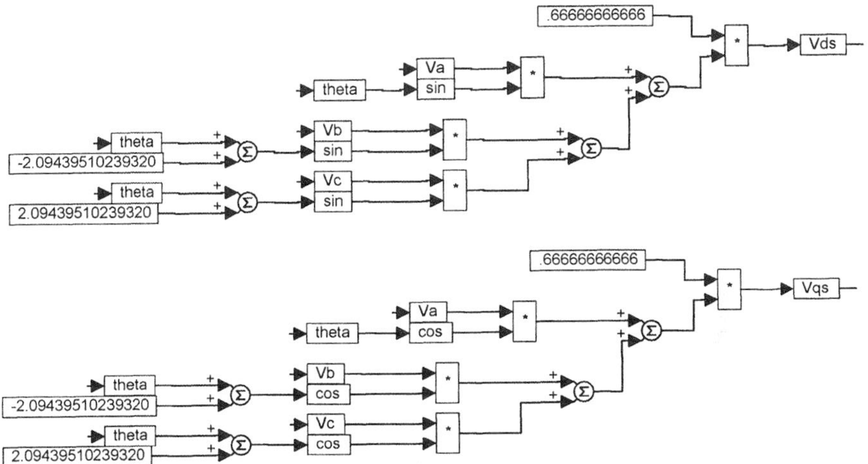

Figure A.15 Inside of the Vabcs →Vdqs block in Fig. A.14.

A.3.5 Feedback Sub-System

The feedback sub-system consists of both the dq←ab block and dq*←dq blocks. Stator current transformation from *dqe* to *dqs* is given in the block diagram below. Equation (A.12) models the *idss* and iqss** variables. The equations are implemented in the block diagram shown in Fig. A.16.

$$\begin{aligned} i_{qs}{}^{s*} &= i_{qs}{}^{e} \cos\rho + i_{ds}{}^{e} \sin\rho \\ i_{ds}{}^{s*} &= -i_{qs}{}^{e} \sin\rho + i_{ds}{}^{e} \cos\rho \end{aligned} \tag{A.12}$$

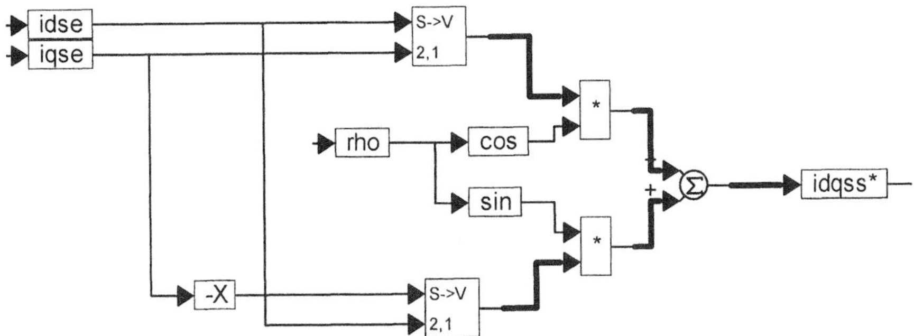

Figure A.16 Command values for the *dq*-axis stator currents (dq_dq* block).

A.3.6 Simulation Results

Speed Response Analysis

The speed response of our designed control system is of the utmost significance to this section of the design. By feeding an input speed command into the system, we observe the speed developed by the rotor of the induction machine over time. The two major criteria that we are trying to fulfill are:

1. the system develops a speed response that closely resembles the input speed command
2. the system is stable because stability is the most important design specification for any control system [2]

Once these two criteria are met, we would also like to achieve a quick transient response, i.e., quick acceleration and a response that is smooth.

Analysis of the Field-Orientated Section of the Design

To increase the speed of the rotor we have to increase the voltage and frequency by the same proportion. This ensures that we maintain a constant magnetic flux in the air-gap of the induction motor. In this section of the design, through the use of scopes placed on nodes of interest, we see that FOC applies this principle of speed operation. FOC transforms the commanded dc signals applied by the user and the feedback signals into sinusoidal signals required by the IM.

FOC is also referred to as vector control because it controls both the magnitude and phase of the voltage signal applied to the stator windings of the IM. The manner in which the phase changes over time for the cos_rho and sin_rho signals is very similar to the way the phase voltage signal develops over time. The phase voltage signal has a similar profile. Thus, the cos_rho and sin_rho signals control the phase, while the iqs and ids signals control the magnitude. It is through these signals, all of which are produced in the Rotor_FO_controller block, that FOC achieves vector control. Furthermore, the speed is linearly increased when the frequency and amplitude of the phase voltage signal is increased by the same proportion.

The principle of increasing the voltage magnitude and the frequency by the same proportion in order to maintain a constant rotor flux is the basis for applying scalar volts per hertz speed control method. However, the equations used to execute FOC consider the parameters of the motor. Therefore, we achieve better control over torque variations. FOC also has the advantage of extending the speed range of operation through field weakening. Finally, FOC is a torque control method. This suits the requirements of the induction motor control design. Figure A.17 shows the electromagnetic torque developed for a ramp speed command. The actual speed is depicted in Fig. A.18.

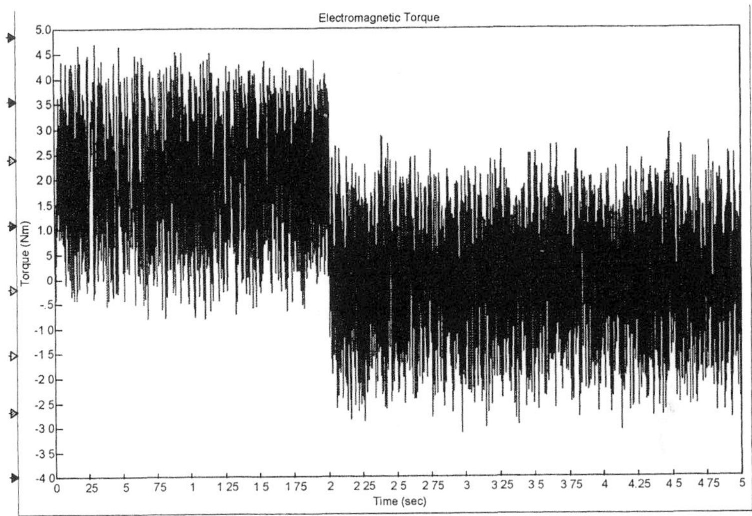

Figure A.17 Electromagnetic torque result from VisSim™ simulation.

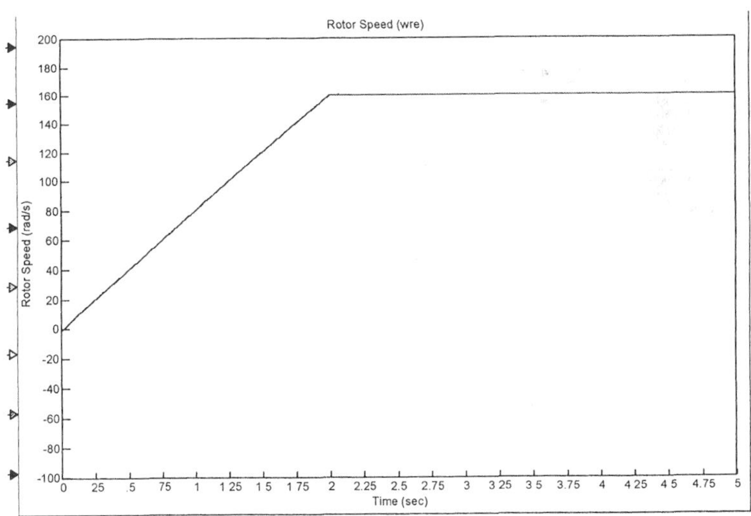

Figure A.18 Rotor speed result from VisSim™ simulation.

A.4 Summary and Improvements

In this appendix, the simulation of a field-oriented IM using VisSim™ was presented. The electromagnetic torque and rotor speed were given. We presented the concept of FOC to be the control technique because it produced controlled results that had better dynamic response to torque variations in a wider speed range compared to other scalar methods. It was shown through the VisSim™ analysis that

FOC is an effective means of driving an induction motor. Also, a speed profile with an adequate transient response and steady-state error was generated.

By simulating the control system, we obtained an acceptable speed response and found there was adequate stability. It is also worthy to note that increasing the proportional gain of the torque controller introduced or amplified noise and harmonics. This affected the short-term and long-term performance of the induction machine. Also, by increasing the proportional gain of the torque controller, the system approached unstable regions of operation.

An improvement on the system could include the implementation of a self-tuning FOC module. In the self-tuning scheme, any change in resistance of the stator and rotor windings due to heat could be accounted for.

References

1. P. Vas, *Vector Control of ac Machines*, Oxford University Press, New York, 1990.
2. D.W. Novotny and T.A. Lipo, *Vector Control and Dynamics of AC Drives*, Oxford University Press, New York, 1996.
3. N. Mohan, T.M. Undeland, and W.P. Robbins, *Power Electronics: Circuits, Devices, and Applications* - 2nd ed., John Wiley & Sons, New York, 1995.
4. B. K. Bose, *Adjustable Speed ac Drive Systems*, IEEE Press, New York, 1987.
5. C-M. Ong, *Dynamic Simulation of Electric Machinery using MATLAB /SIMULINK*, Prentice Hall, New Jersey, 1998.
6. Texas Instruments Inc., Field Orientated Control of 3-Phase AC-Motors, www.ti.com, Texas Instruments Literature number BPRA073, Europe, 1998.
7. Texas Instruments Inc., Clarke and Park Transforms on the TMS320C2xx, www.ti.com, Texas Instruments Literature number BPRA048, Europe, 1998.
8. Texas Instruments Inc., Digital Signal Processing Solution for AC Induction Motor, www.ti.com, Texas Instruments Literature number BPRA043, Europe, 1997.
9. Texas Instruments Inc., Implementation of a Speed Field Orientated Control of a Three-Phase AC Induction Motor, www.ti.com, Texas Instruments Literature number BPRA076, Europe, 2000.
10. Visual Solutions, Inc., VisSim™ 4.5g Fixed-Point Datasheet.
11. Visual Solutions, Inc., VisSim™ 4.5/TI C2000 Rapid Prototyper Datasheet.

INDEX

> Much more of a problem than the cost of travel were the gaps between appointments. I could have a 15-minute appointment at, I don't know 7, then another appointment at 9. You couldn't do anything with the time – it was just wasted time and you didn't get paid anything for that time.

Jenny is not alone in highlighting this issue. Unpaid travel time would go on to become the subject of a court case against companies contracted by Haringey Council to provide care services. In that instance, the use of unpaid travel time took down some workers' wages to less than £4 an hour. In Autumn 2020, the workers were awarded £100,000 in back-dated earnings. But even with this precedent, access to the courts is hard for low-paid workers – wins are uncertain, the process drawn out and the innovation in exploitation is often quicker than workers' ability to fight back.

Jenny's stories shocked me most when she talked about being placed in situations for which she had nothing like the knowledge, support, or training to handle. In one case, thanks to staff shortages, Jenny was asked to provide support for a person with dementia. She had never met the client, and had not been scheduled for the appointment at the beginning of the day:

> I was asked to cover for a client after somebody had called in sick. The client was a person suffering with dementia. I'd never worked with a dementia patient before, and I'd had no training – I'd been promised it, but it was months away. To top it all off, she didn't speak English and I didn't speak Spanish, so I was left trying to communicate with someone whose language I didn't understand. At points, she got very distressed, and there was nothing I could do to help that – I was also quite distressed [. . .] I ended up being there about four hours.

In the end, on the back of traumatic experiences like this, Jenny left the care sector altogether. Forty per cent of people in social care roles do so every year, leaving the sector with over 100,000 vacancies at any given time.[40]

> I felt I was being asked to do things I wasn't qualified to do, and like I'd end up hurting someone. I couldn't take the risk, and it definitely wasn't worth it for the pay and progression.

Jenny's reasons for leaving the sector are consistent with the data on why so many other care workers are leaving their roles.[41] The national body Skills for Care identifies that younger people, people who travel further for work, people on zero-hour contracts, those who do not receive regular training or opportunities to advance qualifications, and those who work in settings with poor standards of care are the most likely groups to leave their roles in any given year. Nearly half leave their job within a year, the majority to work in another sector altogether.[42]

Jenny's story came before Covid. But other social care workers shared their experiences of the pandemic, and how it had impacted them. A senior care manager talked about the fear invading their work life, telling me how:

> Staff are frightened of what could happen. PPE is a concern, and they have gone into isolation after completing an NHS assessment . . . [staff] are frightened to come to work.

Fear, stress and fatigue were common themes. A managing director put things in blunt terms:

> Stress has increased, people are scared. It is likely to get worse. People are physically tired as they have to physically cover others who are self-isolating, many of whom could be at work, but the lack of testing means they have to isolate. The lack of PPE is horrific. Lambs to the slaughter.

I heard the same message from almost every tier of the social care workforce. They had been forgotten. Protective equipment wasn't coming; stress was becoming more unmanageable; and they feared for their safety. A registered manager said:

> The government need to do more with regard to support with PPE, most especially hand sanitizer that is out of reach for most providers. We need more support with aprons, gloves, wet goggles and facemasks.

While a home carer told me:

> [I have] concerns over whether PPE is effective. It's difficult to concentrate. I feel emotional and drained whilst trying to stay positive for clients. I feel it's going to get worse.

'Lambs to the slaughter' is a line that, to me, now reads as tragically prophetic. It is indicative of what a sector run with the twin aims of maximum profit and lowest possible state spend, means for its workers.

THE NATIONAL CARE SERVICE

In the late 1990s and early 2000s, the election of a Labour government raised expectations of bold social care reform. Within the first year of Tony Blair's premiership a royal commission was established – chaired by Sir Stuart Sutherland – to explore how social care was run and funded.

The commission reported just over a year later, and recommended free personal care, based on need and funded by general taxation and a more generous means testing of people to fund their living costs and accommodation.[43] Their recommendations were rejected. The government decided free care would be prohibitively expensive.[44] And while they did adjust the means test for social care living and accommodation costs, it only changed in line with inflation – meaning it did not really become more substantial.

When Blair left office, there was widespread disappointment with his record on social care. Jon Glasby of the University of Birmingham called it a 'tale of policy neglect'. Andrea Rowe, the former chief executive of Skills for Care, said that 'Social care under Blair's leadership has continued to be a 'Cinderella service'. And Bob Hudson, at the University of Durham, noted that it seemed that only health (rather than care) mattered politically.[45]

In 2019, the Labour Party outlined plans to rectify this missed opportunity for lasting change. At their Autumn party conference, and ahead of a December election, they pledged to deliver a National Care Service (NCS) – based on free personal care, better standards for workers and a tougher time for large profit-driven providers. The NCS is now the basis for a current leftist consensus. It has support from academics like Allyson M. Pollock, Luke Clements and Louisa Harding Edgar;[46] trade unions like Unite and GMB;[47] and influential campaign groups like We Own It.[48] Ahead of the 2021 Holyrood elections, the Scottish National Party also committed to delivering a NCS – with SNP leader Nicola Sturgeon calling it a 'top priority' for the Scottish government.[49]

If we take the policies in turn, we find some excellent ideas. Free personal care, the headline proposal in England (implemented in

Scotland in 2002), clearly has merit. Most importantly, it provides a robust delivery mechanism for social care – based on the NHS' own popular principle of 'free at the point of delivery, based on need'. In turn, it immediately addresses variation in care, providing a clear and equal entitlement to social care provision. And while not a policy aimed directly at care costs, it would still immediately reduce the number of people facing catastrophic care costs from around 140,000 people to around 80,000 people.[50]

It is also broadly affordable. It would require spending to increase from around £20 billion per year today, to £36 billion by 2030. That's only £2 billion more per year, in 2030, than the cost of the Conservative Party's 2017 pledge on social care – denigrated as a death tax by *The Daily Mail* and now often cited as a key factor in Theresa May's failure to secure a majority in the 2017 general election.

And it is very popular. YouGov polling in 2018 found that almost three in four English adults support free personal care for all who need it, and that almost 70 per cent would be willing to pay more tax to fund it.[51] Other studies have shown strong cross-party consensus for free personal care, with majority support among both Labour and Conservative voters.[52] Of all the major proposals for social care funding reform, it is by far the easiest to explain – helping it generate its popular appeal.

In terms of the other components of the 2019 Labour manifesto, a better deal for social care workers is clearly important, too. The most common idea is a wage guarantee for care workers. If subsidised by government directly, this could achieve better pay for half a million people – doing highly skilled and intense work – at a cost of £1 billion per year.[53] Other proposals to ensure social care workers have access to development, training and progression are equally vital, in the face of the stories of low hope and exploitation I have already detailed above.

A tougher stance on the worst private providers also makes sense. The Labour Party's 2019 vision for a National Care Service suggests a focus on transparency, financial sustainability and tax compliance. As well as a striking back against private equity, and support sector stability, this would help to address what the Centre for Health and the Public Interest has called 'leakage'. Their research estimates that £1.5 billion of the £15 billion spent on independent care sector 'leaks' out the sector in the form of profit before tax, rent payments, director pay and loan repayments.[54]

THE LIMITS OF THE NCS

The benefits promised by an NCS are undermined by problems that stop it being truly transformational. As a policy around which the left has coalesced, it still needs to be taken further.

Specifically, the National Care Service only asks and answers two of three vitally important questions. The two it answers very well are these: how should better, more accessible care be funded? And by whom? However, to at least some extent, it misses an even more important one: what do we want care to look like? This oversight detaches proposals for a NCS from the lived experiences, needs and ambitions of those who use the social care system.[55]

In *Madness and Civilization*, Michel Foucault explores the evolution of madness.[56] He provides a genealogy of the transition from a time when it was seen as an expression of divinity, something to be treated with kindness, or something that shows the truth of humanity – to a time when it became something that created fear and demanded confinement. His research reveals what he dubs 'the Great confinement' – whereby across Europe, 'insane' people were first segregated to the margins of society, and then segregated physically in medicalised spaces: culminating with the rise of the asylums. He shows how this relates to a wider process of institutionalisation and the enactment of state power, which first defined madness as a moral failure and then as a medical one.

The evolution of social care in the twenty-first century is a modern repetition of the 'great confinement'. On this occasion, the basis is not the binary of mad/sane, but rather vulnerable/normal. From this (false) binary has evolved a highly institutionalised and paternalistic care system designed around the twin goals of i) managing needs and ii) hiding 'vulnerability' from wider society. Predictably, the care system this regime has brought into being is the modern equivalent of the asylum – at least in terms of the power relationship that obtains between the institution and the recipients of care.

This comparison can be shored up through examining how care homes and care at home schemes tend to operate. Too often, care at home functions through the instrument of the 'checklist'. That is, people's 'needs' are boiled down to a mechanical list of the daily chores they need help with. Their support becomes little more than a process of checking off these mechanical chores each day. People become defined, for those who provide their care, not by their humanity, their priorities, their relationships, the life they want to lead and the potential they could reach – but through an oversimplified care plan, that only incidentally relates to what they need. It is a case of need being managed and hidden, rather than actually being met.

One of the worst examples of this one-size-fits-all approach is the continued prevalence of 15-minute appointments. According to Freedom of Information request figures, released by disability charity Leonard Cheshire in 2018, 20,000 people receive visits from their carers that only last 15 minutes.[57] As the charity themselves said, in reacting to these numbers, this can leave a person feeling dismissed and stressed. Reviews have found levels of loneliness in care homes as high as 42 per cent.[58]

Confinement in this system is not about locked doors and padded rooms. Rather, it is a simple result of viewing people through the prism of mechanical tasks, rather than as humans and citizens. More than half of social care users reported feeling isolated in 2017 data.[59] This is a result of a system that views people as problems to be managed, rather than citizens that can be empowered by brilliant public services.

The equivalent of this trend within residential care is 'warehousing'. The modern rise of big, faceless providers has created a system of woefully inadequate care 'factories'. Within these, people often find themselves living lives of idleness, boredom and neglect. In the worst cases, they're packed into small bedrooms – without space for visitors, without scope for even small amounts of autonomy and without design features conductive to social interaction. I have heard care home providers talk about supporting people to do their own laundry as an empowering innovation, in a damning indictment of the extent of our national ambition.

Faced with these modern asylums, the ideas currently mainstreamed within the National Care Service would: help reduce the number of people who lose all their wealth to access this system; help reduce the number of people with no support at all; and would probably bring about some improvements in 'outcomes'. But they do not really change what type of care is provided, what kind of lives it enables and what societal purpose it serves. That is, the NCS does not sufficiently change the nature of care, the power relationships that define it, or the level to which institutionalised and paternalistic care dominates provision.

The experience of Scotland – which has already implemented free personal care (2002) and has a government commitment to a National Care Service (as of 2021) – provides a warning that we still need to develop our ideas, if they are to deliver truly radical change.. Survey research by the charity Independent Age shows that, despite almost two decades of free personal care in Scotland, care provision remains inadequate, with 43 per cent of respondents indicating that they did not feel the amount of care they or their family member received was adequate for their needs.[60] The same research also showed how limited the definition of 'free personal care' could be – it covered washing, getting out of bed, meal prep and toileting, but was not perceived to cover shopping, travelling, or support for relationships and social interaction.[61]

THE RIGHT TO A BRILLIANT LIFE

The idea of a service free at the point of delivery, based on need and available to everyone is important, and makes the concept of a National Care Service worth keeping. But it needs to be combined with a stronger vision of what *good care* looks like – that is, what it would deliver, and not just how it would be delivered.

In thinking through the policy implications, the best starting points are the demands that have been developed by those who use social care support – including the disability rights movement. Too often, these are voices left out of social care discourse and debate. The result is policy that, like the NCS, is half formed.

Social Care Future is one organisation with a shared commitment to major transformation in social care – based on focusing less on *who* provides care, and more on *what* care is available and for what *ends*. In June 2019, the group brought together 80 contributors to draft a shared vision in answer to these questions. The first attempt at capturing the vision read:

> We all want to live in the place we call home with the people and things that we love, in communities where we look out for one another, doing the things that matter to us and we're good at. We all want the peace of mind that should we, our families, or neighbours need some support from public services to do so, that it will be there for us quickly and affordably.
>
> Great support offered, how we want and need it, revives our sense of hope and purpose. It helps all of us keep or regain control over our lives, connect and sometimes reconnect with the things that are most important to us and to realise our potential. By doing so, it allows us

> to keep on contributing to our communities, with the benefits rippling out to everyone.
>
> By investing together, we can create a fair, reliable and effective social care support for everyone. By investing in a better social care future, we can invest in us.[62]

Their vision is brilliant because it doesn't ask 'how to make the asylum cheaper, free or state-run' – but rather, how to smash the asylum completely, and move care onto a more empowering, dignified foundation.

Elsewhere, more ambitious conceptions of what care does and achieves have come from within the disabled people's rights movement. In 2019, the Reclaiming Our Futures Alliance of disabled people and disabled people's organisations set out a plan for a National Independent Living Support Service. This detailed a vision for independent living defined not as disabled people doing everything on their own, but rather having choice and control over their lives; being included in the community; and having the same chances to take part as other people.[63] Critically, while it has important overlaps with the NCS, it also calls for more fundamental and ambitious reforms to how social care support works, and what it looks to achieve.

Looking across the deliberative research, there are three changes that need to be better integrated into any conception of, or narrative around, a National Care Service:

1. **Towards Community:** Everyone should have a right to care in their own home and community wherever possible, delivering alongside people they know and places they call home.
2. **Towards Personalisation:** Care should be personalised around an objective to give everyone a brilliant life, rather than being mechanical and one-size-fits-all.
3. **Towards Empowerment:** Care should be undertaken *with* people, not *to* them – with a fundamental recalibration of power between providers, carers, care workers and people receiving support.

If integrated into a future NCS, these changes would combine to form a positive approach to care funding with a more empowering envision of what care looks like and can achieve.

TOWARDS COMMUNITY

The case for delivering more care in the community is well developed. It is in line with what people who use social care services want. It is an excellent way to maintain people's independence and dignity. It avoids people having to pay catastrophic residential costs, as is the norm in care homes.

Despite these prospective benefits, there has been little real progress in bringing more care into the community. A community-led model of care relies on a system that provides support as soon as its needed, with the ambition of maintaining someone's quality of life and capacity to flourish for as long as possible. But severe funding pressures on local authorities have instead left us with a care system that only intervenes at the latest possible moment – often once someone's needs have deteriorated and where intensive support is the only available option.

This can be seen in the international data. The UK has a much higher proportion of people receiving care in residential and nursing homes – and hospitals – than many other, comparable countries.

Table 4.1 Location of Deaths, Selected European Countries and the UK[64]

	Home	*Hospital*	*Care home*
European average	33%	44%	18%
UK	23%	47%	28%
Difference	–10%	+3%	+10%

One third of people in Europe can expect to die in their home, compared to just a quarter of people in the UK – despite people overwhelmingly saying that dying at home, surrounded by loved ones, would be their preference. This is one, indicative example of how the UK care system is designed around acute need, late intervention and managing need rather than ensuring dignity, independence, or flourishing lives.

In fact, the situation is highly variable across the UK. In Summer 2021, analysis by the IPPR and CF – a healthcare management consultancy and analysis company – revealed a stark postcode lottery determining who has access to community care. In some local authorities, the best being Hammersmith and Fulham, more than eight in ten people receiving care support were receiving it at home. In others, the figure was less than half – such as just 46 per cent of people in Barnsley.[65]

Some of this can be put down to demographics – the average age of some places is very different to others in the UK. But the variation even exists once local authorities are grouped by geographic and demographic features. Indeed, even a moderately ambitious attempt to close avoidable variation between similar places would help 80,000 people to receive more appropriate care and generate efficiencies worth over £2.5 billion a year.

Everyone should have the opportunity to receive good support, as soon as they need it – and for that care to be in their community and home by default.

TOWARDS PERSONALISATION

Guaranteeing people care at home, from the point they need it, should be an opportunity to focus on personalised rather than one-size-fits-all models. This would mean commissioning significantly more diverse, relational and personalised social care – rather than the mechanical and institutional 'life and limb' services that too often dominate.

Isolated care 'warehouses' and 15 minutes 'flying visits' by over-stretched and underpaid care workers need not be the norm. Models of care do exist that provide people with more choices – or which better tailor care to the needs and demands of a specific diagnosis (e.g., dementia). Indeed, there are many examples of care models that are not used as widely as they could be in this country, but which highlight opportunities for greater personalisation and choice within community care.

There are 150 Shared Lives schemes in the UK. Shared Lives provide opportunities for adults to live, independently, in their community – with the support of a community network. Within the schemes, Shared Lives carers use their own home and family life, to share it with someone who needs support. Carers earn up to £32,500 per year – and receive training, breaks and a support network. It provides a community-based and relationship focused alternative to residential care.

Evaluations of shared living schemes have been positive. Over 96 per cent of Shared Lives schemes are rated good or outstanding across England, compared to just 65 per cent of large independent nursing homes and 72 per cent of large independent care homes.[66] In 2018, an independent review found that Shared Lives can provide a preferable model of respite services for older people and people living with dementia.[67] Moreover, estimates suggest that Shared Lives could offer

£225million of savings per year, if every area caught up to the best performing – thanks to reductions in A&E and hospital admissions.

Buurtzorg is another example of excellent community care, pioneered in the Netherlands. Buurtzorg translates to 'neighbourhood care' and is an innovative homecare model developed in the Netherlands. It is a system designed to allow people with care needs to live independently, with less dependence on formal support.

The model runs as follows. A district nurse provides coaching for the individual and their family, to build their capacity and confidence to deliver care. The focus is on relationships, with nurses encouraged to spend most of their time with people.

Evaluations have suggested the scheme comes with significant benefits. The cost per hour is more expensive. However, the care provided is much higher quality and orientated towards prevention, meaning only half as much care as is needed on average. People live better, more independent lives – in the place they call home, and with people they know – for 40 per cent less cost than traditional homecare models.

In residential settings, where they are needed, fantastically better care could also be within reach. In the village of Hogewey, just outside Amsterdam is one example of a cutting-edge care facility. Hogewey provides a space for around 152 residents, all with advanced dementia, to lead a normal life. The village has a town square, a supermarket, a hairdressers, theatre, pub and restaurant. Living arrangements are designed around familiarity – with different living areas themed. The logic of Hogeway is simple – living a normal life is much easier in a place people can call home, with tailored support, than it is when receiving a very poor level of unspecialised care in a place you no longer recognise. Moreover, in Holland, it is publicly funded through the country's social security system and costs the state no more than other forms of nursing home in the Netherlands, as of 2019.[68]

Research on dementia villages specifically remains nascent, and it can be hard to obtain reliable gauges of quality of life from residents with a dementia diagnosis. But the studies that do exist are encouraging. Reviews have found the higher staffing levels, increased physical activity[69] and a reduction in the use of antipsychotic medication. Before the village was introduced, 50 per cent of residents needed antipsychotic medication, which has since been reduced to 12 per cent of residents (as of 2019).[70] Studies of similar models, like the Eden Alternative in Texas, found a reduction in pressure ulcers, use of restraint, staff absenteeism and an increase in residents' levels of movement – all highly positive indicators for people living with an advanced dementia diagnosis.[71]

It works because it looks to provide people who use it a place and a community that they can call home.

TOWARDS EMPOWERMENT

The market has provided a very paternalistic model of social care in the UK. The status quo today is that care is 'done to people' rather than with them. However, though they have merits, progressive alternatives have rarely put forward ideas that would genuinely change this power dynamic. Indeed, concepts like a National Care Service would arguably reiterate it.

This is not, however, inevitable. There is growing evidence that some ownership models support more equal power relations. Specifically, models of democratic and local ownership – normally using highly innovative approaches, and delivering care at a smaller scale – that are exciting prospects in social care delivery. There is a growing consensus on a definition for 'democratic and local ownership'. In 2020, the New Economics Foundation – themselves, building on work by Co-operatives UK, the University of Manchester and the Centre for Health and the Public Interest – laid out some simple criteria for what democratic, local ownership looks like in practice:

- Legal binding to a clear social mission
- Accountable to people needing and providing support
- Organised at the smallest appropriate scale[72]

If care organisations displaying these values became dominant, it would constitute a major empowerment of people who receive care and of workers who provide care.

There are some excellent examples of emerging, empowering and democratic care provision. For example, there is much that is innovative about Equal Care co-operative, operating in West Yorkshire's Calder Valley. It is a platform co-op, meaning it uses a digital platform to provide its service. Their collaborative platform allows each person to choose their relationships, and then decide where and how their support takes place. It is funded by the community, rather than by private equity, through a community shares model. This model allows investment in enterprises that benefit a community and investors to then have a democratic say. They have proven themselves a remarkably successful form of capital since 2009. And they are multi-stakeholder in design, which means they are governed through dialogue – in this case, with both the providers and recipients of care held in focus.

These are emerging services, with evaluations pending. But the early indications are very strong. Testimony from people using the care service suggests a sense of ownership and empowerment, not always associated with social care. Ambitions include a target to pay new care workers £20,000 per year at the very least – a full 25 per cent above the average in the industry.[73]

There are three main advantages to Equal Care's model. First, its innovative platform allows it to make significant efficiencies. Second, its democratic ownership model allows those efficiencies to be invested in people and workers. Finally, its platform is designed around empowering those members, care recipients and providers alike – helping address problems with power dynamics that often plague other provider models.

Elsewhere, there are promising examples of social enterprises. PossAbilities is one such case, in Greater Manchester. It has been set up to help people live the life they choose, rather than one imposed on them. It provides self-contained living spaces, has a voluntary commitment to the Driving Up Quality code, exclusively 'good' and 'outstanding' ratings from the inspector, its own 'Happiness Manifesto' and a clear space for families and shared lives. These are features of quality and personalisation that are far from standard in this country's care provision.

The wider evidence of democratic and local ownership models in social care specifically is still inchoate, but promising. Equally promising is the extensive evidence on the link between organisations focused on social value and the quality of service provision. Compared to other forms of ownership, co-operative models are associated with greater job satisfaction, worker wellbeing, lower pay inequality and higher rates of engagement and productivity.[74] Despite this, the UK has disproportionately fewer co-operatives and mutuals than other OECD countries (and four times fewer than Germany).[75]

MAKING THE CHANGE

There are clear routes to achieving these three aims – more community, more empowerment and more personalisation – through ambitious new policies. Each are compatible with the National Care Service model. Combined, they provide both a model for delivery (free at the point of need, as in the NHS), and a model for what is delivered (the best possible care).

First, we need a formal guarantee that everyone can have access to care at home, where they would most benefit from it, and a commitment to bringing in innovative models of care to ensure choice and personalisation – all of which will require investment in community infrastructure. The pandemic has showed that we just don't currently have the capacity, services, or staff to deliver in line with a community-first model of care.

That is not to say there wouldn't be a return. In the medium to long-term, community care is not only a fundamentally better way of delivering care, but also a more efficient one. This is intuitive: community care is all about good principles of prevention and early intervention. It is about maintaining independence and dignity and delaying the point at which someone needs more intensive or acute support. The upfront investment needed by community care would pay itself back, in better lives and better public finances alike.

In thinking about the size of investment needed, we can look internationally for inspiration. In the USA, community care infrastructure forms a vital component of President Joe Biden's post-Covid stimulus fund. His investment will equal \$400 billion over eight years – or \$50 billion per year from his \$3 trillion investment package. In the UK the equivalent amount of investment would be worth around £5 billion per year, over the same eight-year period.

This figure would provide the capacity to do two things. First, it would allow community capacity to increase. Research has indicated a £2 billion per year stimulus for community capacity would be a reasonable post-Covid settlement. The remaining money could be invested in a new, social care transformation fund – providing local authorities with the money they need to incubate, scale and transition to community-led, democratically-owned and social anchored care provision.

This extra funding would give the government an opportunity to set ambitious standards for ethics within the social care sector. It has previously been suggested that the government should introduce a binding set of ethical standards for the sector, including:[76]

- A commitment to high quality care: Contracts should only be awarded where there is confidence in quality and where there is a clear accountability mechanism between providers, social care workers, carers and people using support.

- Workforce standards: Contracts should only be awarded where organisations pay a real living wage, provide adequate training and support, and engage in sectoral bargaining.
- Economy: Preference should be given to local organisations, that can demonstrate value to their local economy.
- Environment: Preference should be given to organisations that can provide evidence of high environmental standards.

To this list, we might suggest additions, based on the content of this chapter so far. Firstly, that there should be a set preference for co-operatives, social enterprises, small social care providers – and other promising, empowering models of ownership and delivery. This preference already exists in Wales, where the 2014 Social Services and Wellbeing Act embedded an expectation that these types of social value-focused providers be preferred when commissioning government services.

This could be taken a step further. To accelerate the sustainable transition from large, private equity funded and corporate social care providers, a new expectation could be implemented by stipulating that local authorities spend a set proportion of their commissioning budget on a) small providers embedded in the community, b) direct provision of care by the authority itself, or c) care provided by co-operatives, social enterprises, mutuals, or similar.

Second, there should be a focus on empowerment within these standards. Providers delivering government services should be required to put in place meaningful consultation with care recipients, carers and workers – and to action their views. Their ability to do so should be measured by the satisfaction of people using their services, and the number and type of complaints received about their provision. Any sign of a lack of accountability to the people using these services should be considered a break of contract, and reason enough for a local authority to change providers.

Next, we must recognise that the ongoing crisis experienced by the social care workforce severely limits the potential for transformation. Workers need the security, tools and training to provide great care – and if we do not do right by them, as presently, they cannot do right by us.

The Labour 2019 proposals for an NCS included a pay rise for workers. But we also need mechanisms for more fundamental change – to better train and professionalise care workers, to make it an attractive career going forward and to ensure there is scope for progression – including into roles with autonomy and professional expertise, as we might find in the nursing and medical professions.

There are several ways this could be achieved. Significant investment in training is an obvious one. The national body Skills for Care estimates that 48 per cent of workers in the social care sector do not have a care qualification.[77] Given the kind of high-pressure people in these roles, like Jenny, are under – this lack of provision for training is alarming. It reflects a government that invests a woefully small amount in social care training overall, and in continuing development for professionals more widely.

Perhaps most important is the sector's ability to negotiate collectively. The health sector is full of infrastructure to collectively negotiate, campaign and advocate for different parts of the workforce. There are bespoke unions, like the British Medical Association; there are an array of Royal Colleges; there is a formal infrastructure for formal negotiation on pay, training and recruitment between representatives of the sector and the government. The same cannot be said of social care. If we really want to evolve social care, we need a body to advocate for the evolution of the social care career. This means embedding sectoral and collective bargaining for workers – through something akin to a Royal College of Social Care Workers, with union status.

While the conditions and levels of exploitation experienced by the social care workforce demands immediate attention, this is not the only workforce challenge. Local authorities have also seen huge cuts in their workforce in recent years. In 1999, local government had a headcount of just over 2.7 million employees and central government of 2.3 million. Today, local government has around 2 million employees and central government 3.1 million – a big centralisation of capacity.[78]

One area where expertise has been particularly lost to local authorities is commissioning. The staff with relationships to prospective local providers, and who can differentiate between unscrupulous providers out to make a quick buck and those who would genuinely provide good care, have been lost.[79] Restoring this expertise is vital if we have any hope of local authorities being able to identify, partner with and sustain the kinds of social care providers who can challenge the unaccountability, one-size-fits-all provision, profiteering and paternalism of care as it currently exists.

A BETTER DEAL FOR CARERS

There is one more issue that needs close consideration – the role and work of informal or unpaid carers. One of the tragedies of the initial welfare state, as proposed by Beveridge and enacted by the 1940s Attlee government, was that it simply took for granted unpaid labour.

This was often particularly bad news for women. Indeed, Beveridge's recommendations referred explicitly to the duty of women to marry and provide the labour of social reproduction. Subsequently, his recommendations did not include any state intervention where unpaid labour, often by women, could be relied upon.

As we reimagine health and care, we must not make the same mistake. But it's a risk the National Care Service concept currently runs. In thinking exclusively about the formal social care system, it might be accused of having far too little to say about the informal care being provided outside it.

Before the Covid-19 pandemic, the amount of adult care being provided through informal, unpaid means – rather than through the state – was significant. Official estimates in 2016 put it at 7.9 billion hours a year, from just over 2 million carers.[80] It is thought it would take 4 million adult social care workers, working every week of the year at their median hours, to replace those providing informal care.

The consequences of providing large amounts of care, without the proper levels of support, can be severe. Providing unpaid care alongside work can be particularly difficult – with new estimates of the number juggling work and care much higher than once thought, at 4.87 million.[81] 2.6 million people gave up work to care during 2019, and 2 million reduced their working hours to facilitate unpaid care.[82] This underpins a strong association with poverty. Research by the New Policy Institute found that among people providing 20 hours or more of care each week, 37 per cent are living in poverty – compared to one in five people in the whole population.[83] High levels of unpaid care are a significant driver of gender inequality, with estimates suggesting women provide 60 per cent more unpaid care than men overall.

There are several trends that suggest the burden of unpaid care on people's lives will increase significantly in the future. The number of older people receiving informal care is predicted to increase by 60 per cent on 2010 levels, by 2030.[84] As the population ages, the average age of an unpaid carer is also likely to increase. Around 60 per cent of informal adult care is already provided by people older than 60, and that figure is increasing – meaning a situation where many could simultaneously have their own support needs and be providing care to someone else.

Covid-19 is a case study in how unprepared we are for rising demands on unpaid care and informal carers. There is clear evidence that caregivers experienced a significant burden during the pandemic.[85] In many cases, this was down to higher needs of the care recipient, the loss of support services, a lack of opportunities for rest and respite and increased hours of care – all of which could become business as usual in the future, even without another pandemic taking place.[86]

While there are lots of areas where more support can be introduced – and would be both welcome and impactful – the key concern for carers today is often financial support. At present, Carers Allowance is a paltry sum at just over £60 per week. This lack of financial support not only leaves carers vulnerable, but incentivises government to neglect funding improvements in the social care system, because unpaid labour can be cheaply relied upon.

According to the Office for National Statistics, unpaid carers provided social care worth £59 billion a year in 2014, representing an almost 50 per cent rise from just 2005.[87] This figure means, ultimately, that it could cost as much as this to replace unpaid carers with paid carers. On the current trajectory, it may be that this becomes inevitable. However, the better and more proactive policy would include providing unpaid carers with a significant dividend for their contribution.

For example, a care benefit-share scheme, which allocated carers 25 per cent of the economic value they create – just for adult care – would cost around £15 billion a year. That would equate to a payment worth £7,500 for every recipient of care, to support the people providing their care – a figure twice as high as the current Carers Allowance benefit. Importantly, this higher level of payment would better encourage direct state provision of care in the future – whereas free care currently incentivises them to avoid increasing care provision.

There is extensive international evidence showing that a much fairer deal between state and carers is both possible and effective. In Germany, cash payments are offered to carers – either to purchase services or support their carers – with payments ranging from £3,500 to £21,000 per year. Evaluations suggest this is more cost effective and more popular than the care home-led and highly institutionalised model currently offered in England. Importantly, the payment suggested here would be delivered alongside state services – as support for unpaid carers, rather than a personal care budget.

Elsewhere, there are other international examples of support for carers that the UK would do well to adopt. For instance, the UK lags behind on compassionate leave policies, providing no specific entitlement to even unpaid leave. In Poland, carers are entitled to two days paid leave per year worked, with financial entitlement based on 80 per cent of average earnings (essentially, a carers furlough).[88] Adopting this scheme would be straightforward, given the infrastructure developed during the pandemic, and would provide much needed space for individuals to grieve. Furthermore, in supporting people to remain in work, it would come with its own gains. Carers UK estimated in 2013 that supporting carers to say in work, through policies like this, could add as much as £5.3 billion to the UK economy.[89]

Our society needs carers, and the vision should not be to replace them entirely. But what is certain is that the state cannot, sustainably, go into the future taking carers for granted. It must invest in their wellbeing, share the value of their care and pick up its fair share through the social care system.

5
The Sustainability Frontier

The UK was meant to be world leading when it came to major health shocks. In the Global Health Security Index's (GSHI's) 2019 assessment of our capacity to respond to a disease outbreak, we were ranked second – only behind the United States.[1] By Spring 2021, the UK and the United States ranked first and fifth respectively for total Covid-19 deaths – and both were in the top 15 globally, when ranked by deaths per 100,000 people.[2] The countries ranked bottom and second from bottom in the GHSI's 2019 table – Turkey and Albania – had both fared markedly better.[3]

It might seem logical to think that this experience will change how we approach health resilience and sustainability. But, in fact, there is a very real risk that we learn entirely the wrong lessons and repeat our past mistakes. There are pitfalls for both the left and the right, which threaten to undermine our ability to translate the experience of Covid-19 into a UK policy that adds to – rather than detracts from – global health sustainability.

One blind alleyway is an emerging deployment of a 'health security' paradigm, including by the Tory government. In the Spring of 2021, they announced a new body designed to deal with infectious disease threats. Originally, this was to be called the National Institute for Health Protection. At the last minute the name was changed, in favour of 'Health Security Agency'.[4] The deadline-day change indicates an intentional shift in emphasis.

Of course, this emphasis is about more than just the names of new Whitehall departments. In one of his first speeches on the post-Covid future, entitled 'Reinvigorating our System for International Health', former Secretary of State for Health and Social Care Matt Hancock said:

> The pandemic has thrust the G7 health agenda to the centre of global affairs. Health policy is the number one economic policy, security policy. Today I want to set out the UK's G7 agenda for the rest of this

year and also some of the actions we're putting in place immediately to deliver on it. The first area is health security – for everyone.[5]

Whether or not it is couched in the language of equality, 'security' is a loaded concept. Rather than looking to identify, understand and solve the structural reasons for the growing global risk from infectious diseases, it implies a focus on borders and control. It denotes less effort to prevent pandemics and major disease outbreaks elsewhere, and greater efforts to insulate the UK and its borders – so we can avoid the damage when such diseases do emerge. In essence, security is a paradigm that allows rich countries to continue concentrating resource in places where the risk of diseases emerging and doing significant damage are lowest. It is therefore, often, a nativist approach to global health.

And while it's framed as a break from the policies that made the UK and the world so vulnerable to a pandemic, it can in fact be seen as a reiteration or deepening of the pre-pandemic policy status quo. Research by Kate Jones of London's Institute of Zoology has shown a rise in emerging diseases, concentrated on new disease 'hotspots' (areas where conditions are right for disease emergence) in parts of the world with lower latitudes.[6] That is, the risk of infectious disease outbreaks before Covid-19 were concentrated in places where public health capacity and investment is lowest, and reporting is weakest. Jones concludes that 'global resources to counter disease emergence are poorly allocated'.[7] The countries with the best early warning systems and surveillance approaches are not the same as those facing the greatest risks.

That is the status quo that an undue focus on security could further entrench. For instance, one worrying indicator in the aftermath of Covid-19 has been a focus in many advanced economies in the West on developing domestic vaccine manufacturing capacity. They have realised the value to self-interest of the ability to hoard vaccines produced at home – an approach that is hugely damaging at a global scale. At the time of writing, the UK government has been clear its ambition is to attract more manufacturing capability from elsewhere.[8]

Nativism is unlikely to be an optimal approach in a world defined by vulnerability to chronic and infectious diseases. In Covid-19, we have proof that a nationalistic and isolationist security agenda is no match for the level of vulnerability a highly interconnected and inter-reliant global health system faces. Focusing on national security in the face of pandemics when they strike, rather than addressing the fundamental reasons we have become so vulnerable to emerging diseases, is entirely the wrong path.

It's not, however, only the current government and the political right who risk learning the wrong lessons from Covid-19. In the aftermath of the pandemic, the left is also at risk of defining their asks more narrowly; becoming more inward looking, or otherwise not quite grasping the scale of the health challenge posed by this new 'era of pandemics'.

Specifically, the experience of Covid-19 risks strengthening the kinds of approaches to health I have questioned throughout this book: NHS-centrism and a purely defensive position. It would be little surprise, for example, if the poor performance of outsourcing – by Deloitte, Serco, Sitel and others – leads our movement to focus ever more on who provides NHS services. Undeniably, this is important. But as a sole focus, it simply isn't in proportion to the scale of thinking needed in a world that has experienced the shock of Covid.

It would take another full book – and perhaps several – to properly understand the global health order, and how it needs to change. Some of these books already exist, and provide compelling reasons why the World Health Organisation, existing norms, vaccine and treatment supply chains, health treaties and international health infrastructure aren't anything like fit for purpose. I do not intend to spend a few thousand words oversimplifying their points. Instead, and consistent with this book's search for a blueprint for UK public health, I explore what domestic levers exist to embed sustainability and global solidarity in this country's approach.

GLOBAL HEALTH IS DESTABILISING RAPIDLY

Covid-19 was not only predictable but predicted. Elected decision makers might like to put it down to bad luck or interpret it as a shock once in a lifetime occurrence – a 'black swan' event. But this would be to give in to a comfortable delusion. The truth is Covid-19 is a symptom of a fundamental rise in global health vulnerability and the era of pandemics in which we now live.

Instances of prescience are easy to find. In 2014, a team led by Professor Katherine Smith analysed a novel dataset, covering the period 1980 to 2013. This covered 12,102 outbreaks of 215 human infectious diseases. Even with stringent controls for disease surveillance, communications, geography and host availability, they found a significant rise in the total number and diversity of disease outbreaks. That indicates not only an increasing risk of major disease outbreaks, but that the risk is increasing more quickly than humans are developing capabilities in surveillance, warning and prevention.[9]

It is not just in technical academic journal papers that we find predictions of rising pandemic risk, either. They can also be found in the very places policy makers and politicians look most readily for insight and ideas. In early 2019, a white paper was released through the World Economic Forum, documenting the results of a new study on World Health Organisation global outbreak data.[10] It showed how the risk of global disease outbreaks had steadily dropped in the late twentieth century. Alarmingly, it also showed how that progress had been reversed in just the last decade. In the twenty years before 2010, the number of countries experiencing 'significant disease outbreaks' halved – from 36 to 18. Between 2010 and 2018, the number more than doubled, from 18 to 37. It is a trend towards exponential growth in infectious disease events with pandemic potential.[11]

Penning the foreword, Peter Sands – research fellow at Harvard's Global Health Institute and former CEO of Standard Chartered – showed his own foresight:

> Neurobiologically conditioned, as we are, to pay attention to stark contrasts and sudden changes, we often overlook slow moving changes in our environments that may herald disastrous consequences. The evolution of infectious disease risk is one such change. As this report explains, the number and diversity of infectious disease outbreaks are gradually but inexorably increasing, as is their capacity to send shocks through our global economic systems. As we travel, trade and communicate across an increasingly hyperconnected global economy, more and more companies will find themselves exposed to the effects of outbreaks that begin thousands of miles away.[12]

As Sands indicates, we had not got to grips with the risk. But we cannot pretend there were not warning shots before Covid-19 – including SARS, swine flu, avian flu and zika virus. A major health shock was very much on the cards, and that means we must contend with Covid as symptomatic of a chronic trend, rather than an acute crisis.

That means recognising and reacting to the fact that the risk of pandemics in the future are not less likely because of Covid. It would be easy to commit a gambler's fallacy: to believe that because an event has happened in the recent past, it is less likely to happen again soon. But unless both domestic and global policy changes, Covid-19 by itself does nothing to reduce the odds of further major health shocks.

As much as we should worry about the increase in frequency of global disease outbreaks, we should also worry about the prospect of an emerging infectious disease that is more infectious or deadly than Covid-19. The World Health Organisation's list of 'top emerging diseases likely to cause major epidemics' contains some lethal prospects:

Table 5.1 Description of a Sample of EIDs of Significant Concern

Name of Emerging Disease	*Description/Symptoms*	*Comparison to Covid-19*
Crimean Congo Haemorrhagic Fever	A viral disease that can cause fever, muscle pains, headache, vomiting, diarrhoea, bleeding, liver failure and death	Case fatality rate of up to 40 per cent[13]
Ebola Virus Disease	A viral disease that can cause fever, weakness, rash, reduced kidney and liver function and death	Case fatality rate of 25 to 90 per cent[14]
Marburg Virus Disease	From the same family as Ebola, causing severe headache, malaise, high fever and rapid debilitation	Case fatality rate of between 24 and 88 per cent[15]
Lassa Fever	Can cause fever, tiredness, weakness, headaches, bleeding gums, vomiting or low blood pressure	Case fatality rate of 1 per cent, rising to 15 per cent for patients hospitalised with severe cases[16]
Nipah Virus Infection	Can cause fever, headache, drowsiness, disorientation and respiratory issues. May lead to coma within 24–48 hours	Case fatality rate of between 40 and 75 per cent[17]

This is just a small sub-section of infectious diseases of global concern. Notably, highly pathogenic Coronaviruses – a genre containing Covid-19 (SARS-CoV-2) – was also listed as a priority by the World Health Organisation in their 2015 list, indicating that these are not simply distant threats to our health.

Another temper for complacency is the fact that emerging infectious diseases could pose a greater scientific challenge – in terms of producing a vaccine or a cure – than Covid. While violent political apathy delayed the development of effective treatments for HIV/AIDS – itself a case study of an emerging zoonotic disease – the process was further complicated by the biology of the virus. AIDs is a disease that enters the human genome, making a total cure very difficult (though suppression has proven possible).[18] It is a remarkable thing that we ended the first year of the Covid-19 pandemic with viable new treatments and vaccines – albeit with huge challenges around distributing those, but within countries and globally – but there is no guarantee our science will provide an absolute safety net, particularly given how little pre-emptive work there is on emerging infectious diseases.

As well as the lesser-known viruses in *Table 5.1*, influenzas remain a genus of disease with the potential to start a pandemic of significant destructive power. The scale of flu as a threat should not be understated. When it comes to historic disease outbreaks, Spanish Flu (H1N1) is perhaps only less well known than the Black Death and killed somewhere between 50 and 100 million people (c.5 per cent of the global population).[19] In 2009, a strain of H1N1 ('Swine Flu') killed 12,000 people in the United States including over 1,000 children. In the UK, where 392 people died, worst case scenario planning showed potential outcomes where 30 per cent of the UK population were infected and 65,000 people died. Despite this, many struggled to take the risk seriously, with an indicative *Daily Mail* headline from the time reading 'How £300 Million Was Squandered on Swine Flu Jabs that We Didn't Need'.[20] Until now, the right media has been less than favourable on governments who spend money on resilience.

While we are unable to predict the next flu epidemic, we do have a good indication of the biggest risks. Studies note a variety of subtypes of swine and avian-originating flu have high pandemic potential – including some for which the human population currently lacks the antibodies that would provide some degree of resistance.[21]

Elsewhere, the threat is not emerging diseases, but re-emerging illnesses. There are some shocking mechanisms by which we might find ourselves facing diseases otherwise lost to history. Among the most alarming sources of emerging diseases could be the reduction in the Arctic's permafrost. Studies of glacier ice cores have shown active viruses in samples up to 15,000 years old. Some of these viruses are known, whilst some were entirely new to modern science.[22]

It might seem intuitively unlikely that this could lead to outbreaks, but there have already been warnings. A 2016 outbreak of Anthrax in Siberia – which led to the culling of 200,000 reindeer and killed one human – was linked to thawing permafrost.[23] Any individual outbreak linked to permafrost remains unlikely, but it reiterates a worrying point: a huge number of factors are combining to substantially increase the chances, in any given period, of a lethal and disruptive health shock. This is the mechanism of growing global health vulnerability.

THE BOUNDARIES OF GLOBAL PUBLIC HEALTH

I opened this chapter by noting that the UK had come second in the Global Health Security Index. This league table now provides a historical document of the flaws in how we'd defined 'good' when it came to preparedness. Because it wasn't just the UK who scored well in the league tables and then did badly during Covid. A 2020 research paper led by Enoch Abbey, from the Johns Hopkins School of Medicine, found that the Global Health Security Index was not predictive of response or performance in light of the coronavirus outbreak.[24] If this indicates anything, it is most of all how flawed our definitions of 'good' were.

As important as the political failure to recognise the threat pandemics pose has been the failure to develop compelling conceptual frameworks, through which a good approach to global health can be embedded.[25] Accepting that pandemics and disease outbreaks are a chronic risk of the twenty-first century is not enough. We will only come to the right solutions if better able to account for the structural reasons behind rising vulnerability.

Left and grassroots movements have developed compelling concepts and models to better understand and inform action on other existential-level threats. This is particularly true in work around our climate and nature emergency. Kate Raworth has developed the Doughnut model, an alternative to traditional economic figures like supply and demand graphs, with sustainability built in. She builds on earlier work by Rockström et al., which established the 'safe operating space for humanity'.[26] Global public health needs a similar framework – a model that understands the domains of risk that moderate our exposure to emerging and re-emerging diseases, around the world. We need to be clear what the limits of tolerance are for global public health, just as our movements are ever clearer on the limits of tolerance for activities that impact planetary health.

Table 5.2 The Safe Operating Space for Humanity, Global Public Health

Global Public Health Boundaries
Climate change
Changing human/animal interaction
Insensitivity of agriculture
Population growth
Urbanisation
Conflict and displacement
Global poverty and inequality
Global trade
Global travel

The challenge of restoring global health sustainability cannot be understated. Nevertheless, there is much here we still have the power to change. Every global health boundary explored above is amenable to policy – but we must act quickly, and we must act decisively.

a. Climate Change

The link between climate warming and emerging infectious diseases has long been evidenced. Through the late 1990s and into the 2000s, research by Jonathan Patz highlighted a link between changing trends in average temperature and both the incidence and distribution of vector-borne diseases.[27] As he even noted then, his conclusions shouldn't be particularly surprising. In fact, humans have had evidence that climatic conditions impact on epidemic infections for centuries.

While the impact of climate variables is never unrelated to modifying influences – levels of poverty, population density, connectivity, farming – there are some clear and worrying trends. Firstly, small changes in temperature can expand the areas in which a pathogen or vector can thrive – or otherwise push them to unprepared geographies. Second, gradual changes in temperature drive evolutionary responses in pathogens and vectors, changing their characteristics. Or slightly warmer environments can increase the incubation period of pathogens, simply making them last longer.

Models have looked to outline the impact of future climate scenarios on infectious diseases. Increasing temperature and precipitation has been linked to increasing malaria transmission in the African highlands.[28] All reasonable models of climate change have been linked to better conditions for malaria transmission in Zimbabwe.[29] Research has come to similar conclusions when looking at arboviruses in south Florida,[30] the West Nile virus,[31] the consequences of El Niño on water-borne diseases in Peru[32] and the transmission of dengue fever.[33] More recent research has linked greenhouse gas emissions and the resulting change in global temperatures, with infectious disease risks linked to ticks, mosquitos, West Nile virus and an increased number of foodborne illnesses.[34]

Against this, the progression of climate warming – driven by the burning of fossil fuels and greenhouse gasses – shows little sign of abating. Between 1990 and 2019, global CO_2 emissions rose by 62 per cent.[35] In 2018, levels of greenhouse gasses in the atmosphere reached a new peak – at 407.8 parts per million.[36] Worryingly, even the break in economic and human activity enforced by Covid-19 did not stop the trajectory of climate change. Global carbon emissions fell by just 6.4 per cent – far below what many researchers had predicted, and in line with a bounce back by 2022.[37] On current trajectories, warming is likely to exceed 3 degrees Celsius – an outcome described as catastrophic in a signed letter from more than 15,000 scientists from 184 countries, published in 2017.[38]

The UK makes a disproportionate contribution to greenhouse gas emissions – meaning we make a disproportionate contribution to our own global health vulnerability. In 2018, the University of Leeds compared UK emissions to sustainable output. They showed that the UK should have 1.6 tonnes of CO_2 emissions per year, per capita, at most; it should use no more than 0.9 kilograms of phosphorous in agricultural soils per year; and it should use no more than 8.9 kilograms of nitrogen from industrial and international biological fixation. Yet, the current performance of the UK is 12.1 tonnes CO_2; 5.2 kilograms of phosphorus and 72.9 kilograms of nitrogen per capita – representing a 756 per cent; 578 per cent and 819 per cent use of allocation respectively.[39]

The impact in global warming is not felt equally across the world. The UK's land temperature is set to increase by less than in many other countries – but this is no cause for relief. In fact, the countries with the highest vulnerability to emerging and re-emerging infectious diseases are often the same as those that will be hit hardest by the impacts of the climate emergency.[40] This means climate emergency could increase vulnerability most in the places where global shock events are almost most likely to originate, amplifying their impact.

b. Human Animal Interaction and Intensive Agriculture

Within the category of emerging and re-emerging infectious diseases, the biggest threat is posed by zoonotic diseases transmitted to humans by animals. As such, one of the clearest risks is increased human encroachment into animal habitats, and the rise in intensive animal-based agriculture. These two factors constitute a bridge through which pathogens transmissible between humans and animals can emerge, evolve and take hold.[41]

The threat of this happening had been observed in practice many times before Covid-19. In 2001, a paper led by Dr Peter Daszak explored several emerging diseases and explored whether anthropogenic actions impacting animal habitats were increasing the risks posed by zoonotic diseases. His results were worrying. Habitat loss helped disease emergence in several cases. For two diseases in the study, it was the prime driver. In the case of the Nipah virus – a disease with a case-fatality rate of between 40 and 75 per cent – intensive farming brought together fruit bats and pigs, enabling transmission from the former to the latter. In 1998, the virus transmitted from pigs to humans, via the pig-farm workers, resulting in the first ever human fatalities from the virus.[42]

Since the early 2000s, the link between agricultural practices and emerging diseases has only become clearer. In 2019, research led by Professor Jason Rohr, of the University of Notre Dame, concluded that agricultural drivers were associated with more than a quarter of all emerging infectious diseases since 1940 – and 50 per cent of all emerging zoonotic infectious diseases.[43] This finding has now been repeated in systematic reviews.[44]

There are several causal mechanisms behind it. Perhaps most obviously, increases in agriculture often lead to increased interaction between human and animals. The trend of encroaching into previously wild land then brings human and livestock into closer contact with wildlife. As Bryony Jones and others put it in a review conducted for the Department of International Development:

> [. . .]the interaction of humans or livestock with wildlife exposes them to sylvatic disease cycles and the risk of spill over of potential pathogens. Livestock may become intermediate or amplifier hosts, in which pathogens can evolve and spill over into humans, or humans be infected directly from wildlife vectors.[45]

In other words, our food production practices have put in place the conditions for new infectious diseases to emerge, evolve and spread.

Despite this body of evidence, established over decades, intensive agriculture continues to expand. Today, over half the world's habitable land is used for agriculture.[46] Of that, 77 per cent is used for livestock, and 23 per cent for crops – despite the fact the former makes up only 18 per cent of the world's calories (normally, in the diets of the most affluent).[47]

c. Population Growth and Urbanisation

The 1700s and 1800s saw modest annual rises in the global human population, normally of a few tenths of a per cent. At the start of the eighteenth century, the world's population was 600 million and by the end of the nineteenth it was just short of 2 billion. In the following century, population growth exploded. A combination of rising birth rates and declining mortality saw the population hit 2 billion in 1928 and 5 billion in 1987. 1968 marked 'peak population growth' – with the global population increasing 2.1 per cent that year.

Population growth is expected to continue to rise less quickly in the next century – though, by historical records, it will mostly remain rapid. By 2050, the world's population is expected to hit 9.7 billion, and by 2100, it will be almost 11 billion. At that point, it is likely to plateau, or drop.[48] These numbers come with huge uncertainty. The human impact of war, famine, climate change, medical advance, disease, or government policy could yet disrupt these trends. Nonetheless, planning must take account of the fact that, in all likelihood, the number of humans on Earth will only continue to grow throughout most of our lifetimes.

Population growth provides a predictor of emerging infectious diseases. In general studies, population growth as a variable has been linked to zoonotic EID events.[49] In some ways, there are obvious mechanisms behind this – more humans mean more of the interactions (e.g., with animals) that can lead to a disease outbreak. In other cases, the precise ways the world has changed and adapted to a larger population has created specific and avoidable risks.

While we may have fewer resources at our disposal to address population growth, we do have options about how the world changes as it looks to cope with a larger, total population. For example, the spatial consequences of population growth have both helped drive recent global health vulnerability and offered opportunities to address that vulnerability in the future.

One of the key dimensions has been urbanisation. 2007 marked the first time in history that more people in the world lived in urban rather than rural areas, according to the UN's estimates. By 2050, it is predicted that that figure will increase to two-thirds of the world's population.[50] And while urban areas tend to have higher living standards at an aggregate level – high density living has brought about challenges like over-crowding. Globally, almost one in three people living in an urban area live in a slum.[51]

In some ways, twenty-first century urbanisation has elements that increase the vulnerability of urban spaces (and thus the world) – distinct from processes of urbanisation seen in previous centuries. One paper led by the University of Lincoln's Creighton Connolly concludes that 'Contemporary patterns of extended urbanisation fundamentally shift the vulnerability of cities to infectious diseases in ways that differ from those that have historically been associated with urbanisation'.[52] There are few better conditions for infectious diseases to spread than through high-density, highly connected urban spaces. As is shown through examining the association between urbanisation and the impact of infectious diseases since the turn of the millennium.

d. Global Poverty and Inequality

One of the clearest drivers of emerging and re-emerging infectious diseases is global poverty. In many cases, poverty and inequality underpin the emergence of infectious disease by driving many of the trends discussed in this section already. The combination of poverty and urbanisation can lead to squalid living conditions, where waterborne diseases thrive, evolve and are exposed to new forms of pollution.[53] Poverty has a well-established link to deforestation as well – as people expand into forests and other biospheres to sustain a living.

In 2014, a team led by Almudena Marí Saéz explored the conditions behind the severe Ebola outbreak in West Africa.[54] Their evidence implicated bats in the outbreak – specifically, insectivorous free-tailed bats. Contextually, the authors note that the district that the index case for the outbreak – Meliandou – had been known forest region, but had lost approximately 80 per cent of its trees and most of its vegetation in preceding decades. This, in turn, facilitated the closer living and interaction between people and infected species. The first person infected was a two-year-old child playing in a hollow tree.

As well as supporting the conditions for the emergence or re-emergence of infectious diseases, global poverty and inequality accelerate the spread and exacerbate the impact of pathogens. Poverty cannot be disaggregated from levels of sanitation, close proximity of living and lack of clean water. There is compelling evidence, from Covid-19 and prior outbreaks, that urban centres with higher levels of poverty experience the most brutal impacts and largest infection spikes from infectious disease outbreaks.[55]

While genuine global equality, and an end to all forms of poverty, have of course not been realised – there has been progress. Between 1990 and 2015, the number of people living on less than $1.90 a day decreased from 35 per cent to 10 per cent of the world's population.[56] Decreases have also been observed on higher lines of $3.20 and $5.50 a day.[57] Nonetheless, this progress remains very tenuous. Much of the decrease is attributable to China and India, meaning extreme poverty has become concentrated in certain regions, including sub-Saharan Africa.[58] Equally, many recent gains in extreme poverty stand to be undone by the experience of Covid-19 itself – with the World Bank estimating that as many as a further 150 million people will be pushed into extreme poverty – the first rise since 1998.[59]

e. Global Trade and Travel

One of the remarkable traits of the Black Death was its ability to spread. The first introduction to Europe occurred in the Crimea. By mid-1348, the virus had spread through Italy, East Spain, most of France and Greece. By 1349, major outbreaks had hit England and Germany. By 1350, the spread reached Scotland, Northern Ireland and much of Scandinavia.

For a long time, the common knowledge was that fleas and rats spread the Black Death. More recent experimental data challenges this hypothesis. In 2019, Katharine Dean and her colleagues developed a conceptual model to establish whether rats/flea transmission or human transmission were more likely. They concluded with a strong likelihood that it was in fact human to human transmission that spread and sustained the Black Death.[60]

The reason the Black Death thrived was down to its remarkable biology. Several traits of the virus underpinned its success. Some studies have indicated a very long gestation period – with estimates suggesting more than a month between infection and symptoms.[61] Its infectiousness peaks before a person is symptomatic – making it incredibly hard to contain. Combined with a long incubation period, this made quarantine policies incredibly difficult.

Today, it does not take a virus with simply remarkable biology to wreak havoc. Rampant globalisation means a far wider range of diseases have the capacity to survive and spread across borders, where they wouldn't have been able to do so in most of human history. This is down to the significant increase in cross-border connectivity between people and places – driven by a combination of global trade and travel, resulting from wider globalisation.

Cross-border interactions are far more frequent and diverse than ever before. According to the World Tourism Organisation – part of the United Nations – tourism has increased exponentially. The 1950s saw just tens of millions of international tourism arrivals. By 1970, the figure had risen to nearly 200 million. By 1990, it had more than doubled, to over 400 million. And in 2018, the figure reached a new record: 1.4 billion international arrivals.[62]

We need not just look at global travel, either – global trade is also on the rise. Analysis of trade data from 1800 to 2014 shows global trade rising exponentially to 2008 – before levelling out in the years following the financial crash.[63] Compared to 1913, international trade had grown 40 times larger by 2014.[64] This exponential growth represents a meteoric rise in interactions – whether the infrastructure needed to transport goods (haulage, freight, air transport), or the infrastructure needed to govern such a sprawling trade system (international offices, business travel, government departments). That is, it indicates a more connected world, with ever more intensive cross-border interaction.

The right to movement is, correctly, defended by many within the broad church of progressive thought. The implication here is not that people lose the right to flee prosecution, to travel, or to migrate for economic or other reasons. Instead, we should question the power dynamics that allow the benefits of trade and travel to predominantly accrue to wealthier people and countries, and for those activities to take place beyond what is sustainable for the world.

The right response, then, is for a redirection of the rules that govern travel and trade. As campaigner Kierra Box has put it: 'simply put, trade rules exist to make it easier to trade – and increase the amount of all trade globally. Their primary aim is not to end poverty or lower carbon emissions'.[65] The right response is to reshape national and international rules to make global health sustainability a core part of how trade and travel systems work. That might mean fundamental changes to the World Trade Organisation, giving domestic law primacy over international trade, or raising shared production values. It might mean bringing the concept of a Public Net Zero, discussed in Chapter 3, to bear on activities that harm global health – not just domestic health.

f. Conflict and Displacement

Whilst trade and tourism represent indulgences of commercial globalisation, they are not the only ways that human movement can increase the potency and potential impacts of pandemics. This can also come down to population movement resulting from conflict and displacement.

Refugees constitute a major area of growth in the intercontinental movement of people – whether fleeing war, violent conflict, environmental disaster, or political discrimination. Many elements of the current refugee experience can lend themselves to creating and sustaining outbreaks of infectious diseases, including emerging infectious diseases – notably, malnutrition, overcrowding, unhygienic conditions and a lack of basic medical services. All of which are common in formal refugee camps.[66]

In 2007, a paper in the journal *Emerging Infectious Diseases* superimposed outbreaks of recent emerging or re-emerging infectious disease outbreaks onto countries affected by conflict between 1990 and 2006. The overlap was significant, with sites of significant conflict overlapping with major outbreaks of Ebola, H5N1 influenza, monkeypox, plague and malaria.[67] This included cases where otherwise successful eradication of disease broke down – malaria had been virtually eliminated in Tajikistan in the 1960s – but following civil war in 1992/3, and a huge displacement of people to neighbouring countries, malaria parasites returned in 1994. By 1997, cases hit 30,000 per year.[68]

Problems can range. Conflict often undermines surveillance and early warning systems, meaning public health interventions come too late. People are forced to move, whether as refugees or because conflict and emergency demands a redeployment of people. Healthcare settings are often stretched. Supply chains – for both resources and information – are difficult to maintain. And due to medical shortages, inappropriate and outdated medicines can be used – with poor treatment compliance – supporting the development of drug resistance.[69]

In short, environments that lead to large numbers of refugees also create the conditions for infectious diseases to emerge, spread, or otherwise be reintroduced to places where they'd formerly been eradicated – often with huge human cost. It should not be lost on us, of course, that the Black Death is often thought to have been introduced to Europe during a siege (of Caffa) – while the Spanish flu was amplified, significantly, by the movement of people after World War One.[70]

This should accentuate alarm at the projected increase in refugee populations – driven by the consequences of climate change, global inequality and political tension. One estimate from the Institute for Economics and Peace – a thinktank that produces global terrorism and peace indexes – predicted 1.2 billion people from 31 countries were at risk of displacement from ecological threats alone.[71] By way of comparison, the UN's human rights agency had capacity to resettle just over 60,000 people in 2019.[72]

A HEALTHY WORLD FOR FUTURE GENERATIONS

How can we reflect the implications of this understanding of global health vulnerability in UK health policy?

One of the most important interventions in the aftershock of Covid would be a legal guarantee to protect future generations from the impact of pandemics, and other consequences from growing global health insecurity. As it stands, younger and future generations are not just set to be economically worse off than their parents, they face what has been called a 'toxic inheritance'.[73] This includes climate change, species extinction, destabilised oceans and – now, as the pandemic has proven – the growing plight of infectious diseases.

If it is often noted that children today are to be worse off than their parents economically – then we should also note that we are close to the first generation in the modern medical era that will have worse health than its parents. Research already shows stagnating life expectancy in countries like England – and actively declining life expectancy in the poorest parts of our country.[74]

Worse, many of the drivers of the remarkable health gains seen in the twenty-first century are far from guaranteed into the future. The fact we are exceeding the boundaries of global public health is also driving alarming rates of resistance to anti-microbials, meaning we face going into an uncertain future without anti-biotics and a range of other vital medicines.

The global and national consequences of anti-microbial resistance (AMR) are already plain. Today, 123 countries report extensive multi-drug resistant tuberculosis. Up to 2 billion people lack effective access to antimicrobials globally, and 700,000 die at the hands of drug resistant infections. That figure is predicted to rise to 10 million, with a cumulative cost of $100 trillion, if by 2050 is no action is taken.[75] Like the emerging diseases explored in this chapter so far, AMR is sustained and accelerated by instances where we exceed the boundaries of sustainable global health. Its links to poverty, hunger, irresponsible agricultural practice and climate change are well known.[76]

It is not uncommon to see news articles predicting that we'll reach new heights in human longevity. Scientists can almost guarantee a front page by making bold predictions on how long humans will live in the future. As recently as May 2021, a prediction humans would reach 150 years old made a splash in *The Telegraph*.[77] Various predictions by Dr Aubrey de Grey that the first person to live to 1,000 – or even, that the first immortal child – has already been born have been widely covered.[78]

But in reality, our future is far more uncertain than the steady, linear progress we've become used to in the last seventy years. A combination of health shocks, chronic illness, rising socio-economic inequality and resistant microbes could wipe out decades of health progress. Which is to say that the relatively healthy 100 years we've experienced is at genuine risk of being little more than a remarkable blip in history.

It is vital we protect future generations from the damage to health sustainability we are facing today. One of the best ways of doing this would be the introduction of a Future Generations Act, providing formal legal recognition of the right of people to live in a stable public health environment. The intent would be to embed the kind of health sustainability definitions outlined here – which are significantly preferable to the limited definitions of 'security' that ranked the UK second on global health in 2019 – and to shift the UK's culture of policymaking towards long-term thinking and participatory democracy.

Conceptions of a Future Generations Act have often seen it as a mechanism to deliver the kind of civil and political rights embedded in the Human Rights Act – a constitutional protection should the government fail to meet targets around both global planetary health and global public health.[79] To this end, the legal duty should come alongside a range of government and accountability infrastructure – some of which can be adopted from Wales, a country that has pioneered this kind of legislation:

- An Office for Future Generations (OFG): To sit within government, with membership from Cabinet level ministers and accountable to the Prime Minister. The office should have an advisory role across local and national government, and there should be a duty on ministers to align policy to independently published OHG advice. The office should provide a counterbalance to the Treasury, which currently embeds an overly explicit focus on output growth, and policy short-termism.[80]
- Future Generations Assessments: Much as all policies make assessments on their economic and fiscal impact, a new future generations impact assessment should be introduced for all government spending decisions. This should pay reference to global planetary and health boundaries.
- A Future Generations Select Committee: The powerful role of select committees – otherwise, denigrated as stiff and dull meetings – has been seen during the pandemic. Their role in holding the health and social care secretaries to account, or in pulling Cameron over the coals after Greensill, has shown their role in ensuring accountability.[81] A new select committee should be set up to hold the government to account on its future generations obligations.
- A Permanent Health Sustainability Citizens Assembly: Future generations thinking means involving citizens from all generations. There is very little democratic participation on the UK's approach to health, whether local, national, or global. A standing citizens jury, drawn from representative samples of the country and

rotating annually, should work to both produce a steady stream of new ideas of global health sustainability, and to challenge the government on bad policy or inaction.

As with many policies on global health sustainability, these will need to be aligned with action on the highly related climate and nature emergencies. Done properly, they can help to ensure a healthy future for generations to come – and ensure the UK does its duty in extending that beyond its own, limited borders.

REGULATE FINANCE

Focusing on boundaries brings in scope finance. This is central to other agendas that focus on sustainability, such as climate and economic justice, and should also be a key consideration within public health sustainability.

In many cases, the relevant numbers are available from work focused on the climate emergency – but take on further urgency when also put in the context of global health sustainability. For instance, it remains the case, despite the rhetoric of financial companies from major banks and investment fund managers, that there is significant investment in fossil fuel companies. Research by the Rainforest Action Network has shown that the world's biggest banks have provided $3.8tn of financing for fossil fuel companies since the Paris Climate Accords (2016).[82] The evidence in this chapter shows these investments don't just constitute a risk to planetary health, but also to global public health.

The financial system provides scale and sustainability for sectors that drive other variables within my list of global health boundaries – beyond emissions. The role of the UK's high street banks in funding the arms trade is now well established. 2019 research from the organisation Facing Finance implicated some familiar high-street brands in supporting companies who have been linked to weapon exports in unstable/crisis affected countries in the Middle East and North Africa.[83] These included Deutsche Bank, Lloyds, Barclays and Santander – among others.[84]

If these facts are not new, what is novel is the opportunity to link these factors to infectious disease outbreaks. In Covid, we have proof that rich countries cannot 'safely' make money out of international conflict or climate emergency, without consequences to their own populations – the world is more interlinked than they've built into their cost-benefit analyses. Equally, we now have a case study of the consequences borne by an infectious disease outbreak – in terms of human and economic cost, nationally and internationally. The value in linking these factors to health is a strengthening of political capital, the weight of popularity and the cost-benefit analyses that can encourage political action.

There are a few structural factors that enable the financial system to underpin a breakdown in global health sustainability. Most obviously, there is little reason for banks to quantify and act on the health impacts of their investment. If a product they back causes chronic ill health, sustains foreign conflict and displacement, contributes to climate emergency through intensive emissions, or otherwise enters into activities that help concentrate wealth in Western countries and absolute poverty elsewhere, then neither the immediate cost, nor the long-term increased risk of infectious disease outbreak, need to be borne by banks or equity funds.

It is very difficult for this reason to bring these costs to the foreground. The size and power of banks makes them more difficult to regulate, as they have significant power in the face of states. Equally, it aggravates the size of mistakes. As the UK's initial Green New Deal Group argued in their work on defining the specifics of a Green New Deal, big banks make big mistakes.[85] And as the financial crisis showed, when financial institutions become 'too big to fail', they are more liable to take risks – like maintaining an approach to the sustainability of global public and planetary health that could plunge us into regular human, social and economic chaos.

After the pandemic, there is likely to be a pronounced focus on protecting the financial system from the risks posed by infectious diseases and health shocks. But what we really need is protection for global human health from the risks of the more unscrupulous tendencies of finance. That is, we need to bind the financial system to the boundaries of sustainable health, rather than further insulating its relentless search for growth from considerations of wider human flourishing.

In many cases, the specific policy instruments we need in global health policy can be adapted from those developed by the climate sector – particularly, from the excellent and innovative green finance movement. For instance, a combination of the Bank of England and the Treasury could look to use their lending schemes to leverage commercial banks to invest in projects that are sustainable from a global health perspective. Second, the UK's financial regulators could introduce measures that force financial institutions to account for pandemic risks. There will be no better time to do this, while private financial institutions count the cost of an unexpected health shock.

Elsewhere, tools in this book could be used to create health-specific financial targets. In 2021, the landmark report of the IPPR Environmental Justice Commission concluded that 'UK financial institutions should be legally required to set targets, including interim targets, to align their investments with net zero . . .'[86] The PHN-ZERO makes a similar ambition possible, from a sustainability perspective, with financial firms challenged to ensure their portfolio does not make a net-negative impact on the global health boundaries.

MOBILISE PUBLIC RESEARCH AND DEVELOPMENT

The public sector had a starring role in generating a vaccine, a key achievement in the pandemic so far. While Prime Minister Boris Johnson said he thought the success was due to the 'greed of capitalism', the truer picture points to the contributions of public universities, publicly funded researchers, state investment, clinical researchers from the NHS and the National Health Service's ability to then distribute the vaccine across the country quickly.

Despite this, the public sector's contribution to global sustainability remains broadly untapped – particularly when it comes to generating new knowledge and innovation.

It's most undermined by a systematic lack of public investment. In 1986, the UK invested about 2 per cent of GDP in R&D per year – about the same as the average in other OECD nations. However, this year marked the first in a long-term decline. While OECD average investment in R&D activities has remained around 2 per cent per year, on average, UK investment has dropped to just over 1.5 per cent.[87] That means, over the course of 35 years, UK investment is £222 billion lower than if it had kept up with the average in other advanced economies.[88] This is puzzling, given the public popularity, economic boost and social value of science.

The political blocks to public investment in R&D since the 1980s are ideological. There remains an orthodoxy that the market can do things better (because of its supposedly greater efficiency in allocating resources). This suggests the role of the state is to get out of the way – in this case, it is argued that any unnecessary state involvement will 'crowd out' more efficient private sector activities.

Evidence has now discredited both ideas. First, it is very clear that the market needs help in allocating R&D investment to health priorities. For example, we might expect many diseases to receive a level of investment by the UK broadly in line with their burden on UK health.[89] But this is not the case. Cancer, renal, urogenital and skin conditions are receiving research investment in line with, or above, what their overall impact on health would suggest.[90] In other cases, conditions are missing out on significant research spend – chief amongst them the leading causes of both UK mortality (cardiovascular disease) and UK morbidity (neurological conditions and mental ill health). While an estimated £29 per person is invested in cancer research each year, only half as much goes into mental health research, despite it having a very similar overall burden on health. Just £9 per person goes into cardiovascular disease research. That represents a £650 million deficit for cardiovascular disease and a £350 million deficit for mental health conditions ever year.[91]

Second, research has shown that public investment does not simply replace or 'crowd out' private investment. In fact, the opposite is true. A 2016 study led by Jon Sussex showed that a 1 per cent increase in public sector investment in medical research was associated with a 0.81 per cent increase in private sector investment in research.[92] Where public investment goes, private sector investment follows. That only makes the value proposition for public-led R&D stronger – it means when it invests £1 in research, it can expect much more than £1 of research to take place.[93]

If we accept the case for a major programme of public sector investment and stewardship on health R&D, then a focus on pandemic prevention and sustainability is an obvious priority. Research by the Access to Medicines Foundation has demonstrated a severe lack of R&D on the priority pathogens that have the clearest epidemic potential. Having been struck by Covid-19, there is now lots of research on Covid-19 and coronaviruses, proxied by the 63 products in the R&D pipelines (i.e., future focused projects). Ebola and Zika, which have caused well-known outbreaks, have 5 and 4 items in their R&D pipeline. Chikungunya, Marburg and non-polio enteroviruses have 4, 1 and 1 projects in their pipeline respectively. And the ten other diseases on the list have no active R&D pipeline at all.

The reason is clear. The odds of any one of these diseases causing a shock event are very low – meaning there is little market today, and a very uncertain market for tomorrow. The impact they would have, however, and therefore the social need for research into them, is very high. This is a tailor-made area for the government to step in – and to use its historical underinvestment in R&D to rectify one of the biggest challenges we'll face in the future. This creates a distinct opportunity for the public sector to mobilise its existing resources, to help steer what kinds of research get done and to ensure the focus is on areas of greatest social priority. It would constitute a major and beneficial democratisation of health R&D.

6
The Public Health New Deal

The five frontiers of public health examined in the previous five chapters each proffer the potential for seismic shifts, both in improving the total stock of health in the country and in distributing that stock of health more justly.

Most ambitiously, the five frontiers in this book could be put together into a cohesive package through which to achieve radical advancement for our health. Taken together, this book constitutes a full blueprint for a healthy future: a public health new deal. We need arguments that go beyond the individual case for the policies discussed in the book so far. We need bespoke analysis on how such ideas fit together as a whole – not least, because the individual ideas together constitute an intervention an order of magnitude larger than anything discussed so far.

Three metrics strike me as particularly important. First, whether the blueprint is affordable. Second, whether a Public Health New Deal could be implemented without simply leaving us open to privatisation. And finally, whether we can balance the national nature of the public health new deal with the flexibility to adapt to the different material conditions of the local places and communities people actually live in.

* * *

There's no getting away from the fact that a Public Health New Deal is expensive. This is probably the most significant barrier to generating the political will and popular consensus needed to gain momentum and deliver real change.

Where ambitious and necessary ideas have come without a clear costing, they have sometimes been open to attack. At worst, reactive voices have seized the initiative by associating radical ideas with hyperbolic price tags. Often, it's little more than a cynical play to undermine public support. For example, the American Action Forum – a group suspicious of ambitious climate policy – undercut the US movement for a Green New Deal by attaching a costing of $93 trillion to it in 2019.[1]

While their figure has not stood up to scrutiny, it has nonetheless proved damaging.

So, it's important to explore cost and affordability from the outset. This is simultaneously about policy substance and about communication. I'm aware the left is not in total agreement about how things should be funded – or, indeed, whether affordability is even meaningful. But even as we have these debates, we need to be realistic that a pragmatic price tag and justification for the investment is integral to the willingness of the public to support our suggestions for change. As such, the below constitutes a very safe estimate for costings, and whether they can be paid for – with the aim, at this early stage, of ensuring a concrete case for change.

COSTING THE NHS FRONTIER

In many cases, restoring the NHS from fragile safety net to comprehensive service would ensure a more efficient service in the long-term – either by improving the health of the country (and so boosting the economy), or by reducing the total cost of healthcare delivery.

Indeed, research shows large returns on investment are possible from expanding healthcare universality to the frontier of scientific possibility. Beyond the figures I've already cited in Chapter One, evidence indicates there are huge long-term returns on investment just waiting to be secured by extending universality in the NHS. For instance, estimates suggest that adopting the very best technology and AI could be worth £12.5 billion a year for the NHS – just because of the time it frees up for healthcare professionals.[2] Not only is this a significant sum of money, but healthcare professionals 'time to care' is the most valuable resource in health and care today.

None of this means that upfront investment wouldn't be necessary. The costs are often short-term, whilst the returns are mid–long term. Innovation – from new technology and new software, to digital tools – often has an upfront cost itself. But the more significant cost is often that of 'double running' services. The NHS cannot just end an old service before a new one gets up and running, as more often than not transformation relies on a patient period of running them in tandem.

One way to conceptualise the up-front investment we might need is to look at what comparable health systems with the best health outcomes deem necessary themselves. Despite our status as a G7 nation, and despite the popularity of NHS funding increases with the public, the UK spends significantly less on health than other countries (relative to GDP). Covid-19 emergency funding skews later statistics, but 2019 figures show that the UK spent around 10.25 per cent of GDP on health, compared to a G7 average of 11.5 per cent. Meeting the G7 average would imply increasing investment by an estimated £27.5 billion per year.[3]

This would provide the funds needed to deliver the four shifts outlined in Chapter One. The breakdown for this funding against the four components of extending universality could be:

1. **Bespoke Transformation Funding:** In 2020, IPPR research recommended a £10 billion transformation fund for the National Health Service could make a significant contribution to closing the gap in performance both within the country, and between the UK and comparable countries.[4]
2. **A Resilience Guarantee:** Many aspects of resilience will be achieved by other recommendations in the book – for example, public health advances will increase the population's health, likely putting us in better stead should health shocks (infectious or not) strike. Beds, community care and capital funding are three areas where funding is needed to deliver the 'resilience rules' discussed in Chapter 1. Combined, the evidence suggests that the current funding implications of a step-change in these areas would come to £3.4 billion per year.[5]
3. **Major Workforce Investment:** Pay justice for NHS workers could be achieved at a cost of £5.2 billion per year – sufficient for a 15 per cent pay rise. Outside of pay, a major uplift in mental health services for NHS workers could be achieved at an estimated cost of just £100 million.[6] In July 2020, a new scheme offered to cover childcare costs for GPs to boost recruitment. Expanding universal childcare across the NHS would cost an estimated £400 million.
4. **An Institutional Injustice Fund:** The remaining £8.4 billion per year should be allocated to radically driving forward access for people not currently well served by the NHS. Much of this funding should be focused on ensuring health services do not just sit back and wait for people to come in for care, but actively take care into the community. It should also look to significantly strengthen training and progression opportunities for people working in the NHS, who have

been systematically excluded from progression and leadership roles throughout the institution's history.

Table 6.1 Costing the NHS Frontier

Universalising the Best Expansion	*Estimated Cost*
Transformation fund	£10 billion per year
Resilience fund	£3.4 billion per year
Workforce fund	£5.7 billion per year
Inclusion health fund	£8.4 billion per year
Total	£27.5 billion per year

COSTING THE SOCIAL JUSTICE FRONTIER

The five prescriptions associated with a Universal Public Health Service are based on areas where strong evidence on cost and efficacy exist, often from the wider Universal Basic Services movement.

Inevitably, a first element of setting up the service will be a major expansion in headcount among existing public health teams. The Universal Public Health Service as a concept depends on the ability of public health professionals to build meaningful relationships with other local organisations and with communities themselves – within places, rather than as Whitehall body giving top-down diktats. The UK centralises more power than any other comparable country, and the policy prescriptions in this book should not be an excuse to further continue this centralisation.[7]

As such, the service would likely need a fleet of professionals operating in natural communities – not entirely different to a public health equivalent of the general practitioner, albeit designed around a social-prescription rather than a medical-prescription model. The proportion of existing public health budgets that go towards staffing costs is not systematically reported. Nonetheless we know that the NHS spends approximately 45 per cent of its total budget on staff pay (less during the pandemic years), and local councils a little more. It is reasonable to budget at a similar level here.

Costings for the five initial prescriptions for the Universal Public Health Service can then be estimated as follows:

1. A universal entitlement to a nutritious and sustainable diet has been costed at £1.5 billion per year.[8]
2. Social house building schemes have been costed at £1.2 billion in year 1 rising to £8.1 billion seven years later (and then lasting to year 30) by University College London.[9]
3. The cost of the education prescription can be estimated at £8 billion per year, based on extending existing investment in the pupil premium.[10]
4. Universalizing utilities, excluding digital access, has been costed at £5.1 billion per year elsewhere.[11] The additional inclusion of digital rebates has been costed at £300 million.[12]
5. Adapting existing costings for 'skill's wallet' schemes of £1.8 billion per year would give the Universal Public Health Service a budget equivalent to £10,000 per person/lifetime.
6. Given the unpredictable impact of Covid-19, an estimate for low pay and unemployment levels comes with significant uncertainty. Pre-Covid, giving public health professionals the ability to prescribe up to £500 per month to those receiving less than a minimum income standard would cost £11 billion per year (adjusted to account for elements of minimum income standards, like utilities, captured elsewhere).

These are likely to be long-run returns, both in terms of growth to the economy and reduced spending on other public services. For instance, NHS fiscal pressures might be reduced in the long-term, thanks to better population health and less avoidable illness. There are likely to be economic benefits from better health, on metrics like productivity, reduced absenteeism and reduced presenteeism. And there may also be decreases in the number of people relying on welfare payments as a consequence of poverty, too. As these benefits cannot be calculated with sufficient certainty to present here, and are unlikely to be realised in the short-term, the upfront investment is presented here without discount. As such, an estimated opening budget for the Universal Public Health Service is £39.2 billion.

In many cases, some of these costs will return almost immediately to the chancellor. For instance, a major expansion of jobs – in a period where unemployment is predicted to rise – would reduce reliance on unemployment benefits. Equally, public sector jobs are liable to income tax, national insurance and VAT on extra consumption, meaning much money returns to government. Estimates put the exact amount at between 50 and 80 per cent.[13] These discounts have not been applied here.

Table 6.2 Costing the Social Justice Frontier

Public Health Prescriptions	*Estimated Cost*
Universal nutrition prescription	£1.5 billion per year
Universal housing prescription	£1.2 billion in year 1, rising to 8.1 billion per year in year 8
Universal education prescription	£8 billion per year
Universal utilities prescription	£5.4 billion per year
Universal skills prescription	£1.8 billion per year
Universal work prescription	£11 billion per year
Expansion of public health teams	£13 billion per year
Total	c.£42 billion per year (initial)

COSTING THE ECONOMIC FRONTIER

The evidence shows that health taxes do not have a long-term impact on the business values of impacted companies and that a healthier workforce is likely to have its own economic value. Where taxes are enacted – including on public health harming practices like pollution, unhealthy food production, unscrupulous exploitation of mental health by social media, or discrimination of women and ethnic minorities when it comes to pay and progression – the revenues are likely to be substantial.

An exact measure of revenue potential is very hard to estimate (nor is it fixed). Two confounding factors are at play here. Firstly, as the case study of the soft drinks industry levy shows – in-line with other international health taxes and regulation – companies are often more likely to change their practices rather than pay taxes or pass it on to consumers. Most sugary drink producers did remove sugar from their products to avoid the levy – and, in fact, that was the whole idea. Second, as I point out in chapter three, exact targets and rates of tax would need to be established properly – most adequately by a unit of public health specialists sat either within the Treasury, the Cabinet Office or across Whitehall.

Nonetheless, the possibility of significant revenue – even with corporate behaviour change the primary aim – has a precedent. As of 2019, so-called 'green taxes' equalled 2.26 per cent of UK GDP – for a value of approximately £60 billion per year.[14] Even this is not the limit of potential. Green taxes make up a much bigger percentage of GDP and share of taxation in countries like Serbia, the Netherlands, Denmark and Italy. Undeniably, public health taxes have similar potential: the government commissioned National Food Strategy (2021) highlighted that nutrition taxes alone could raise around £3 billion per year.[15]

Elsewhere, state monopolies could also generate significant revenue. In 2017, estimates suggest Sweden's alcohol monopoly generated 29.4 billion SEK – or around £2.46 billion. That is revenue in a country with a population about one sixth the size of the UKs, and one fifth the size of England.[16] Meanwhile, the legal sale of marijuana is generating billions of pounds in tax revenue alone, but even more in private profit – introducing a state monopoly would generate sizable funds. Applying the joint potential of state sale and levies across a whole public health net zero ambition suggests some potential to fund the whole package of investment outlined elsewhere in the book, or more.

Despite this, I do not put forward an arbitrary figure for what could be raised, given the inevitable uncertainty that figure would rely on. We can be confident that a good proportion – or, even, the whole of the proposal recommended here – could be financed through these methods. But we would not want our public health services to become overly reliant, or worse, actively conditional, on revenue raised from negative corporate health behaviours. The case for affordability is stronger if it can stand up even without reference to the revenue raising potential of PHN-ZERO.

COSTING THE SOCIAL CARE FRONTIER

My proposals for social care begin from the existing consensus on the left around a National Care Service, delivered on the same basis as the National Health Service. The costs of Labour's 2019 social care commitments have been costed at £6 billion per year elsewhere.[17]

My evolution of these ideas combines the National Care Service with ideas to help reform what care looks like – and what role it plays. The recommendations which require investment are service double running and a social care transformation fund. These could be linked to two sources: first, calculations of the efficiencies possible by focusing care on community-settings and independence (£2 billion per year),[18] and the savings possible from the fact the new social care model will replace

NHS Continuing Healthcare funding (worth £3 billion per year).[19] This would imply setting the level of initial investment in transformation at approximately £5 billion per year.

This has a symmetry to President Joe Biden's recent stimulus package for community social care infrastructure, as part of his post-Covid $2 trillion stimulus package. His scheme invests $400 million over eight years to improve, expand and make resilient community and home care. The equivalent of this, once adjusted for the size of the UK economy and inflation, would be a major £40 billion investment in social infrastructure over an eight-year period.

On top of this, I estimate the cost of a carer dividend at £15 billion per year (as per Chapter Four).

Table 6.3 Costing the Social Care Frontier

Item	*Estimated Annual Cost*
Free personal care	£5 billion per year
Pay justice	£1 billion per year
Expansion of new care models	£5 billion per year
Unpaid care 'dividend'	£15 billion per year
Total	£26 billion per year

COSTING THE SUSTAINABILITY FRONTIER

The sustainability frontier makes one recommendation with notable expenditure implications. Namely, major public investment in R&D targeted at global public health challenges. Given the UK's cumulative underinvestment since the 1980s – £222 billion in total – a proportionate approach would be to rectify that over the same period, on top of existing plans. This process would make as much as £5 billion a year available for R&D on the structural drivers of global health and planetary health emergencies. This should be ring-fenced and must be over and above existing government spending ambitions.[20]

Brought together, this suggests the investment needed in setting up a Public Health New Deal would equal £100 billion. The hurdle we need to jump is whether this can be communicated as a reasonable or affordable amount of spending.

Table 6.4 Total Investment for a Public Health New Deal

Measure	*Estimated Extra Cost (Annually)*
NHS frontier	£27.5 billion per year
Social justice frontier	£42 billion in year 1, rising to 48.9 billion
Economic frontier	Revenue raising
Care frontier	£26 billion per year
Sustainability frontier	£5 billion per year
The Public Health New Deal	**c. £100–110 billion per year**

There are a few ways to communicate why this large amount of investment is reasonable.

First, we might point out that £100 billion doesn't even match the amount taken out of public finances during austerity. Research from 2019 shows that UK spending, as a percentage of GDP, was 8.1 per cent lower than the European average – which comes to approximately £230 billion.[21] The supposed upside of this policy was meant to be a stronger economy. However, other research, using figures produced by the Office for Budget Responsibility, has shown that the economy was £100bn smaller in 2018 than it would have been without public spending cuts.[22]

Second, we might point out how ambitious other countries have been on investment – compared to the UK – in the aftermath of the immediate Covid-19 shock. In the US, Joe Biden has led that agenda – with his $1.9 trillion stimulus package. The UK equivalent to a US stimulus would come to £190 billion (8.6 per cent of GDP).[23] Even then, the UK would not be the most ambitious country amongst those with similarly advanced economies. As early as December 2020, Japan announced a stimulus worth over 13 per cent of GDP. Converted to the UK, that would translate to a much bigger budget for economic stimulus still – somewhere between £250 and £300 billion. Such a sum could cover a Public Health New Deal, a UK Green New Deal and a significant amount of sustained support for people impacted[24] by the ongoing effects of the Covid-19 pandemic.[25]

Both these indicate the affordability of a Public Health New Deal. In one sense, it can be seen as a way to use public health to truly end austerity. In another, it can be a package through which the UK reacts proportionately to Covid, strengthens the economy, delivers growth, and prepares public services for an uncertain decade ahead. That is the economic and the health case both suggest the time is now ripe for a bold new deal style stimulus framed around, and justified through, public health.

ACHIEVABLE

In throwing our movement behind a radical new approach to health, we need to be confident that a shift away from a defensive, rear-guard defence on privatisation won't simply give a free hand to those who would destroy our NHS.

Having studied several private incursions into health and social care service in detail, there is a clear pattern.[26] The conditions for privatisation are most likely to emerge when the public sector faces false constraints which limit its ability to provide the best. That is, private health and care providers capitalise upon the failure to Universalise the Best.

In Winter and Spring, there was (rightly) an outcry from the left when Operose Health took over 37 more GP practices in London. Operose is a subsidiary of the US company Centene, and at the time the company already ran 21 GP practices in England. This new incursion took its total to 58. This means, at the time of writing, this large, for-profit company controlled the care of 500,000 patients. It is the biggest single primary care provider in the country.

The unified position on the left was the correct one: Operose must be stopped and immediately barred from delivering healthcare services for profit. Just as I have shown in social care, large operators are poorly suited to provide effective, efficient, and ethical health and care services.

But once a private provider has managed to infiltrate the NHS, it is far harder to get rid of them. The fact that we only realised the Operose threat *after* they had taken a significant market share makes our campaigns far more difficult. And while we could put this down to the sneakiness of American corporate interests, or on backdoor dealing within government, this might lack some necessary self-reflection. The signs of trouble – and the conditions that lead to privatisation – were visible long before most on the left had ever heard of Operose.

When privatisation happens, we on the left tend to ask the same three questions:

1. How did a private company manage to infiltrate the NHS?
2. What are the downsides of this privatisation?
3. How can we can reverse it?

We are less good at working out the conditions that lead to privatisation, and the right strategies to prevent it from ever happening – rather than just reacting to it.

Therefore the most important question is the one asked the least often: what conditions allowed Operose Health to successfully buy so many GP practices? While the answer doesn't help us get them out of our healthcare system now, it does teach us vital lessons about how their incursion was allowed to occur. And that is vital in preventing it from happening again.

To answer the question adequately we need to begin with a little primary care history. GPs have always worked on a slightly different basis to the rest of the NHS. Whereas other medical professionals are usually employed on an NHS contract, GPs are often self-employed. Traditionally, practices are run by GP partnerships. They are run as a business: employing their own staff, managing their premises and delivering services for the NHS on an independent contractor basis.

In 1948, the model was seen to have advantages worth keeping. Specifically, it was felt that leaving general practitioners to work as partners in small practices would have two benefits: i) innovation – partners with a financial stake would have skin in the game, and an incentive to adopt best practices; ii) localism – in a very centralised system, it was hoped that GPs would understand their patients and tailor the NHS to meet their needs.

It was also useful from a political perspective. Bevan faced opposition to his NHS Bill from the medical unions and the opposition parties. Maintaining the GP status quo helped ensure the legislation passed.

Seventy years later, our system of small, business-owning GPs has reached breaking point. Even before Covid-19, general practice was facing a crisis of workload. Surveys by the British Medical Association regularly showed the immense pressure facing GPs in particular – with 63 per cent of clinicians indicating the workload in their practices as 'unmanageable' in January 2020.[27] By 2021, Pulse found GPs were working eleven-hour days on average.[28]

This is bad for patients. Such sharp rises in workload, predictably, have consequences for access. In 2020, my research showed that 36 per cent of people in England were waiting more than a week for a GP appointment. And 3 per cent of people were not able to book an appointment at all, the last time they needed one. That won't immediately hit anyone as a massive number, but it's nonetheless the equivalent of 1.7 million people in England going without access to their GP.[29] Given General Practice provides the gateway to the rest of the NHS, that means severely limited access to the whole gamut of healthcare.

Working in a broken system is severely harming GP numbers. Despite government efforts to recruit more people into general practice – including a golden handshake policy, where new GPs were rewarded with an immediate £10,000 – the number of GPs in England per 100,000 people has reached its lowest level since 2003. Compared to 2003, the population has much more chronic, complicated health needs – meaning the same numbers per person translates to significantly less GP capacity.

The simple arithmetic is this: people are retiring from general practitioner roles more quickly than people are entering the profession. Not enough prospective doctors want to become GPs. It has become a profession associated with poor reward, linear careers, overwork and stress. It perfectly epitomises the worst of the worrying workforce trends I worked through in Chapter One.

Those left in the profession are taking proactive steps to manage their own workload and burnout. In 2015, the average GP worked the equivalent of 0.84 FTE (full-time equivalent). In 2019, that had dropped to 0.76 FTE.[30] That drop is the equivalent of losing a whole 3,000 GPs – in just four years. Faced with a major increase in work, GPs are choosing to work part-time to manage their work/life balance. That is entirely understandable.

The decline in GP numbers comes with a hidden sting in the tail. It means that there are less and less people willing to take over a GP practice, when one of the existing 'partners' comes to sell. In the worst cases, this leads to a 'last person standing' problem – whereby a skeleton crew of GP partners who want to stop working simply cannot, because they are financially bound to their practice, and unable to pass it on. They are locked in.

This brings us back to Operose. Having done extensive research on primary care, I've had the chance to talk candidly to many people working in senior Operose roles. I often hear the same remarks. They tell me that General Practitioners are coming to them, sometimes in quite desperate and unhappy circumstances, and pleading with them to take over the practice. They might be doing it so they can retire. They might be doing it because the workload is simply too much – and they want to pass on the practice management to someone else, while continuing to focus on clinical duties. Without Operose, their only way out may very well be financial ruin.

So, a big corporate entity has become the only route of escape for doctors desperate to get out of the sector with their health and finances intact. A sub-optimal model of organising general practice, a workforce crisis and limited patient access to services are the determinants of privatisation. That is, privatisation slips into the gaps where state provision and best practice separate – in the growing areas where Universalise the Best is not secure.

The more proactive strategy for preventing privatisation would have been a simple expansion of Bevan's fundamentals: nationalisation of GPs. In early 2020, just months before the Operose takeover, I proposed this very policy – delivered through a right to NHS (i.e., public sector) employment for all current general practitioners and compulsory NHS employment for all new general practitioners. This would have immediately ensured several improvements:

- Pay: GPs earn less in this country than elsewhere. Average pay should be aligned to international averages.
- Flexibility: GPs currently have poor access to flexible working, with anecdotal evidence indicating many struggles not only to take lunch breaks, but toilet breaks. Flexible working should be key to a new role, with aspirations to use any boost in recruitment to cut the GP working week.
- Variation: GPs today do not want to just rush patient to patient. They want to do research, leadership, strategy and a host of other valuable activities. All GPs should have the right to combine their role with another activity, as part of a portfolio career.

These were proposals developed after extensive consultation with GPs, looking to understand their aspirations, needs and experiences of the system today. It was oriented towards benefitting workers and, in turn, alleviating the difficulties with access emerging from dire recruitment and retention in general practice. That is, it was developed from the realisation that if the state didn't act, it was only a matter of time before the private sector did.

A frontier approach to health, based on universalising the best, would allow us to implement the policies necessary to avoid privatisation. A defensive, rear-guard action forces us to wait until it happens, and only then react.

In fact, our movement can sometimes be its own worst enemy.[31] We could have prevented privatisation by nationalising GPs. This would be a major change to the status quo. And yet, despite the fact it is a move so in-line with the evidence and with Bevan's principles, the idea of changing the model of general practice is fiercely opposed by the key institutions of the health labour movement: royal colleges and trade unions. Both are staunch defenders of the status quo in general practice, as broken as it is.[32] Both put out media statements dismissing the nationalisation policy – six months later, Centene was the biggest GP operator in the country.

We need our own compelling vision for the future. Otherwise, we will simply find ourselves defending the way things are, even as they break. And it is those breaks through which private companies can slip in. The Public Health New Deal is about avoiding this, at scale, across public health as a whole.

PLACE

Clement Attlee's Labour government did not all agree on what format the National Health Service should take. One side of the debate was taken up by Nye Bevan, who was convinced of the merits of a national system – famously declaring that the sound of a dropped bedpan in Tredegar should reverberate around Whitehall. The case for a much more locally led health service was championed by the Leader of the House of Commons, Lord President of the Council and future Deputy Leader of the Opposition: Herbert Morrison.

A former leader of London Country Council, Morrison set out on a quest to convince colleagues that the health service should be run from local councils. His case relied on a few arguments. First, he said that taking cherished and long-standing functions like health away from local authorities would severely diminish their power. Second, he argued that it would install a one-size-fits-all system, making it much harder to meet the different health needs of different places.

Bevan, as we know, won out. He convinced colleagues that a centrally funded National Health Service should equally be controlled by central government. And he won the argument that a *national* service would be more equal and more efficient than a more locally led NHS, which would allow a postcode lottery to emerge.

In thinking about how the Public Health New Deal is implemented, it is important to consider the level at which it will be led. I believe there is a balance to be struck here. In principle, the Public Health New Deal is a national proposition. Like the NHS, it is driven by the idea that a central proposition applicable to the whole country is key to both the popularity of the policy, and its ability to deliver justice. Throughout, the principles driving a new blueprint are that the right to good health, for free, can be expanded within and beyond the NHS.

However, we cannot ignore that our health is defined by the material conditions of the places in which we live. We can see this at different levels of geographical analysis. From a regional perspective, the data shows us that health and care needs cluster in the North and the Midlands. Take, for example, the rate at which people die of ill health before they reach their seventy-fifth birthday – a concept used to indicate 'premature' mortality in health statistics (that is, mortality that should be prioritised as avoidable with the best preventative and medical practices).

Table 6.5 Under 75 Mortality Rate per 100,000 Population, England, By Region[33]

	All Causes	*CVD*	*Cancer*	*Respiratory Disease*	*Stroke*	*Liver Disease*
North East	394.74	82.80	152.60	44.20	16.31	25.5
North West	388.39	86.60	145.60	46.30	15.20	25.7
Yorkshire and the Humber	363.18	82.00	141.20	41.20	14.29	19
East Midlands	334.42	73.50	133.40	34.90	12.62	17.5
West Midlands	354.41	78.40	138.30	36.30	13.28	21.3
East	302.12	63.40	126.00	29.80	11.56	15
London	303.32	70.50	120.10	30.30	13.13	16
South East	292.31	59.00	123.60	28.70	10.72	15.7
South West	301.46	61.90	125.60	28.10	11.19	15.9
England	330.49	71.7	132.3	34.7	12.83	18.5

The shading in Table 6.5 show where a region has a mortality rate before age 75 per 100,000 people above the England average. If each shaded box counted as one point, the North and Midlands would have a score of 28 out 30 – representing the fact that the East Midlands has *just* lower than average mortality from stroke and liver disease. By contrast, the East and South of the country would score just one point out of a possible 24 – with London having above average premature mortality from stroke.

Underneath this regional analysis is a clustering of ill health in smaller places. At local authority level, of the ten places with the highest rates of mortality considered preventable from cardiovascular, respiratory disease, liver disease, stroke and cancer, six areas are in the top ten for all five conditions and a further two for four conditions.[34] These places are overwhelmingly outside the more affluent South East of the country, and places where socio-economic deprivation is highest. It strongly suggests a one-size-fits-all approach to health won't work, if only because it does not address the systemic injustices of the country as we find it today.

The interaction between place and health was made very clear by Covid. Almost exactly two years after Boris Johnson introduced 'levelling up' as a key rhetorical theme of his leadership – during the Conservative leadership election in 2019 – Professor Sir Michael Marmot released new analysis exploring the disproportionate impact Covid-19 had had across the country in 2020. His analysis showed that mortality had been 25 per cent higher in Greater Manchester than in England as a whole. As a result, life expectancy had fallen faster in the North West that year than in the rest of England. And a close relationship between deprivation and Covid mortality meant that a fall in life expectancy had been focused on both the poorest parts of England and the poorest parts of the North West region.

This is all to say that we need to have a much stronger conception of the role of place – in designing, delivering and tailoring this major new public health plan – than there was when the NHS was formed.

The key to a workable place-based approach to the Public Health New Deal is two-fold. First, we need a plan that genuinely redistributes power over health – to people and local leaders, to co-create services in places together. Second, we need a firm appreciation of the material conditions implicit in place and a plan for ensuring that our investment in health provides an anchor and a source of community development for the places that need it most. That is to say, the level of investment implicit in a Public Health New Deal can be a public health intervention itself – above and beyond the specific services it funds – because of its capacity to catalyse regeneration in the parts of the country worst hit by Thatcherism, deindustrialisation and austerity.

We do not need to begin from scratch. There are a range of models that can provide a basis for implementing the Public Health New Deal – nationally – in a way that is cognisant of the different material conditions of place.

For example, the last decade saw Wigan embark on a major process of change, towards a model of 'asset-based working at scale'.[35] The subsequently named 'Wigan Deal' was based on a new relationship between public services and local people. First the council and its partners engaged in intensive, local interviews over a week-long period. They mapped out its services, its voluntary sector, its citizens' capabilities and its leadership. It developed a system where public sector workers had permission to innovate, where a dedicated community investment fund could fund community development and where council service commissioning was more collaborative.[36] It has been remarkably successful, with health and social care improved, financial sustainability maintained despite the severity of austerity's cuts and improved wellbeing outcomes for citizens.[37]

Outside of health, other place-based schemes provide a basis for using the investment implicit in a Public Health New Deal to deliver community regeneration. In Preston, the now famous 'Preston Model' was developed under the leadership of councillor Matthew Brown, who heads the council. The Preston model is an initiative formed on building community wealth through a different type of economics of place. Having found that 56 per cent of spend from 'anchor institutions' – vital public services, with big procurement, delivery and employment footprints – had left the region, Preston acted. It began a large-scale influencing project, pushing their public services to procure from local and socially responsible suppliers. Procurement processes and documents were overhauled, to ensure these suppliers could win contracts.

It is a model of regeneration that keeps money in the local economy, and ensures it reaches local people. City council staff receive at least a living wage and play a role in advocating for its adoption across employers. The amount of procurement spend staying in Preston has more than doubled since 2012/13. The co-operative sector has grown. Major housing and commercial real estate developments are required to provide skills and employment plans for local people.[38]

Preston is the most famous example of a social-value approach to regeneration and a local-retention approach to spending. But it is not the only such example. Salford's mayor Paul Dennett has pioneered a model of 'sensible socialism' that is also reaping dividends. This has included a council-owned development company to drive social housing; a similar focus on anchor institutions and social value standards to Preston; and a regime of borrowing at low interest rates to invest in the local economy, infrastructure and carbon footprint.[39]

These kinds of schemes suggest several lessons in how place can feature in the implementation and value-maximisation of a major investment in public health:

- Investment should be kept within places, particularly those where socio-economic deprivation is the highest
- Regulation associated with the Public Health Net Zero and services associated with the Universal Public Health Service should be based on established principles of community wealth building
- New public health, care and health services should be done *with* people, through sustained engagement – rather than done *to* them, in a paternalistic fashion
- In implementing the deal, there should be a focus on local employment, supporting socially valuable enterprises, developing and involving community assets and building partnerships – things that are good for public health, above and beyond what new services are launched
- The £100 billion should not be controlled exclusively by Whitehall, but used to empower local leaders as much as possible – in line with the well-established principles of devolution in health and care

While the entitlement to good health is national, and while the Public Health New Deal is a national solution to a problem we all face, the value of the deal will only be realised in full if it works with places, in the context of the material conditions of places, and uses its investment to create thriving communities of people above and beyond the specific services it implements.

A HEALTHY FUTURE IS POSSIBLE

When I talk to people about the future of health, there is often a sense of pessimism. They wonder how much longer general taxation can support the NHS in the face of growing health need. They wonder how we'll catch-up with the health outcomes of our international peers. They fret about the trajectory of a social care support system mired by crisis. They predict widening inequality, a second bout of austerity and continued reversals in healthy life expectancy and longevity.

Public perceptions research shows that this pessimism isn't unique to my friends and colleagues. As much as the country love the NHS, they are fearful about the future. 81 per cent of the public think our health system is overstretched. In other countries, the average number who think their own national health system is stretched is just 55 per cent.[40] Polling has also shown that the public think health services are getting worse, and a similar amount think the general standard of care will get worse in the near future.[41] There is little optimism about health and care.

But if I've argued for anything over the last few hundred pages, it is that a healthy future is possible. Good health remains one of the greatest, untapped resources for a just society and a just economy – locally and nationally. It is possible to come up with a positive vision for how we can unleash that potential following the greatest health shock the country has experienced in its modern history.

It was from the destruction of war that the NHS was born. The charge for such a radical policy was led by popular writers and radicals, George Orwell among them. From the ashes of Covid-19, we need the same level of radical change. Once again, we face the stark realities of sub-optimal health and rife health injustice.

This decade sees us at a crossroad. Two futures are possible. One will see the status quo sustained. The NHS will be strained, inequalities will widen, life expectancy will fall back and elderly people will continue to suffer in care homes. The other offers a radical advancement of public health, a reinvention of Bevan's democratic socialism over five frontiers. With it will come a fairer economy, social justice and greater sustainability.

It's no longer enough to maintain the status quo. In this desperately uncertain moment, we must be bold in reaching out and grasping it. Only then is a healthier, fairer future possible.

Epilogue
Labour's Medicine

I have not spoken much about Labour Party strategy in this book. I have not wanted to suggest the Parliamentary Labour Party are the only route to change, nor that they are synonymous with the left and its movements. Nonetheless, as the UK's biggest progressive political party there is no doubt that they're important. It is worth considering how the ideas outlined in this book fit with their agenda and electoral prospects.

Health has long been a foundation of the Labour Party's electoral strategy. Under some very different leaders – from a full array of the coalition that makes up the 'Parliamentary Labour Party' – it has campaigned on its ability to steward the nation's health service and improve the nation's health. In the same way the Conservatives have dined on their reputation in governing 'the economy', Labour have presented themselves as 'the party of the NHS'.

Historically, this approach has served the Labour Party well. Winning on health has allowed it to capitalise on an issue at the very heart of what people care about, and that genuinely impacts on how they have cast their vote.

The market research company Ipsos Mori have created monthly 'issues indexes' now for decades. My analysis of their polling shows that, when the polls were first carried out, the NHS was considered something of a fringe issue. In the first month, it was selected as most important by just 3 per cent of the public. The next month, just 2. The importance of the NHS rose steadily through the next decade. By 1988, it was the most important issue for a massive 64 per cent of voters. In the three decades since, health has been the most consistent priority issue for the public.

Health may well be *the* defining issue in deciding modern elections.

In recent times, Labour have tended to do well at the ballot box when they've been successful at putting the NHS at the top of the agenda. In his own analysis of Ipsos polling data, health and care commentator Richard Sloggett suggests two things have predicted Labour election victories recently. First, the NHS being listed as important by more than 40 per cent of voters. Second, the NHS rising in importance during the election campaign period.[1]

In 2010, the NHS fell in importance for voters during the election campaign (- 2 per cent) and was listed as important by only a quarter of voters. Throughout Gordon Brown's premiership, voter prioritisation of healthcare hit lows not seen since the early 1980s. Labour recorded a historic loss. In 1997 and 2017, where Labour broke 40 per cent of the national vote, the NHS recorded massive increases in salience during the election campaign window: 14 and 13 per cent respectively.

The implication is that the Labour Party have not been able to win without health and, unless something changes quite dramatically, are significantly less likely to win without health in the future. And so, it should worry their strategists deeply that there have been signs that the Labour Party are beginning to lose their advantage on health. Despite an exception in 2017 – where Labour's narrative was far more nuanced than in 2019 – the party's story isn't resonating quite as well or widely as it used to (*Table E.1*).

Table E.1 Labour Lead on Healthcare Ahead of Election[2]

Year	*Labour lead on healthcare among all voters*	*% Mentioning healthcare as important*	*Increase (%) from month before*	*Labour vote share*
2019 (December)	+12	55	+ 1	32.2
2017 (June)	+14	61	+ 13	40.0
2015 (May)	+13	44	+ 3	30.4
2010 (May)	+9	22	+ 2	29.0
2005 (May)	+12	36	+ 8	35.2
2001 (June)	+28	58	+ 17	40.7
1997 (May)	+32	63	+ 24	43.2
1992[3] (April)	+23	41	+ 8	34.4

These figures suggest performance on health in the lead up to elections is a strong predictor of election outcome. The trend seems to be that the manner in which health, as an agenda, supports a Labour election victory is a product of three things: a large Labour lead; lots of public salience; and a decent rise in salience in the general election campaigning period (when people who normally don't obsess about politics are tuning in more intensely).

We might group the election results in Table E.1 into three broad groups.

- Group One (Over 40% vote share): Labour manage to mobilise the health and care as a key issue through the campaign, people list health and care as a key issue, and Labour are rewarded with a big lead on the issue. (1997, 2001, 2017)
- Group Two (30-40% vote share): Healthcare is seen as important and Labour obtain a moderate lead, but do not manage to increase salience of the agenda in the election campaign window (2015, 2005, 1992)
- Group Three (below 30% vote share): Labour don't manage to mobilise health as a key issue, people don't list health as a key issue, and Labour don't have a big lead on the issue (2010)

The low-tens and high single-digit figure leads on healthcare that Labour saw in the 2010s are reminiscent of the 1980s – Thatcher's decade of dominance, and a decade of misery for the Labour Party. Then, Thatcher was very good at constraining Labour's lead on health – and in reducing their lead on health during general election campaigns. She managed to cut it by 18 points from previous peaks, in the period directly preceding the 1983 and 1987 elections.

The political right's strategists evidently sense that Labour's strength on health is waning. There is a real belief at Conservative Party HQ and in Number 10 that they can win on health consistently – and in doing so negate one of their historic problem areas. They have demanded that Conservative MPs and officials get in line with the idea of a publicly funded health service, even if they don't particularly like it. Where the NHS was once considered a toxic topic for Conservatives, it has now become one of their key lines. Prime Minister Boris Johnson has shown a particular predilection for putting health at the heart of his campaigning – including during the referendum on Britain's membership of the EU (£350 million/week) and in the 2019 general election (where the NHS was positioned as *the* thing we could turn to once the country 'got Brexit done').

In tandem, the right has begun to more strongly challenge the narrative that they are a risk to the NHS, namely by highlighting how they've been in charge for most of its existence (now, 43 of its 73 years).[4] There have even been incursions to Labour's 'home turf'. One of the 2019 TV leader debates saw Boris Johnson successfully attack Jeremy Corbyn on the role of Private Finance in the NHS despite Jeremy Corbyn being a staunch critique of PFI and Johnson having been a minister in governments that had actively used the scheme.

The recent years of Conservative Party dominance have been defined by remarkable ideological agility. They have positioned themselves as the party of common sense, of generational equality and the party of the working class. Now they are positioning themselves, with some success, as the party of the NHS and health more widely. And just as presenting Labour as a party of good economic governance helped Blair win elections, so could managing to present themselves as a party of good health governance help the right create further electoral success.

The canary in the coalmine for the success of their approach is the willingness of NHS workers to vote Conservative. According to polling undertaken by Nursing Notes, 42 per cent of healthcare workers voted for the Conservative Party during the local elections in 2021.[5] That figure is despite the run up to the election being dominated by government failures on Covid-19, and a scandal around a proposed post-Covid pay rise for NHS workers of just 1 per cent.

The challenging reality is that the Labour Party need a far more compelling vision and story when it comes to health, and to create new political divides and a vision unique to leftist politics. They need to speak far more about why right-wing politics undermines resilience in the NHS. They need to speak of how the crisis in care stems from political ideology, not just incompetence. Most of all, they need to speak about how the conditions of social injustice, and the exploitations of capital and structure of our economy, make us all sick – and disproportionately kill the poorest and least powerful in our society.

In short, they need to do far better at meeting people where they are and engaging with the lived reality of health. This is how they can reinvigorate the health agenda, and present good health as something only Labour can provide – at a time when both parties are willing to pump large sounding funding numbers into healthcare.

The Labour Party cannot survive if they lose health. It's time for them to adopt a radical new blueprint for the future of public health.

Notes

All articles last accessed 8 September 2021.

PREFACE

1. Clare Wenham, 'What is the future of UK leadership in global health security post Covid-19?', *Progressive Review*, 27/2 (2020), onlinelibrary.wiley.com/doi/full/10.1111/newe.12201.
2. Jacqueline Rose, 'Pointing the finger', *London Review of Books*, 7 May 2020, www.lrb.co.uk/the-paper/v42/n09/jacqueline-rose/pointing-the-finger.
3. See, for example Grace Blakeley, 'Capitalism's covid recovery will deepen social inequality', *Tribune*, 13 April 2021, www.tribunemag.co.uk/2021/04/capitalisms-covid-recovery-will-deepen-social-inequality.

INTRODUCTION

1. The point at which he adopted his more famous penname, George Orwell.
2. A hospital that still exists today, albeit with a much-enhanced reputation.
3. George Orwell, 'How the Poor Die', (1946). Available at: www.orwellfoundation.com/the-orwell-foundation/orwell/essays-and-other-works/how-the-poor-die/.
4. Over 39.4 degrees Celsius, well above the 38 degree threshold for a fever.
5. Orwell, 'How the Poor Die'.
6. Ibid.
7. Notably, 'How the Poor Die' was first published in the same year Bevan brought his NHS bill before parliament (1946).
8. *The Sunday Express*, 'We're all in this together and can beat coronavirus', editorial, 15 March 2020, www.express.co.uk/comment/expresscomment/1255323/coronavirus-outbreak-UK-Government-measures-latest-death-toll.
9. John Harris, 'Coronavirus means we really are, finally, all in this together', *Guardian*, 29 March 2020, www.theguardian.com/commentisfree/2020/mar/29/coronavirus-means-we-really-are-finally-all-in-this-together.
10. *Evening Standard*, 'A rainbow of hope as London stands together', editorial, 25 March 2020, www.standard.co.uk/comment/comment/evening-standard-comment-a-rainbow-of-hope-as-london-stands-together-a4397296.html.
11. See Toyin Owoseje, 'Coronavirus is "the great equalizer," Madonna tells fans from her bathtub', *CNN*, 23 March 2020, www.edition.cnn.com/2020/03/23/entertainment/madonna-coronavirus-video-intl-scli/index.html.

12. Public Health England, *Disparities in the risk and outcomes of COVID-19*, August 2020, https://assets.publishing.service.gov.uk/government/uploads/system/uploads/attachment_data/file/908434/Disparities_in_the_risk_and_outcomes_of_COVID_August_2020_update.pdf.
13. Ibid. p. 39.
14. Author's analysis of Office for National Statistics, *Coronavirus (COVID-19) related deaths by occupation, England and Wales*, dataset, 2021, www.ons.gov.uk/peoplepopulationandcommunity/healthandsocialcare/causesofdeath/datasets/coronaviruscovid19relateddeathsbyoccupationenglandandwales. Figures given for men, with data for deaths among women involving Covid-19 by occupation far more limited. The data that does exist coheres to this trend.
15. Ibid.
16. Hannah E. Davis, Gina S. Assaf, Lisa McCorkell et al., 'Characterising long COVID in an international cohort: 7 months of symptoms and their impact', *EClinicalMedicine*, 38/101019 (August 1, 2021), https://doi.org/10.1016/j.eclinm.2021.101019.
17. Office for National Statistics, *Prevalence of ongoing symptoms following coronavirus (COVID-19) infection in the UK: 1 July 2021*, 2021, www.ons.gov.uk/peoplepopulationandcommunity/healthandsocialcare/conditionsanddiseases/bulletins/prevalenceofongoingsymptomsfollowingcoronaviruscovid19infectionintheuk/1july2021.
18. Ibid.
19. See Mick Tucker and Paul Kenyon, 'Long Covid pay decision for key workers could take a year', *BBC*, 18 May 2021, www.bbc.co.uk/news/uk-57146120.
20. Healthy life expectancy is an official measure of life lived in 'reasonable' health. It takes into account quality of life as well as longevity.
21. Office for National Statistics, *Health state life expectancies by national deprivation deciles, England and Wales: 2017 to 2019*, data release, 22 March 2021, www.ons.gov.uk/peoplepopulationandcommunity/healthandsocialcare/healthinequalities/bulletins/healthstatelifeexpectanciesbyindexofmultipledeprivationimd/2017to2019.
22. Theresa A. Hestert et al., 'Disparities in cancer incidence and mortality by area-level socioeconomic status: a multilevel analysis', *Journal of Epidemiology and Community Health*, 69/2, (2015): 168–176. Doi: 10.1136/jech-2014-204417.
23. Author's calculations. As this book enters production in July 2021, where most adults have been vaccinated, the figure is 1 in every 33 Covid deaths globally. But this has only been enabled by a vaccination inequality that has allowed many rich countries to double-jab their least vulnerable before less wealthy countries have really even begun their own programmes. Data analysis uses official population estimates and John Hopkin's Covid-19 tracker: https://coronavirus.jhu.edu/data/cumulative-cases. (Accessed 28 February 2021 and 12 July 2021).
24. Harry Quilter-Pinner, 'The hidden cost of Covid-19 on the NHS – and how to "build back better"', Institute for Public Policy Research, 16 August 2020, www.ippr.org/blog/the-hidden-cost-of-covid-19-on-the-nhs.

25. Parth Patel, Chris Thomas and Harry Quilter-Pinner, *Without skipping a beat*, Institute for Public Policy Research, 2021, www.ippr.org/research/publications/without-skipping-a-beat.
26. Parth Patel, Chris Thomas and Harry Quilter-Pinner, *The state of health and care: The NHS long term plan after Covid-19*, Institute for Public Policy Research, 2021, www.ippr.org/research/publications/state-of-health-and-care
27. Andrew Hood, Thomas Jemmmett and Victor Yu, 'Mental health surge model', *The Strategy Unit*, November 2020, www.strategyunitwm.nhs.uk/mental-health-surge-model
28. UCL, 'Majority feel they comply with Covid-19 Rules better than others', 4 December 2020, www.ucl.ac.uk/news/2020/dec/majority-feel-they-comply-covid-19-rules-better-others; University College London, 'Lockdown compliance improving but low take up of Covid tests "worrying"', 13 January 2021, www.ucl.ac.uk/news/2021/jan/lockdown-compliance-improving-low-take-covid-tests-worrying.
29. Marina Sitrin (ed.), *Pandemic solidarity: Mutual aid during the Covid-19 crisis*, Pluto Press, 2020.
30. Friedrich Engels and Karl Marx, *The communist manifesto*, translated by Samuel Moore, Penguin Books, 2002 [1848], p. 225.
31. Milton Friedman, *Capitalism and freedom*, University of Chicago Press, p. xiv.
32. Centre for Systems Science, *Covid-19 dashboard*, www.arcgis.com/apps/opsdashboard/index.html#/bda7594740fd40299423467b48e9ecf6.
33. Office for National Statistics, 'Coronavirus and the impact on output in the UK Economy: December 2020', 12 February 2021, www.ons.gov.uk/economy/grossdomesticproductgdp/articles/coronavirusandtheimpactonoutputintheukeconomy/december2020#:~:text=6.,The%20UK%20economy%20during%20the%20coronavirus%20(COVID%2D19)%20pandemic,to%20the%20measurement%20of%20GDP.
34. Gyongyosi Gyozo and EmilVerner, 'Financial crisis, creditor-debtor conflict and political extremism', *Beltrage zur Jahrestagung des Vereains fur Socialpolitik, Digitale Wirtschaft* – Session: International Financial Markets, (April 2021), http://dx.doi.org/10.2139/ssrn.3289741.
35. More recently, the party have changed their name to National Rally.
36. Manuel Funke, Moritz Schularick and Christoph Trebesch, 'Going to extremes: Politics after financial crises, 1870 – 2014', *European Economic Review*, 88/1 (2016): 227–260.
37. Kristian Blickle, 'Pandemics change cities: Municipal spending and voter extremism in Germany, 1918-1933', *Federal Reserve Bank of New York Staff Reports*, 921 (May 2020), www.newyorkfed.org/medialibrary/media/research/staff_reports/sr921.pdf
38. Mark A. Senn, 'English life and law in the time of the Black Death', *Real Property, Probate and Trust Journal*, 38/3 (2003): 507–588; Elizabeth Bruenig, 'If Anti-Vaxxers want a revolution, they just might get one', *New*

Republic, 4 February 2015, www.newrepublic.com/article/120960/disease-epidemics-have-long-history-causing-political-upheaval.

39. R. J. Evans, 'Epidemics and revolutions: Cholera in nineteenth-century Europe', *Past & Present*, 120 (August 2020): 123–146.
40. Reginald E. Zelnik, *Labor and society in Tsarist Russia*, Stanford University Press, 1971.
41. Documented in leaked emails from June.
42. Neena Modi, 'Health systems should be publicly funded and publicly provided', *British Medical Journal*, 10/362 (2018), https://doi.org/10.1136/bmj.k3580.
43. Florien M. Kruse et al., 'Do private hospitals outperform public hospitals regarding efficiency, accessibility and quality of care in the European Union? A literature Review', *The International Journal of Health Planning and Management*, 33/2 (2018): 434–453, onlinelibrary.wiley.com/doi/full/10.1002/hpm.2502.
44. Sonia Adesara and Chris Thomas, 'Environmental collapse means more crises are coming – how can the NHS resist?', *IPPR Progressive Review*, 27/3 (2020), onlinelibrary.wiley.com/doi/abs/10.1111/newe.12208; Neena Modi, 'Health systems should be publicly funded and publicly provided'.
45. Ben Clover and Katherine Hignett, 'NHS braces for £10bn spend on outsourcing work to private hospitals', *The Health Service Journal*, 2020, www.hsj.co.uk/finance-and-efficiency/nhs-braces-for-10bn-spend-on-outsourcing-work-to-private-hospitals/7028255.article.
46. Mark Dayan and Helen Buckingham, 'Will the new Health and Care Bill privatise the NHS?', *Nuffield Trust*, 15 July 2021, www.nuffieldtrust.org.uk/news-item/will-the-new-health-and-care-bill-privatise-the-nhs.
47. It probably was in the 1980s, when there were very lively discussions among the Thatcher government about making denationalisation of the NHS official government policy.
48. House of Commons, *National Health Service Bill: Order for second reading*, Hansard, 30 April 1946, api.parliament.uk/historic-hansard/commons/1946/apr/30/national-health-service-bill.
49. This type of insurance is now being actively offered by major mortgage brokers, some of whom make the argument that those who can afford it should not risk their health relying on a less than comprehensive NHS.
50. LaingBuisson, *Private healthcare: Self-Pay UK market report*, 2020, www.laingbuisson.com/category/private-hospitals/.
51. LaingBuisson, *Private healthcare: Self-Pay UK market report*, 2021, 3rd edn., www.laingbuisson.com/shop/private-healthcare-self-pay-uk-market-report-3ed/.
52. As reported in Denis Campbell, 'NHS Waiting Times "driving people to turn to private treatment"', *Guardian*, 11 September 2017, www.theguardian.com/society/2017/sep/11/nhs-waiting-times-driving-people-to-turn-to-private-treatment.
53. See Jessica Murray, 'Fears of "two-tier" system as NHS waiting lists prompt more people to go private', *Guardian*, 27 October 2020, www.theguardian.

com/society/2020/oct/27/fears-of-two-tier-system-as-nhs-waiting-lists-prompt-more-people-to-go-private.

54. 'Out of pocket' expenditure figures available through the OECD 'Health Spending' dataset, data.oecd.org/healthres/health-spending.htm. 'Out of pocket' is a term that refers to health services purchased directly by private individuals, out of private savings.
55. Ibid. GDP figures available through the OECD 'Gross Domestic Product' dataset, data.oecd.org/gdp/gross-domestic-product-gdp.html.
56. £114 billion in 2018/19. Rachel Harker, *NHS expenditure*, UK Parliament research briefing, 17 January 2020, commonslibrary.parliament.uk/research-briefings/sn00724/#:~:text=In%202018%2F19%2C%20NHS%20England%20population%20and%20needs%2Dbased%20formula.
57. A move seen as so egregious that it prompted Nye Bevan to resign from the government. He had moved from minister of health to minister of labour and national service just months earlier – weakening the party ahead of a 1951 election it would go on to lose to Winston Churchill's Conservatives.
58. In the 1952 National Health Service Act. See, for example, House of Lords, *National Health Service Bill*, House of Lords, Hansard, May 1952, api.parliament.uk/historic-hansard/lords/1952/may/06/national-health-service-bill.
59. See Veronica Dale et al., *NHS dental charges and the effect of increases on access: An exploration*, University of York, the King's Fund/Prepare, 2021, www.york.ac.uk/media/healthsciences/images/research/prepare/NHS%20dental%20charges%20and%20the%20effect%20of%20increases%20on%20access%20-%20an%20exploration.pdf.
60. Stephen Armstrong, *The new poverty*, Verso, 2018.
61. John Curtice, 'Will Covid-19 change attitudes towards the welfare state?', *Progressive Review*, 27/1 (2020): 93–104.
62. Office for National Statistics, *NHS70: Marking 70 years of the National Health Service*, statistical briefing, 2018, www.ons.gov.uk/peoplepopulationandcommunity/healthandsocialcare/healthandwellbeing/articles/nhs70marking70yearsofthenationalhealthservice/2018-06-26.
63. British Heart Foundation. *Trends in coronary heart disease 1961-2011*, February 2011, www.bhf.org.uk/~/media/files/research/heart-statistics/bhf-trends-in-coronary-heart-disease01.pdf.
64. Macmillan Cancer Support, *Living after diagnosis: Median cancer survival times*, 2018, www.macmillan.org.uk/documents/aboutus/newsroom/livingaftercancermediancancersurvivaltimes.pdf.
65. The Richmond Group of Charities, *Just one thing after another*, 2018, richmondgroupofcharities.org.uk/sites/default/files/final_just_one_thing_after_another_report_-_singles.pdf.
66. Mai Stafford, Adam Steventon, Ruth Thorlby et al., *Understanding the health needs of people with multiple health conditions*, The Health Foundation, November 2018, www.health.org.uk/publications/understanding-the-health-care-needs-of-people-with-multiple-health-conditions.
67. Ibid. p. 1.

68. 16 years for women and 19 years for men – often, the whole of the average person's non-working life. Office for National Statistics, *Health State Life Expectancy, all ages, UK*, dataset, January 2021, www.ons.gov.uk/peoplepopulationandcommunity/healthandsocialcare/healthandlifeexpectancies/datasets/healthstatelifeexpectancyallagesuk.
69. Predictably, during the pandemic, it went down fast. A study by Michael Marmot's Institute for Health Equity found that life expectancy fell by about a year in England (though disproportionately in some areas, normally where socio-economic deprivation is higher). Michael Marmot, Jessica Allen et al., *Build back fairer in Greater Manchester: Health equity and dignified lives*, Institute of Health Equity, June 2021, www.instituteofhealthequity.org/resources-reports/build-back-fairer-in-greater-manchester-health-equity-and-dignified-lives/build-back-fairer-in-greater-manchester-main-report.pdf.
70. Michael Marmot et al., *Health equity in England: The Marmot Review 10 years on*, The Health Foundation/Institute of Health Equity, 2020, www.health.org.uk/publications/reports/the-marmot-review-10-years-on.
71. Veena S. Raleigh, 'Trends in life expectancy in EU and other OECD countries: Why are improvements slowing', *OECD Health Working Paper*, 108 (2019), https://doi.org/10.1787/223159ab-en.
72. Anne Case and Angus Deaton, 'Life expectancy in adulthood is falling for those without a BA degree, but as educational gaps have widened, racial gaps have narrowed', *PNAS*, 118/11 (2021), https://doi.org/10.1073/pnas.2024777118.
73. From the website: www.nhsnewdeal.org/.

CHAPTER 1

1. Nick Thomas-Symonds MP replaced Diane Abbott as Shadow Home Secretary when Keir Starmer took over the Labour Party leadership.
2. Nicklaus Thomas-Symonds, *Nye: The political life of Aneurin Bevan*, I. B. Taurus, 2014, p. 2.
3. The Labour Party, *It's time for real change: The Labour Party manifesto 2019*, labour.org.uk/wp-content/uploads/2019/11/Real-Change-Labour-Manifesto-2019.pdf.
4. The Labour Party, '24 Hours to vote Labour and save the NHS', press release, 7 June 2017, labour.org.uk/press/24-hours-to-vote-labour-and-save-the-nhs/.
5. Team GB, 'Togetherness…', tweet, 5 July 2020, www.twitter.com/teamgb/status/1279686703725326338.
6. Steve Richards, 'Don't laugh, politicians deserve pity not ridicule', *The Independent*, 1 August 2012, www.independent.co.uk/voices/commentators/steve-richards/steve-richards-don-t-laugh-politicians-deserve-pity-not-ridicule-7999205.html.
7. For more on this bill see Nicholas Timmins, *Never again? The story of the Health and Social Care Act 2012*, The King's Fund, Institute for Government,

2012, www.kingsfund.org.uk/sites/default/files/field/field_publication_file/never-again-story-health-social-care-nicholas-timmins-jul12.pdf.

8. This was a cynical and dangerous language, and could justify its own section – but such critiques are well established and the focus here is what we can learn about leftist health discourse from the pandemic.
9. The Labour Party, 'Labour brands pay cut for the 225,000 NHS heroes in the Midlands the "ultimate insult"', press release, 18 March 2021, www.labour.org.uk/press/labour-brands-pay-cut-for-over-225000-nhs-heroes-in-the-midlands-the-ultimate-insult/.
10. Liberal Democrats, '1% pay rise is an insult to our NHS heroes says Lib Dems', press release, 1 March 2021, www.cheltlibdems.org.uk/give_nhs_heroes_a_pay_rise.
11. Ben Glaze, 'NHS heroes take to the streets to demand a pay rise from Tory Ministers', *The Mirror*, 11 September 2020, www.mirror.co.uk/news/politics/nhs-heroes-take-streets-demand-22663431.
12. Samantha Batt-Rawden, 'We need to stop calling NHS staff heroes – for a very important reason', *The Independent*, 1 May 2021, www.independent.co.uk/voices/nhs-covid-stress-burnout-heroes-b1840683.html.
13. Available at Wilfred Owen, 'Dulce et Decorum Est', 2021 [1920], *Poetry Foundation*, www.poetryfoundation.org/poems/46560/dulce-et-decorum-est.
14. YouGov, *The NHS vs. European healthcare*, tracker, 2021, www.yougov.co.uk/topics/travel/trackers/the-nhs-vs-european-healthcare.
15. Ibid. The rest either thought it was about the same (27 per cent) or don't know (19 per cent).
16. Eric C. Schneider et al., *Mirror, mirror 2017: International comparison reflects flaws and opportunities for better U.S. health care*, Commonwealth Fund, 2017, interactives.commonwealthfund.org/2017/uly/mirror-mirror/.
17. Our third place ranking on administrative efficiency will cause snide smiles among any healthcare workers reading – I have yet to meet anyone within the NHS who considers its administration efficient.
18. The 2021 follow-up to the rankings saw the UK slide to fourth place. This was put down to the impact of Covid-19, which impacted access to care and care processes.
19. At best, it sometimes sneaks into the middle of the pack.
20. Melina Arnold et al., 'Progress in cancer survival, mortality, and incidence in seven high-income countries 1995-2014 (ICBP SURVMARK-2): A population-based study', *The Lancet*, 20/11 (2019): 1493–1505, www.thelancet.com/journals/lanonc/article/PIIS1470-2045(19)30456-5/fulltext.
21. Irene Papanicolas et al., 'Performance of UK National Health Service compared with other high income countries: observational study', *British Medical Journal*, 367/I6326 (2019), www.bmj.com/content/367/bmj.l6326.
22. Rates of death from conditions that shouldn't prove fatal, with best practice healthcare in place. Christopher Murray et al., 'Healthcare access and quality index based on mortality from causes amenable to personal health care in

195 countries and territories, 1990 – 2015: A novel analysis from the global burden of disease study 2015', *The Lancet*, 390/10091 (July 2017): 231–266.

23. Russell Viney et al., 'Deaths in young people aged 0-24 years in the UK Compared with the EU15+ countries, 1970-2008: Analysis of the WHO Mortality Database', *The Lancet*, 384/9946 (2014): 880–892, www.thelancet.com/journals/lancet/article/PIIS0140-6736(14)60485-2/fulltext.
24. OECD, *Health at a glance: Europe 2016*, 2017, www.oecd.org/health/health-systems/Health-at-a-Glance-EUROPE-2016-Briefing-Note-UNITED%20KINGDOM.pdf.
25. OECD, *United Kingdom: Raising standards*, 2016, read.oecd-ilibrary.org/social-issues-migration-health/oecd-reviews-of-health-care-quality-united-kingdom-2016_9789264239487-en#page29.
26. Annabel Denham, 'Covid-19 has exposed the faults within the broken NHS System', *City A.M.*, 1 October 2020, www.cityam.com/our-broken-nhs-system-needs-replacing/.
27. See Kristian Niemietz, 'Viral myths: Why we risk learning the wrong lessons from the pandemic', *Institute of Economic Affairs*, 9 February 2021, iea.org.uk/wp-content/uploads/2016/12/Niemietz-NHS-Interactive.pdf. For a wider account of the author's arguments see Kristian Niemietz, *Universal healthcare without the NHS*, Institute of Economic Affairs, 2016, iea.org.uk/wp-content/uploads/2016/12/Niemietz-NHS-Interactive.pdf.
28. TaxPayers' Alliance, 'We need to find a new way to pay for healthcare', 22 January 2015, www.taxpayersalliance.com/we_need_to_find_a_new_way_to_pay_for_healthcare.
29. Benedict Spence, 'The NHS is the closest thing we have to a religion – and that's why it must be privatised', *The Independent*, 7 February 2017, www.independent.co.uk/voices/nhs-crisis-jeremy-hunt-health-service-religion-privatise-save-it-a7567056.html; Fraser Coppin, 'The Big Debate: We need to privatise the NHS', *The Badger*, 1 March 2017, thebadgeronline.com/2017/03/big-debate-need-privatise-nhs/; Elizabeth Truss, 'The NHS is wrong to demonise privatisation', *The Telegraph*, 4 June 2008, www.telegraph.co.uk/comment/personal-view/3559102/The-NHS-is-foolish-to-demonise-privatisation.html.
30. Martin Gorsky, '"Searching for the people in charge": Appraising the 1983 Griffiths NHS Management Inquiry', *Medical History*, 57/1 (2013): 87–107.
31. Richard Lewis, 'Thirty years on, the Griffiths report makes interesting reading', *Health Service Journal*, 21 July 2014, www.hsj.co.uk/future-of-nhs-leadership/thirty-years-on-the-griffiths-report-makes-interesting-reading/5072885.article.
32. Paul Joyce, 'Governmentality and risk: Setting priorities in the new NHS', *Sociology of Health and Illness*, 23/5 (2001): 594–614, onlinelibrary.wiley.com/doi/pdf/10.1111/1467-9566.00267.
33. Alternatively, they could reduce the cost of delivering care to those patients.
34. Integration would later be championed again by one of Labour's health advisors, Simon Stevens, upon becoming chief executive of the NHS in England (2014–2021). For some, the Johnson government's championing of

integration has associated it with privatisation, but conceptually it remains a potential collaboration-focused alternative to competition and markets.

35. 10 Downing Street Archives, *Speech by the Prime Minister Tony Blair about the NHS*, speech, 2017, web.archive.org/web/20080609053134/http://www.pm.gov.uk/output/Page1089.asp.
36. Polly Toynbee, 'NHS: The Blair years', *British Medical Journal*, 334/7602 (2001): 1030–1031, www.ncbi.nlm.nih.gov/pmc/articles/PMC1871752/.
37. NHS funding as a per cent of GDP had stagnated for several years under John Major and dropped slightly in the year before Blair was elected. It then rose from around 4.5 per cent to over 7.5 per cent during the Blair and Brown premierships. This is a much faster funding increase than seen in any other period of the NHS's history.
38. Documented in Toynbee, 'NHS: The Blair years'.
39. Robert Francis QC, *Report of the Mid Staffordshire HS Foundation Trust Public Inquiry*, 2013, assets.publishing.service.gov.uk/government/uploads/system/uploads/attachment_data/file/279124/0947.pdf.
40. Qtd in Kiran Stacey, 'Cameron defends "efficiency" drive', *Financial Times*, 8 May 2021, www.ft.com/content/8b5333ce-9939-11e1-948a-00144feabdco.
41. Slightly unfairly, given that his role as an official was to implement the policy – but the design and direction was decided by politicians and ministers.
42. Institute for Fiscal Studies, *UK health spending*, 8 November 2019, https://ifs.org.uk/uploads/R165-UK-health-spending2.pdf.
43. NHS Confederation and NHS Providers, *A reckoning: the continuing cost of Covid-19*, 2 September 2021, https://nhsproviders.org/resource-library/briefings/a-reckoning-the-continuing-cost-of-covid-19
44. A pilot hosted by the World Economic Forum, and on which I was a member of the England team.
45. Chris Thomas, *Resilient health and care*, Institute of Public Policy Research, July 2020, www.ippr.org/files/2020-07/resilient-health-and-care-july20.pdf.
46. Ibid, p. 13.
47. OECD, *Hospital Beds*, dataset, 2020, data.oecd.org/healtheqt/hospital-beds.htm.
48. A. Bagust, 'Dynamics of bed use in accommodating emergency admissions: Stochastic simulation model', *British Medical Journal*, 319/7203 (2013): 155–158, emj.bmj.com/ content/21/5/575.long.
49. Chris Thomas, *Resilient health and care.*
50. Including bad flu seasons.
51. The Organisation for Economic Co-operation and Development – an international organisation with membership from the 'advanced economies'. Amongst other things, it provides data that shows how the UK performs against countries with broadly similar economic power – though, realistically, the UK remains a bigger economy that the vast majority of members. OECD, *Nurses*, dataset, 2020, data.oecd.org/healthres/nurses.htm; OECD, *Doctors*, dataset, 2019, data.oecd.org/healthres/doctors.htm.

52. OECD, *Long-Term care workforce: Caring for the ageing population with dignity*, 2017, www.oecd.org/health/health-systems/long-term-care-workforce.htm.
53. As per 2019, Office for National Statistics, *Employment in the UK: August 2019*, 2019, www.ons.gov.uk/employmentandlabourmarket/peopleinwork/employmentandemployeetypes/bulletins/employmentintheuk/august2019#:~:text=Download%20this%20chart,-Image%20.csv%20.xls&text=Estimates%20for%20April%20to%20June,year%20to%20reach%2024.11%20million.
54. NHS Confederation, *System under strain*, 30 April 2018, www.nhsconfed.org/sites/default/files/media/System-under-strain-report_0.pdf.
55. The Institute for Health Metrics and Evaluation, *Global burden of disease*, 2021, https://vizhub.healthdata.org/gbd-compare/.
56. British Heart Foundation. *Heart Failure: A blueprint for change*, 2021. www.bhf.org.uk/-/media/files/health-intelligence/heart-failure-a-blueprint-for-change.pdf?la=en&rev=f89dedb7c933452e8086cc063ff98c26&hash=98E3BEADD6A46974EF0AAD86044B8DC5314F4E2B.
57. A campaign idea most regularly found in Unite the Union.
58. See NHS Bill Now, *The bill*, www.nhsbillnow.org/the-bill/.
59. Chris Thomas, Harry Quilter-Pinner and Prashant Verma, *The innovation lottery*, Institute for Public Policy Research, 2020, www.ippr.org/files/2020-06/the-innovation-lottery-june20.pdf.
60. Ibid.
61. Taken from David Oliver, 'The pandemic has delivered clinical service innovations worth keeping', *British Medical Journal*, 373 (2020), https://doi.org/10.1136/bmj.n1306.
62. Wellcome Collection. *Medics, migration and the NHS*, 2018, https://wellcomecollection.org/articles/WyjPPScAALyZnoX7.
63. Bellinda Finlayson et al., 'Mind the gap: the extent of the NHS nursing shortage', *British Medical Journal*, 325/7363 (2002), doi: 10.1136/bmj.325.7363.538.
64. James Buchan et al., *Falling short*, The Health Foundation, November 2019, www.health.org.uk/publications/reports/falling-short-the-nhs-workforce-challenge.
65. See Rachel Worsley, 'We know the public love the NHS, but do they think it is a good employer', *The King's Fund*, 18 November 2019, www.kingsfund.org.uk/blog/2019/11/do-public-think-nhs-good-employer.
66. After the announcement and before the implementation of a 3 per cent pay rise for NHS workers covering 2021/2.
67. These figures vary due to the multitude way of calculating the figure. For this method, see Chris Thomas, 'Unjustifiable snub to exclude nurses and carers from pay boost, says IPPR', press release, Institute for Public Policy Research, 2020.
68. ITV, 'Revealed: Nurses forced to use food banks as Covid financial pressures drive many to brink of quitting', 17 December 2020, www.itv.com/

news/2020-12-17/revealed-nurses-forced-to-use-food-banks-as-covid-and-financial-pressures-drive-many-to-brink-of-quitting.

69. Hanna Wheatley, *Still no homes for nurses*, New Economics Foundation, 2019, neweconomics.org/uploads/files/StillNoHomes_May2019.pdf.
70. Parth Patel and Chris Thomas, *Recover, reward, renew: A post-pandemic plan for the healthcare workforce*, Institute of Public Policy Research, March 2021, www.ippr.org/research/publications/recover-reward-renew.
71. NHS Digital, *NHS workforce statistics*, NHS Digital, 2021. Available at: digital.nhs.uk/data-and-information/publications/statistical/nhswork force-statistics.
72. Mind, 'Two in five GPs have a mental health problem', 2018, www.mind.org.uk/news-campaigns/news/two-in-five-gps-have-a-mental-health-problem/.
73. Clare Gerada, 'Doctors and Suicide', *British Journal of General Practice*, 68/669 (2018): 168–169, bjgp.org/content/68/669/168#ref-2.
74. Lisa Elliott, Jonathon Tan and Sarah Norris, *The mental health of doctors: A systematic literature review*, Beyond Blue, 2010, das.bluestaronline.com.au/api/BEYONDBLUE/document?token=BL/0823
75. National Audit Office, *Discharging older patients from hospital*, Department of Health, 2016, www.nao.org.uk/report/discharging-older-patients-from-hospital/.
76. Thomas, *Resilient health and care*.
77. As a country average and over the course of a year. There will naturally be some oscillation – and having capacity to deal with that is the whole point.
78. Sometimes called an 'asset-based approach'.
79. Macmillan et al., *Improving the cancer journey: More than the sum of its parts, second report from a five-year evaluation by Edinburgh Napier University*, 30 September 2020, www.macmillan.org.uk/_images/Glasgow-improving-cancer-journey-programme-full-evaluation-2017_tcm9-324593.pdf.
80. For instance, Rebecca Thomas, 'Spending on Community Services cut by £300m', *Health Service Journal*, 19 July 2018, Hsj.co.uk/finance-and-efficiency/spending-on-community-services-cut-by-300m/7022954.article. See also Chris Thomas, 'Hitting the poorest worst? How public health cuts have been experienced in England's most deprived communities', *Institute of Public Policy Research*, 5 November 2011, IPPR.org/blog/public-health-cuts.
81. Indicatively, the creation of the NHS created an immediate shortage of 54,000 nurses – just through immediate increase in health provision.
82. Roberta Bivins, *The Windrush Generation and the NHS: By the Numbers*, People's History of the NHS, 2019, peopleshistorynhs.org/the-windrush-generation-and-the-nhs-by-the-numbers/.
83. Emma Jones and Stephanie Snow, 'Immigration and the National Health Service: Putting history to the forefront', *History and Policy*, 8 March 20211, www.historyandpolicy.org/policy-papers/papers/immigration-and-the-national-health-service-putting-history-to-the-forefron.

84. Emma Pitchforth, Michael Anderson, Chris Thomas et al., *Sustainability and resilience in the English health system*, London School of Economics, Partnership for Health System Sustainability and Resilience, www.3.weforum.org/docs/WEF_PHSSR_England_Report.pdf.
85. Matt Hancock, *NHS EI CNO Bulletin*, 2021, www.linkedin.com/pulse/secretary-state-blog-international-nurse-recruitment-matt-hancock/.
86. Jonathan Ashworth, *Jonathan Ashworth Labour Party conference speech*, 28 September 2018, keepournhspublic.com/ashworth-conference-speech/.
87. MedAct, *Patients not passports: Challenging healthcare charging in the NHS*, October 2020, www.medact.org/wp-content/uploads/2020/10/Patients-Not-Passports-Challenging-healthcare-charging-in-the-NHS-October-2020-Update.pdf
88. James Kirkup and Robert Winnett,'Theresa May interview: "We're going to give illegal immigrants a really hostile reception"', *The Telegraph*, www.telegraph.co.uk/news/0/theresa-may-interview-going-give-illegal-migrants-really-hostile/.
89. World Health Organisation, *Health workforce*, www.who.int/health-topics/health-workforce#tab=tab_1.
90. My analysis of the NHS England, *Cancer patient experience survey*, 2020, www.england.nhs.uk/statistics/statistical-work-areas/cancer-patient-experience-survey/.
91. NHS England, *NHS workforce race equality standard*, 2019, www.england.nhs.uk/wp-content/uploads/2020/01/wres-2019-data-report.pdf.
92. Ibid, p. 13.
93. Patel and Thomas, *Recovery, reward, renew*.
94. Cabinet Office, *Race disparity audit: Summary findings from the Ethnicity Facts and Figures website*, October 2017, www.ethnicity-facts-figures.service.gov.uk/static/race-disparity-audit-summary-findings.pdf.
95. Ibid.
96. Marie Curie, *Palliative and end of life care for Black, Asian and Minority Ethnic Groups in the UK*, 2013, www.mariecurie.org.uk/globalassets/media/documents/policy/policy-publications/june-2013/palliative-and-end-of-life-care-for-black-asian-and-minority-ethnic-groups-in-the-uk.pdf.

CHAPTER 2

1. D. W. Lewis, 'What was wrong with Tiny Tim', *American Journal of Diseases of Children*, 146/12 (1992): 1403–1407, pubmed.ncbi.nlm.nih.gov/1340779/.
2. R. W. Chesney, 'Environmental factors in Tiny Tim's near-fatal illness' *Archives of Paediatric and Adolescent Medicine*, 166/3 (2012): 271–275, doi:10.1001/archpediatrics.2011.852.
3. At this point in history modern treatments for TB would not have been available, contextualising the author's choice further.
4. Ibid.
5. Quoted in David Olusanga and Melanie Backe-Hansen, *A house through time*, Picador, 2020.

6. Ministry of Housing, Communities and Local Government, *English housing survey: Headline report*, 2020, www.wassets.publishing.service.gov.uk/government/uploads/system/uploads/attachment_data/file/945013/2019-20_EHS_Headline_Report.pdf.
7. Department for Business, Energy and Industrial Strategy, *Annual fuel poverty statistics in England, 2020*, 30 April 2020, www.assets.publishing.service.gov.uk/government/uploads/system/uploads/attachment_data/file/882404/annual-fuel-poverty-statistics-report-2020-2018-data.pdf.
8. Lucie Middlemiss and Ross Gilard,'Fuel poverty from the bottom-up: Characterising household energy vulnerability through the lived experience of fuel poor', *Energy Research and Social Science*, 6/1 (2015): 146–154, www.reader.elsevier.com/reader/sd/pii/S2214629615000213?token=87E3C870E732CDCF8CD91B414B8759F99423085F2A868B24EED205A738FAB30DD5514EBA15225B4C5F67AC16108696A3.
9. Sorcha Daly, *Later life in the UK*, Institute of Health Equity, University College London, 2016, www.institutehealthequity.blogspot.com/.
10. See Emer O'Connell, 'How your body copes with cold weather', *Public Health England*, 21 October 2019, https://publichealthmatters.blog.gov.uk/2019/01/16/how-your-body-copes-with-cold-weather/.
11. Ministry of Housing, Communities and Local Government, English Housing Survey, Headline Report, 2018–2019, www.assets.publishing.service.gov.uk/government/uploads/system/uploads/attachment_data/file/860076/2018-19_EHS_Headline_Report.pdf.
12. Ibid.
13. Gangtani Pnnumam et al., 'Socioeconomic Deprivation and Notification Rates for Tuberculosis in London during 1982-91', *British Medical Journal*, 350 (1995): 963–966, www.ncbi.nlm.nih.gov/pmc/articles/PMC2549356/pdf/bmj00588-0017.pdf.
14. For primary evidence see: Khansa Ahmad et al., 'Association of poor housing conditions with Covid-19 incidence and mortality across us counties', *Plos One*, 2020, https://doi.org/10.1371/journal.pone.0241327 N; J. A. Patel et al.,'Poverty, inequality and Covid-19: The forgotten vulnerable', *Public Health*, 183 (2020): 110–111, journals.plos.org/plosone/article?id=10.1371/journal.pone.0241327.
15. Rebecca Roberts-Hughes, *The case for space*, Royal Institute of British Architects, Homewise, 2011, www.brand-newhomes.co.uk/RIBA-Case-for-space-2011.pdf.
16. National Housing Federation, '1 in 7 people in England directly hit by the housing crisis', 2011, www.housing.org.uk/news-and-blogs/news/1-in-7-people-in-england-directly-hit-by-the-housing-crisis/.
17. I recognise these cannot be distinguished from class, but rather add a different, intersectional dynamic to experiences of health injustice.
18. D. D. Reid et al., 'Cardiorespiratory disease and diabetes among middle-aged male civil servants: a study of screening and intervention', *The Lancet*, 303/7854 (1974): 469–473, www.thelancet.com/journals/lancet/article/PIIS0140-6736(74)92783-4/fulltext.

19. Michael G. Marmot et al. 'Health inequalities among British Civil Servants; The Whitehall II Study', *The Lancet*, 337/8754 (1991): 1387–1393, www.thelancet.com/journals/lancet/article/PII0140-6736(91)93068-K/fulltext.
20. Equality and Human Rights Commission, *Is Britain fairer: Key facts and findings on ethnicity*, 2015, www.equalityhumanrights.com/sites/default/files/is-britain-fairer-findings-factsheet-ethnicity.pdf; Institute for Government, *Why does ethnic diversity in the civil service matter?*, www.instituteforgovernment.org.uk/explainers/ethnicity-civil-service.
21. His half day off at Christmas is very hard won.
22. HM Government, *Good work: The Taylor review of modern working practices*, 2017, assets.publishing.service.gov.uk/government/uploads/system/uploads/attachment_data/file/627671/good-work-taylor-review-modern-working-practices-rg.pdf.
23. Trade Union Congress, *Living on the margins: Black workers and casualisation*, 2015, www.tuc.org.uk/sites/default/files/LivingontheMargins.pdf.
24. M.E. Davies and E. Hoyt, 'A longtitudinal study of piece rate and health: Evidence and implications for workers in the US gig economy', *Public Health*, 180 (2020): 1–9, www.sciencedirect.com/science/article/pii/S0033350619303415
25. Ursula Huws, Neil H. Spencer et al., *Work in the European gig economy: Research results from the UK, Sweden, Germany, Austria, The Netherlands, Switzerland, and Italy*, Foundation for European Progressive Studies (FEPS), 2017, uhra.herts.ac.uk/bitstream/handle/2299/19922/Huws_U._Spencer_N.H._Syrdal_D.S._Holt_K._2017_.pdf.
26. Back pain is often not taken seriously, but when measuring its consequences in terms of 'disability adjusted life years' – a measure that looks to account for quality of life, not just quantity – it has a larger aggregate impact on health in England than lung cancer or ischaemic heart disease. Nicholas Steel et al., 'Changes in health in the countries of the UK and 150 English Local Authority areas 1990–2016: a systematic analysis for the Global Burden of Disease Study 2016', *The Lancet*, 392/10158 (2018): 1647–1661, https://doi.org/10.1016/S0140-6736(18)32207-4.
27. Ibid.
28. Ibid.
29. Department for Business, Energy, and Industrial Strategy, *The characteristics of those in the gig economy*, 2018, www.gov.uk/government/publications/gig-economy-research.
30. Elena Cottini, 'Health at Work and Low Pay: A European Perspective', *The Manchester School*, 80/1 (2011): 75–98, onlinelibrary.wiley.com/doi/abs/10.1111/j.1467-9957.2011.02250; Elena Cottini, and Claudio Lucifora, *Inequalities at work: Job quality, health and low pay in European workplaces*, GINI Discussion Paper 86, August 2013, http://archive.uva-aias.net/uploaded_files/publications/86-4-3-4.pdf.

31. Tracy Shildrick et al., *The low-pay, no-pay cycle*, Joseph Rowntree Foundation, November 2010, www.basw.co.uk/system/files/resources/basw_123014-5_0.pdf.
32. And the striver/shirker binary it relied on. Indicatively, see Patrick Wintour, 'Welfare reforms: We will make work pay, says George Osborne', *Guardian*, 2 April 2013, www.theguardian.com/politics/2013/apr/02/george-osborne-work-welfare-tax.
33. Andrew Smith, Emma Wadsworth and Christina Shaw, *Ethnicity, work characteristics, stress and health*, Cardiff University and Queen Mary, University of London, Health & Safety Executive, 2005, www.hse.gov.uk/research/rrpdf/rr308.pdf.
34. HM Government, *Race in the workplace: The McGregor-Smith Review*, Department for Business, Energy and Industrial Strategy, February 2017, assets.publishing.service.gov.uk/government/uploads/system/uploads/attachment_data/file/594336/race-in-workplace-mcgregor-smith-review.pdf.
35. And was roundly rejected by the Thatcher government.
36. Michael Marmot et al., *Fair society healthy lives (the Marmot review)*, February 2012, www.instituteofhealthequity.org/resources-reports/fair-society-healthy-lives-the-marmot-review.
37. HM Government, *Ministry of Health Act 1919*, 1919, www.legislation.gov.uk/ukpga/Geo5/9-10/21/enacted.
38. Department of Health and Social Care, *Prevention is better than cure: Our vision to help you live well for longer*, white paper, 2018, www.gov.uk/government/publications/prevention-is-better-than-cure-our-vision-to-help-you-live-well-for-longer.
39. Boris Johnson, 'Face-it, it's all your own fat fault', *The Telegraph*, 27 May 2004, www.telegraph.co.uk/politics/0/face-fat-fault/.
40. Patrick Wintour and Colin Blackstock, 'Let the poor smoke, says health secretary', *Guardian*, 9 June 2004, www.theguardian.com/uk/2004/jun/09/smoking.politics.
41. Ben Bradley MP, 'Tweet 6 September 2020, 8.05pm', *Twitter*, twitter.com/bbradley_mans/status/1302684380327030790.
42. Dolly R.Z. Theis and Martin White, 'Is Obesity Policy in England Fit for Purpose? Analysis of Government Strategies and Policies, 1992-2020', *The Milbank Quarterly*, 99/1 (2021): 126–170, https://doi.org/10.1111/1468-0009.12498.
43. All author's analysis of Public Health England, *Local authority health profiles*, data, 2021, fingertips.phe.org.uk/profile/health-profiles.
44. Joint Strategic Needs Assessment Blackpool, *Housing and homelessness*, 2020, www.blackpooljsna.org.uk/People-and-Places/Wider-determinants-of-health/Housing.aspx#HousingQualityinBlackpool.
45. Public Health England, *Local authority health profiles: Blackpool*, 2021, fingertips.phe.org.uk/profile/health-profiles/data#page/13/gid/1938132696/pat/6/par/E12000002/ati/202/are/E06000009/cid/4.
46. escholarship.org/content/qt4sz4s617/qt4sz4s617.pdf.

47. Ibid.
48. Arguably NHS provision also stops before the end of the pathway. The hospice movement delivers a great deal of end of life and palliative care, but it is funded by the voluntary sector, and does not have nearly enough beds to provide good quality care for all. That means the NHS covers three of the seven steps in my model pathway.
49. The King's Fund, *Making the case for public health interventions*, 2014, www.kingsfund.org.uk/audio-video/public-health-spending-roi.
50. S. Martin et al., *Is an ounce of prevention worth a pound of cure?*, University of York, 2019, www.york.ac.uk/media/che/documents/papers/researchpapers/CHERP166_Impact_Public_Health_Mortality_Morbidity.pdf.
51. Thomas, 'Hitting the poorest worst?'.
52. Chris Thomas, Anna Round and Sarah Longlands, *Levelling-up health for Prosperity*, Institute for Public Policy Research, 2020, www.ippr.org/research/publications/levelling-up-health-for-prosperity
53. Chris Thomas, *Progress at last*, Institute for Public Policy Research, forthcoming.
54. Outside of certain exceptional circumstances.
55. Michael Marmot et al., *Fair Society, Healthy Lives*, 2010, www.parliament.uk/globalassets/documents/fair-society-healthy-lives-full-report.pdf.
56. Jo Bibby, *How do our education and skills influence our health?*, The Health Foundation, 2017, www.health.org.uk/infographics/how-do-our-education-and-skills-influence-our-health.
57. Department for Education, *Permanent and fixed-period exclusions in England*, data, 2020, explore-education-statistics.service.gov.uk/find-statistics/permanent-and-fixed-period-exclusions-in-england.
58. Ibid.
59. T. J. Ford et al., 'The relationship between exclusion from school and mental health: A secondary analysis of the British Child and Adolescent Mental Health Surveys, 2004-2007', University of Exeter, 2017, ore.exeter.ac.uk/repository/bitstream/handle/10871/28337/Psychological%20medicine%20revision%2023%20June%202017.pdf?sequence=1.
60. Zahra Bei et al., 'How do we progress racial justice in education', *Progressive Review*, 28/1 (2021), https://doi.org/10.1111/newe.12242.
61. Education Policy Institute, *Unexplained Pupil Exits from Schools: A Growing Problem?*, 2019, epi.org.uk/publications-and-research/unexplained-pupil-exits/.
62. Sally Weale, 'Attainment gap between poor pupils and their peers in England is widening', *Guardian*, 26 August 2020, www.theguardian.com/education/2020/aug/26/attainment-gap-between-poor-pupils-and-their-peers-widening.
63. Hilary Stewart et al., 'The cost of school holidays for children from low income families', *Childhood*, 25/4 (2018): 516–529, https://doi.org/10.1177/0907568218779130.

64. As also concluded by the Sutton Trust. See Rebecca Montacute, *Social Mobility and Covid-19*, Sutton Trust, 2020, www.suttontrust.com/wp-content/uploads/2020/04/COVID-19-and-Social-Mobility-1.pdf.
65. Rates available at Education and Skills Funding Agency, *Pupil Premium: Conditions of grant 2021 to 2022 for local authorities*, 2021, www.gov.uk/government/publications/pupil-premium-allocations-and-conditions-of-grant-2021-to-2022/pupil-premium-conditions-of-grant-2021-to-2022-for-local-authorities.
66. Luke Sibieta, *2020 Annual report on education spending in England*, Institute for Fiscal Studies, 18 September 2020, www.ifs.org.uk/publications/15025.
67. Independent Schools Council, *ISC annual census*, 7 May 2021, www.isc.co.uk/research/annual-census/.
68. Nicholas W. Papageorge and Kevin Thom, 'Genes education and labor market outcomes: Evidence from the Health and Retirement Study', *National Bureau of Economic Research*, 18/3 (2020): 1351–1399, https://doi.org/10.1093/jeea/jvz072.
69. That is not to say that private schools should continue – while outside the remit of the public health service, repurposing these assets for public use would be worthwhile from a health perspective.
70. Of course, it would not just be the children receiving the extra investment that benefit – there would be a knock-on positive effect for everyone.
71. Anna Ambrose and Harry Quilter-Pinner, *The new normal: The future of education after Covid-19*, Institute of Public Policy Research, October 2020, www.ippr.org/research/publications/the-new-normal.
72. Eva Orberle, 'Screen time and extracurricular activities as risk and protective factors for mental health in adolescence: a population-level study', *Journal of Preventative Medicine*, 141 (December 2020), https://doi.org/10.1016/j.ypmed.2020.106291.
73. Nicola Wright et al., 'Interplay between long-term vulnerability and new risk: Young adolescent and maternal mental health immediately before and during the COVID-19 pandemic', *JCPP Advances*, 1/1 (2021), https://doi.org/10.1177/0907568218779130.
74. Feeding Britain, *Ending hunger in the holidays*, 2017, feedingbritain.org/wp-content/uploads/2019/01/Ending_Hunger_in_the_Holiday_Report_Dec_2017-3.pdf.
75. Public Health England, *Obesity profile*, NHS Digital, 2021, digital.nhs.uk/services/national-child-measurement-programme/.
76. Corinna Hawkes et al., 'Smart food policies for obesity prevention', *The Lancet*, 345 (2015), http://repositorio.uchile.cl/bitstream/handle/2250/133565/Smart-food-policies-forobesity-prevention.pdf?sequence=1.
77. Dean Hochlaf and Chris Thomas, *The whole society approach*, Institute of Public Policy Research, 27 August 2020, www.ippr.org/research/publications/the-whole-society-approach.
78. Susan Lloyd, *Rose vouchers for fruit and veg – an evaluation report*, Lambeth Project Final Evaluation, 2017, www.alexandrarose.org.uk/Handlers/Download.ashx?IDMF=d22cf114-923d-4dec-84a2-22fbed94f7f7.

79. Paul Coleman et al., *Building a food system that works for everyone*, Institute of Public Policy Research, April 2021, www.ippr.org/research/publications/building-a-food-system-that-works-for-everyone.
80. Luke Murphy et al., *Fairness and opportunity*, Institute of Public Policy Research, July 2021, www.ippr.org/research/publications/fairness-and-opportunity.
81. Stuart Adam et al., *Social housing in England: A survey*, Institute for Fiscal Studies, November 2015, www.ifs.org.uk/uploads/publications/bns/BN178.pdf#page=9.
82. Trust for London, *London's poverty profile*, 2020, www.trustforlondon.org.uk/data/rent-affordability-borough/#:~:text=In%20every%20London%20borough%20the,gross%2Dmedian%20pay%20in%20London.
83. As per Clare et al., *No longer managing*, Institute of Public Policy Research, 2021, www.ippr.org/research/publications/no-longer-managing-the-rise-of-working-poverty-and-fixing-britain-s-broken-social-settlement.
84. Social Prosperity Network, *Social prosperity for the future: A proposal for universal basic services*, UCL Institute for Global Prosperity, 2017, www.ucl.ac.uk/bartlett/igp/sites/bartlett/files/universal_basic_services_-_the_institute_for_global_prosperity_.pdf.
85. Crisis, *Homelessness monitor*, 2019, www.crisis.org.uk/ending-homelessness/about-homelessness/#:~:text=There%20is%20no%20national%20figure,are%20homeless%20across%20the%20UK.&text=For%20the%20last%20five%20years,at%20the%20end%20of%202019.
86. Many of these with overlap with the kind of private rented sector homes that come with serious health hazards. HM Government, *Overcrowded households*, 2020, www.ethnicity-facts-figures.service.gov.uk/housing/housing-conditions/overcrowded-households/latest. This varies significantly by ethnicity, with 24 per cent of Bangladeshi households living in over-crowded homes, compared to 2 per cent of white British households.
87. Centre for Public Impact, *Eradicating homelessness in Finland: The housing first programme*, 2019, www.centreforpublicimpact.org/case-study/eradicating-homelessness-finland-housing-first-programme.
88. Ibid.
89. Lynne Chester and Alan Morris, 'A New Form of energy poverty is the hallmark of liberalised electricity sectors', *Australian Journal of Social Issues*, 46/4 (2016), https://doi.org/10.1002/j.1839-4655.2011.tb00228.x.
90. Office for National Statistics, *Exploring the UK Digital Divide*, 2019, statistical briefing, www.ons.gov.uk/peoplepopulationandcommunity/householdcharacteristics/homeinternetandsocialmediausage/articles/exploringtheuksdigitaldivide/2019-03-04.
91. Centre for Economic and Business Research, *The economic impact of basic digital skills and inclusion in the UK*, 2015, www.goodthingsfoundation.org/insights/economic-impact-basic-digital-skills/.
92. Joseph Rowntree Foundation, *How does money influence health?*, 2015, www.jrf.org.uk/report/how-does-money-influence-health.

93. Siobhan Palmer 'Homelessness a "major issue" for employers, new figures suggest', 2019, www.peoplemanagement.co.uk/news/articles/homelessness-a-major-issue-for-employers#gref
94. Joseph Rowntree Foundation, *UK Poverty 2019/20*, 2019, www.jrf.org.uk/file/54566/download?token=nBjYDICV&filetype=full-report.
95. Trussell Trust, 'Do working people need food banks', FAQ, 6 November 2019, www.trusselltrust.org/2019/11/06/working-people-at-food-banks/.
96. The Liberal Democrats, *But what is a skills wallet?*, 2019, www.libdems.org.uk/skills-wallet.
97. Kela, 'Results of Finland's basic income experiment: small employment effects, better perceived economic security and mental wellbeing', press release, 2020, www.kela.fi/web/en/news-archive/-/asset_publisher/lNo8GY2nIrZo/content/results-of-the-basic-income-experiment-small-employment-effects-better-perceived-economic-security-and-mental-wellbeing.
98. Kela, *Basic income experiment*, 2020, www.kela.fi/web/en/basic-income-experiment.
99. A. C. Logan et al., 'Golden Age of Medicine 2.0: Lifestyle Medicine and Planetary Health Prioritised' *Journal of Lifestyle Medicine*, 9/2 (2019): 75–91, www.ncbi.nlm.nih.gov/pmc/articles/PMC6894443/.

CHAPTER 3

1. It later emerged this was only true because they'd launched a massive campaign of free samples for doctors, just before the survey period.
2. Tracy A. Ruegg, 'Historical perspectives of the causation of lung cancer', *Global Quality of Nursing Research*, 2, www.ncbi.nlm.nih.gov/pmc/articles/PMC5342645/.
3. Robert Proctor, 'The History of the discovery of the cigarette-lung cancer link: evidentiary traditions, corporate denial, global toll', *Tobacco Control*, 21/2(2012), tobaccocontrol.bmj.com/content/21/2/87; Centers for Disease Control and Prevention, 'Achievements in public health, 1900-1999: Tobacco use – United States, 1900-1999', *Morbidity and Mortality Weekly Report*, 48/43 (1999), www.cdc.gov/mmwr/preview/mmwrhtml/mm4843a2.htm.
4. An ingenious cohort to choose because they are registered and can be easily followed up with.
5. Richard Doll et al., 'Mortality in relation to smoking: 50 years' observations on male British doctors', *British Medical Journal*, 328/7455 (2014); 1519, www.ncbi.nlm.nih.gov/pmc/articles/PMC437139/#ref11.
6. Action on Smoking and Health, *Key Dates in Tobacco Regulation 1962 – 2020*, 16 April 2020, ash.org.uk/wp-content/uploads/2020/04/Key-Dates.pdf.
7. British American Tobacco, *The global market*, 2020, www.bat.com/group/sites/UK__9D9KCY.nsf/vwPagesWebLive/DO9DCKFM#:~:text=The%20most%20recent%20estimates%20for,19%25%20of%20the%20world's%20

population; The Tobacco Atlas, *Consumption*, 2016, tobaccoatlas.org/topic/consumption/#:~:text=About%205.7%20trillion%20(5%2C700%2C000%2C000%2C000)%20cigarettes%20were%20smoked%20worldwide%20in%202016.

8. Tobacco Manufacturers Association, *Fast facts*, http://the-tma.org.uk/fast-facts/; Tobacco Manufacturers Association, *Tobacco industry in the UK*, http://resource.download.wjec.co.uk.s3.amazonaws.com/vtc/2014-15/15_07_Economics/pdf/cbac/uned7/25.%20Tobacco%20Industry%20Data.pdf.
9. Nicholas Lyett, 'Imperial Brands – Dividend Increased', Hargreaves Lansdown, financial results, 18 May 2021, www.hl.co.uk/shares/share-research/202105/imperial-brands-dividend-increase.
10. BAT, *Building the enterprise of the future: Annual report and form 20-f 2020*, 2020, www.bat.com/annualreport.
11. Philip Morris International, *Together. forward: 2020 annual report*, 2021, www.pmi.com/resources/docs/default-source/investor_relation/pmi_2020_annualreport.pdf?sfvrsn=402b8eb4_2.
12. Office for National Statistics, *Adult smoking habits in the UK: 2019*, statistical bulletin, 2020, www.ons.gov.uk/peoplepopulationandcommunity/healthandsocialcare/healthandlifeexpectancies/bulletins/adultsmokinghabitsingreatbritain/2019#:~:text=In%202019%2C%20the%20proportion%20of,falling%20smoking%20prevalence%20since%202011; Public Health England, *Smoking and tobacco: Applying all our health*, 16 June 2020, www.gov.uk/government/publications/smoking-and-tobacco-applying-all-our-health/smoking-and-tobacco-applying-all-our-health.
13. Office for National Statistics, *Coronavirus (COVID-19): 2020 in charts*, 2020, www.ons.gov.uk/peoplepopulationandcommunity/healthandsocialcare/conditionsanddiseases/articles/coronaviruscovid192020incharts/2020-12-18.
14. National Health Service, *What are the health risks of smoking?*, www.nhs.uk/common-health-questions/lifestyle/what-are-the-health-risks-of-smoking/.
15. I.e., deaths that could be prevented with the right intervention.
16. Sanjay Agrawal, *Health inequalities and tobacco*, Royal College of Physicians, 2020, www.rcplondon.ac.uk/news/health-inequalities-and-tobacco#:~:text=In%20the%20UK%2C%20tobacco%20use,and%20the%20most%20deprived%20communities.&text=Currently%20a%20quarter%20of%20a,due%20to%20expenditure%20on%20tobacco.
17. CDC (undated), *Smoking and tobacco use*, fact sheet, www.cdc.gov/tobacco/data_statistics/fact_sheets/fast_facts/index.htm.
18. Sarah E. Jackson et al. 'Moderators of changes in smoking, drinking and quitting behaviour associated with the first COVID-19 lockdown in England', *Addiction*, 25 August 2021, https://doi.org/10.1111/add.15656.
19. Dereck Yach and Douglas Bettcher, 'Globalisation of tobacco industry influence and new global responses', *Tobacco Control*, 9/2 (2000): 206–216, tobaccocontrol.bmj.com/content/9/2/206.

20. Gianna Gayle Herrera Amul et al., 'A systematic review of tobacco industry tactics in Southeast Asia: Lessons for other low-and-middle-income regions', *International Journal of Health Policy and Management*, 10/6 (2021): 1–14, www.ijhpm.com/article_3834_3be135653f75d1a0228559cfdaad2103.pdf.
21. Luk Joossens, 'Vietnam: Smuggling adds Value', *Tobacco Control*, 12/2 (2003): 119–120, tobaccocontrol.bmj.com/content/12/2/119.
22. J. Collin et al., 'Complicity in contraband: British American tobacco and cigarette smuggling in Asia', *Tobacco Control*, 13/2 (2004), tobaccocontrol.bmj.com/content/12/2/119.
23. And also, beyond it, which sit outside my reflections here.
24. The focus on companies that cost the NHS money is another example of NHS-centrism limiting the scope and scale of the progressive public health mainstream.
25. The Labour Party, *Medicines for the many: Public health before private profit*, 2019, labour.org.uk/wp-content/uploads/2019/09/Medicines-For-The-Many.pdf.
26. Imperial College London, 'Obesity has doubled since 1980, major global analysis of risk factors reveals', 4 February 2011, www.imperial.ac.uk/news/96402/obesity-doubled-since-1980-major-global/#:~:text=BMI%3A,cent%20for%20women%20in%201980.&text=The%20UK%20has%20the%20sixth,around%2027%20kg%2Fm2; Hochlaf and Thomas, *The whole society approach*.
27. Anne Case and Angus Deaton, *Deaths of Despair and the Future of Capitalism*, Princeton University Press, 2020.
28. GambleAware, 'Nearly half of all people with gambling disorder have not accessed treatment or support according to in-depth research into the demand for treatment of gambling harms', press release, 2020, www.begambleaware.org/sites/default/files/2020-12/2020-05-19-treatment-needs-gap-analysis-press-release.pdf.
29. Beveridge won a Berwick-upon-Tweed by-election in 1944, only to be defeated in the 1945 general election and subsequently made a peer.
30. Margaret Jones and Rodney Lowe, *From Beveridge to Blair*, Manchester University Press, 2002.
31. William Beveridge, *Social insurance and allied services*, HM Government, November 1942, www.ncbi.nlm.nih.gov/pmc/articles/PMC2560775/pdf/10916922.pdf.
32. Indeed, it is often noted that the NHS is a treatment not a whole health service – which is the basis for the Universal Public Health Service outlined in Chapter 2.
33. Average retirement age is 64 years old, and average healthy life expectancy is 63 years old. ONS, *Healthy state life expectancies, UK: 2017-2019*, 2021, www.ons.gov.uk/peoplepopulationandcommunity/healthandsocialcare/healthandlifeexpectancies/bulletins/healthstatelifeexpectanciesuk/2017to2019.
34. David Cameron, *PM Speech on Wellbeing*, 25 November 2010, www.gov.uk/government/speeches/pm-speech-on-wellbeing.

35. David Cameron, *Transforming the British Economy: Coalition strategy for economic growth*, 2010, www.gov.uk/government/speeches/transforming-the-british-economy-coalition-strategy-for-economic-growth.
36. Institute for Fiscal Studies, *Recent cuts to public spending*, 2015 www.ifs.org.uk/tools_and_resources/fiscal_facts/public_spending_survey/cuts_to_public_spending.
37. Harry Quilter-Pinner, *The end of austerity*, Institute of Public Policy Research, 18 April 2019, www.ippr.org/blog/austerity-there-is-an-alternative-and-the-uk-can-afford-to-deliver-it.
38. Kate Raworth, *Doughnut economics: Seven ways to think like a 21st century economist*, Penguin Random House, 2017. See Chapter One, available at doughnuteconomics.org/tools-and-stories/18; see also Kate Raworth, 'Why it's time for Doughnut Economics'. *Progressive Review*, 24/3 (2017): 216–222, onlinelibrary.wiley.com/doi/abs/10.1111/newe.12058.
39. Kate Raworth, 'Seven ways to think like a 21st century economist', *OpenDemocracy*, 5 April 2017, www.opendemocracy.net/en/transformation/seven-ways-to-think-like-21st-century-economist/.
40. Robert F. Kennedy, *Remarks at the University of Kansas*, 18 March 1968, www.jfklibrary.org/learn/about-jfk/the-kennedy-family/robert-f-kennedy/robert-f-kennedy-speeches/remarks-at-the-university-of-kansas-march-18-1968.
41. Jacqueline McGlade et al., 'Towards a sustainable wellbeing economy', April 2019 www.researchgate.net/profile/Paul-Sutton/publication/332233189_Towards_a_sustainable_economy/links/5cacb5b3458515cd2b0beofc/Towards-a-sustainable-economy.pdf; B. F. Giannetti et al. 'A review of limitations of GDP and alternative indices to monitor human wellbeing and to manage eco-system functionality', *Journal of Cleaner Production*, 87/1(2015): 11–25, www.advancesincleanerproduction.net/papers/journals/2014/2014_jcp.pdf; Catherine Colebook, *Measuring what matters: Improving the indicators of economic performance*, Institute of Public Policy Research, 2018, www.ippr.org/publications/measuring-what-matters.
42. At least in the short-term – and almost all governments working to a five-year election cycle significantly over-value the short-term.
43. E. Stamatakis et al., 'Overweight and obesity trends from 1974 to 2003 in English children: what is the role of socioeconomic factors?', *Archives of Disease in Childhood*, 90/10 (2004), adc.bmj.com/content/90/10/999; Public Health England, *Obesity profile for England*, fingertips.phe.org.uk/profile/national-child-measurement-programme/data.
44. Matthew Limb, 'Deaths from alcohol hit record high during 2020, show figures', *British Medical Journal*, 327 (2021), https://doi.org/10.1136/bmj.n317.
45. House of Lords Select Committee on the Social and Economic Impact of the Gambling Industry, *Gambling harm – Time for action*, 2020, https://publications.parliament.uk/pa/ld5801/ldselect/ldgamb/79/79.pdf.

46. Office for National Statistics, *Human health and social work activities (£M):CP*, dataset, 2020, www.ons.gov.uk/economy/grossdomesticproductgdp/timeseries/kkn7/bb.
47. There is some evidence from the Covid-19 pandemic that suggests the UK measures GDP in a way that is particularly susceptible to this problem. Office for National Statistics, *Comparisons of GDP during the Coronavirus (COVID-19) pandemic*, 2021, www.ons.gov.uk/economy/grossdomesticproductgdp/articles/internationalcomparisonsofgdpduringthecoronaviruscovid19pandemic/2021-02-01.
48. Simon Stevens, *Simon Stevens speech to Institute of Directors Annual Convention*, 7 October 2015, www.england.nhs.uk/2015/10/directors-convention/.
49. Hochlaf and Thomas, *The whole society approach*.
50. See for instance, the work of Gillian Pascall – including Gillian Pascall, *Social policy: A feminist analysis*, Tavistock Press, 1986; Gillian Pascall, *Social policy: A new feminist analysis*, Routledge, 1997; and Gillian Pascall, *Gender equality in the welfare state*, Policy Press, 2012.
51. See for example Jameel Hampton, *Disability and the welfare state in Britain: Changes in perception and policy 1948-1979*, Policy Press, 2016.
52. Johnathan Watkins et al., 'Effects of health and social care spending constraints on mortality in England: A time trend analysis', *BMJ Open*, 2017, bmjopen.bmj.com/content/7/11/e017722.citation-tools; For a similar figure from a different method, see also: Hochlaf et al., *Ending the blame game*, Institute of Public Policy Research, 2019, www.ippr.org/research/publications/ending-the-blame-game.
53. Clare Bambra et al., *Health for wealth*, Northern Health and Science Alliance,2018, www.thenhsa.co.uk/app/uploads/2018/11/NHSA-REPORT-FINAL.pdf.
54. Chris Thomas et al., *Levelling-up health for prosperity*, Institute of Public Policy Research, 2021, www.ippr.org/research/publications/levelling-up-health-for-prosperity.
55. 2019 estimates. See Statista, 'Gross value added (GVA) of agriculture in the United Kingdom (UK) from 2003 to 2019', data, 2021, www.statista.com/statistics/315804/agriculture-gross-value-added-in-the-united-kingdom-uk/.
56. Public Health England, 'Third year of industry progress to reduce sugar published', 2020, www.gov.uk/government/news/third-year-of-industry-progress-to-reduce-sugar-published.
57. Mark Petticrew, 'Dark nudges and sludge in big alcohol: Behavioural economics, cognitive biases, and alcohol industry corporate social responsibility', *The Milbank Quarterly*, 98/4 (2020): 1290–1328, onlinelibrary.wiley.com/doi/full/10.1111/1468-0009.12475.
58. Ibid.
59. Drinkaware, *Alcohol and cancer*, www.resourcesorg.co.uk/assets/pdfs/Alcohol-and-cancer.pdf.

60. As per D. M. Parkin, 'Cancers attributable to consumption of alcohol in the UK in 2010' *British Journal of Cancer*, 105/.Suppl 2 (2011): S14–18, 10.1038/bjc.2011.476.
61. With exceptions for some high-sugar drinks that may arguably be healthy (e.g., fresh orange juice).
62. HM Revenue and Customs, *Soft drinks industry levy statistics commentary 2020*, 2021, www.gov.uk/government/statistics/soft-drinks-industry-levy-statistics/soft-drinks-industry-levy-statistics-commentary-2020#:~:text=Soft%20Drinks%20Industry%20Levy%20Receipts,-Headlines&text=Total%20SDIL%20receipts%20for%20financial,not%20due%20until%20July%202018.
63. See Vera Zakharov et al., 'Refreshing investment in children's health: Using the Sugary Drinks Tax to improve healthy food access in schools', School Food Matters, 2020, www.schoolfoodmatters.org/sites/default/files/Refreshing-Investment-in-Childrens-Health.pdf.
64. Action on Sugar, *Sugar reduction: Report on progress between 2015 and 2019*, 2020, www.actiononsugar.org/news-centre/sugar-in-the-news/2020/2020-stories/sugar-reduction-report-on-progress-between-2015-and-2019.html.
65. Cherry Law, 'An analysis of the stock market reaction to the announcements of the UK Soft Drinks Industry Levy', *Economics and Human Biology*, 38/1 (2020), www.sciencedirect.com/science/article/pii/S1570677X19302096?via%3Dihub.
66. E. Martos, *Assessment of the impact of a public health product tax: Final report*, World Health Organisation, 2015, www.euro.who.int/__data/assets/pdf_file/0008/332882/ assessment-impact-PH-tax-report.pdf.
67. Emma Smith et al., 'Should we tax unhealthy food and drink?', *Proceedings of the Nutritional Society* 77/3 (2019): 314–320, www.ncbi.nlm.nih.gov/pmc/articles/PMC5912513/.
68. Office for National Statistics, *Gender pay gap in the UK: 2020*, 2020, www.ons.gov.uk/employmentandlabourmarket/peopleinwork/earningsandworkinghours/bulletins/genderpaygapintheuk/2020.
69. Clara Guibourg and Eleanor Lawrie, 'Gender pay gap grows at hundreds of big firms', *BBC*, 20 February 2019, www.bbc.co.uk/news/business-47252848#:~:text=The%20BBC%20looked%20at%20a,and%20the%20middle%2Dranking%20man.&text=Of%20those%201%2C146%20companies%2C%20the,improvement%20from%209.7%25%20last%20year.
70. Jonathan Platt, 'Unequal depression for equal work? How the wage gap explains gendered disparities in mood disorders', *Social Science and Medicine*, 149 (2016): 1–8, www.sciencedirect.com/science/article/abs/pii/S0277953615302616.
71. Chartered Insurance Institute, *Securing the financial future of the next generation*, Insuring Women's Futures, 2018, www.cii.co.uk/media/9224351/iwf_momentsthatmatter_full.pdf.
72. Office for National Statistics, *Human Capital Estimates in the UK: 2004 to 2018*, 2019, www.ons.gov.uk/peoplepopulationandcommunity/wellbeing/articles/humancapitalestimates/2004to2018#human-capital-by-sex.

73. British Heart Foundation, *How has the smoking ban changed our health?*, 2017, www.bhf.org.uk/informationsupport/heart-matters-magazine/news/smoking-ban; See also Michelle Sims et al., 'Short term impact of smoke-free legislation in England: retrospective analysis of hospital admissions for myocardial infarction', *British Medical Journal*, 340 (2010), https://doi.org/10.1136/bmj.c2161.
74. Action on Smoking and Health, *Smokefree: The First Ten Years*, 2017, ash.org.uk/wp-content/uploads/2017/06/170107-Smokefree-the-first-ten-years-FINAL.pdf.
75. See James Meadway,'Creating the digital commons after COVID-19', *OpenDemocracy*, 2020, www.opendemocracy.net/en/oureconomy/creating-digital-commons-after-covid-19/.
76. Nominet, *Digital Access for All launches to solve problem of digital exclusion*, press release, 2019, www.nominet.uk/digital-access-for-all-launches-to-help-solve-problem-of-digital-exclusion/.
77. For example, Harisa Mardiana, 'The impact of teenagers' digital literacy on the use of social media', *The 3rd International Conference of Advance & Scientific Innovation*, 20 June 2020, www.researchgate.net/publication/345783914_The_Impact_of_Teenagers'_Digital_Literacy_on_the_Use_of_Social_Media.
78. See Sonia Livingstone, 'Developing social media literacy: how children learn to interpret risky opportunities on social network sites', *Communications*, 39/3 (2014): 283–303, http://eprints.lse.ac.uk/62129/1/Developing%20social%20media%20literacy.pdf.
79. Department for Education, *Bullying in England April 2013 to March 2018*, 2018, assets.publishing.service.gov.uk/government/uploads/system/uploads/attachment_data/file/754959/Bullying_in_England_2013-2018.pdf
80. Gwen Schurgin O'Keeffe and Kathleen Clarke-Pearson, 'The impact of social media on children, adolescents, and families', *Pediatrics*, 127/4 (2011): 800–804, www.cooperativa.cl/noticias/site/artic/20110329/asocfile/20110329173752/reporte_facebook.PDF.
81. Ofcom, ICO, *Internet users' concerns about and experience of potential online harms*, 2019, www.ofcom.org.uk/__data/assets/pdf_file/0028/149068/online-harms-chart-pack.pdf.
82. S. Boniface, C. Thomas, J. Vohra et al., 'Underage adolescents' reactions to adverts for beer and spirit brands and associations with higher risk drinking and susceptibility to drink: a cross-sectional study in the UK', *Alcohol and Alcoholism*, agab018 (2021), https://doi.org/10.1093/alcalc/agab018; Nathan Critchlow et al., 'Awareness of marketing for high fat, salt or sugar foods and the association with higher weekly consumption among adolescents: a rejoinder to the UK government's consultations on marketing regulation', *Public Health Nutrition*, 23/14 (2020): 2637–2646, www.stir.ac.uk/research/hub/publication/1500476; Nathan Critchlow et al., 'Social media, higher risk consumption, and brand identification', *Addiction Research and Theory*, 27/6 (2021): 515–526, www.stir.ac.uk/research/hub/publication/1074907;

Chris Thomas et al., *Under pressure*, Cancer Research UK, 2020, www.cancerresearchuk.org/sites/default/files/under_pressure.pdf.

83. See Rhianna Schmunk, 'B.C. pub company partially at fault for drunk-driving crash, judge rules', *CBC*, 10 March 2017, www.cbc.ca/news/canada/british-columbia/pub-liability-drunk-driver-1.4020142.
84. See Center for Democracy and Technology, *Overview of the NetzDG Network Enforcement Law*, 2017, https://cdt.org/insights/overview-of-the-netzdg-network-enforcement-law/.
85. Alcohol control is often an area where public health policies are most advanced, if only thanks for the long impact of the temperance movement. In Britain, it emerged from the 'gin craze'.
86. Government Offices of Sweden, *Swedish alcohol retailing monopoly (Systembolaget Aktiebolag)*, www.government.se/government-agencies/swedish-alcohol-retailing-monopoly--systembolaget-aktiebolag/.
87. Tea Olsen, 'Explaining the strong support for the Swedish alcohol retail monopoly', *Institute of Alcohol Studies*, 24 May 2019, www.ias.org.uk/2019/05/24/explaining-the-strong-support-for-the-swedish-alcohol-retail-monopoly/.
88. Tim Stockwell et al., 'Estimating the public health impact of disbanding a government alcohol monopoly: Application of new methods to the case of Sweden', *BMC Public Health*, 18/1400 (2018), bmcpublichealth.biomedcentral.com/articles/10.1186/s12889-018-6312-x. Corroborating evidence from 2010. See Thor Norström et al., 'Potential consequences of replacing a retail alcohol monopoly with a private licence system: results from Sweden', *Addiction*, 105/12 (2010): 2113–2119.
89. Medieakademin, *Förtroendebarometern*, 2019, medieakademin.se/wp-content/uploads/2019/03/F%C3%B6rtroendebarometern-2019-1.pdf.
90. Robin Room, 'Alcohol monopolies', *Shaap*, 12 February 2021, www.shaap.org.uk/blog/332-alcohol-monopolies.html.
91. Helping to explain why this chapter has not advocated for any public health bans.
92. Robin Room and Jenny C. Örnberg, 'Government monopoly as an instrument for public health and welfare: Lessons for cannabis from experience with alcohol monopolies', *International Journal of Drug Policy*, 74 (2019), doi: 10.1016/j.drugpo.2019.10.008.
93. For example Jonathan P. Caulkins and Beau Kilmer 'The US as an example of how *not* to legalize marijuana', *Addiction*, 111/12 (2016): 2095–2096, https://onlinelibrary.wiley.com/doi/10.1111/add.13498; Rachel Barry and Stanton Glantz, 'Marijuana regulatory frameworks in four US states: an analysis against a public health standard', *American Journal of Public Health*, 108/7 (2018): 914–923, www.ncbi.nlm.nih.gov/pmc/articles/PMC5993386/; Rachel Barry and Stanton Glantz, 'A public health framework for legalized retail marijuana based on the US experience: avoiding a new tobacco industry', *Plos Medicine*, 13/9 (2016), https://journals.plos.org/plosmedicine/article?id=10.1371/journal.pmed.1002131.

94. Kojo Koram, *Cannabis and capitalism*, Common Wealth, 2019, www.common-wealth.co.uk/interactive-digital-projects/cannabis-and-capitalism.
95. Estimate from Buzzfeed Amanda Chicago Lewis, 'America's whites-only weed boom', *Buzzfeed News*, 16 March 2020, www.buzzfeednews.com/article/amandachicagolewis/americas-white-only-weed-boom.
96. James Meadaway, *Creating a digital commons*, Institute of Public Policy Research, 2020, www.ippr.org/files/2020-08/creating-a-digital-commons-august20.pdf; British Academy and The Royal Society, *Data management and use: governance in the 21st century*, June 2017, https://royalsociety.org/-/media/policy/projects/data-governance/data-management-governance.pdf.
97. European Parliament, *Is data the new oil?*, 2020, www.europarl.europa.eu/RegData/etudes/BRIE/2020/646117/EPRS_BRI(2020)646117_EN.pdf .

CHAPTER 4

1. Parth Patel et al., *State of health and care*, Institute of Public Policy Research, 2021, www.ippr.org/files/2021-03/state-of-health-and-care-mar21.pdf.
2. HM Government, *National Assistance Act*, 1948, www.legislation.gov.uk/ukpga/Geo6/11-12/29.
3. The King's Fund 'The origins and development of social care', undated, www.kingsfund.org.uk/sites/default/files/Securing_Good_Care_Chapter_1.pdf.
4. Stewart Player and Allyson Pollock, 'Long-term care: from public responsibility to private good, *Critical Social Policy*, 21/2 (2001): 231–255.
5. The same Roy Griffiths who reviewed the NHS in Chapter One.
6. Roy Griffiths, *Community care: Agenda for action*, HMSO, 1988.
7. Ray Jones, 'A Journey through the Years: Ageing and Social Care', *Ageing Horizons*, 6 (2007): 42–51.
8. Player and Pollock, 'Long-term care'; p. 235.
9. Jones, Ray, 'A Journey through the Years'.
10. Not enacted until 1993.
11. HM Government, *National Health Service and Community Care Act 1990*, 1990, www.legislation.gov.uk/ukpga/1990/19/contents
12. Player and Pollock, 'Long-term care'. https://allysonpollock.com/wp-content/uploads/2013/04/CriticalSocialPolicy_2001_Player_LongTermCare.pdf
13. Bob Hudson, 'The unsuccessful privatisation of social care: why it matters and how to curb it', *London School of Economics*, 4 April 2016, blogs.lse.ac.uk/politicsandpolicy/why-social-care-privatisation-is-unsuccessful/?utm_content=buffer4a900&utm_medium=social&utm_source=twitter.com&utm_campaign=buffer.
14. Grace Blakeley and Harry Quilter-Pinner, *Who cares? The financialisaton of adult social care*, Institute of Public Policy Research, 2019, www.ippr.org/files/2019-09/who-cares-financialisation-in-social-care-2-.pdf.
15. Ibid, p. 6.
16. Hudson, 'The unsuccessful privatisation of social care'.
17. At the time of writing, one is in administration.
18. Blakeley and Quilter-Pinner, *Who cares?*.

19. Joe Dromey and Dean Hochlaf, *Fair care: A workforce strategy for social care*, Institute for Public Policy Research, 25 November 2018, www.ippr.org/research/publications/fair-care.
20. Care Quality Commission, *The state of adult social care services 2014 to 2017: Findings from the CQCs initial programme of comprehensive inspections in adult social care*, 2018, www.cqc.org.uk/sites/default/files/20170703_ASC_end_of_programme_FINAL2.pdf; Unison, 'Employers in the care sector are hiding behind complex and incomplete pay slips to break wage laws', press release, 2018, www.unison.org.uk/news/2018/04/employers-care-sector-hiding-behind-complex-incomplete-pay-slips-break-wage-laws/; Susan C. Eaton, 'Beyond "unloving care": linking human resource management and patient care quality in nursing homes', *The International Journal of Human Resource Management*, 11/3 (2000): 591–616.
21. Care Quality Commission, *The state of adult social care services 2014 to 2017*.
22. Care Quality Commission, *The state of health care and adult social care in England 2019/20*, 2020, www.cqc.org.uk/sites/default/files/20201016_stateofcare1920_fullreport.pdf.
23. Ara Darzi, *Better health and care for all*, Institute of Public Policy Research, 2018, www.ippr.org/files/2018-06/better-health-and-care-for-all-june2018.pdf.
24. The King's Fund, *How serious are the pressures in social care?*, 2015, www.kingsfund.org.uk/projects/verdict/how-serious-are-pressures-social-care.
25. Christopher Thomas, *Health inequalities: Time to talk*, Macmillan Cancer Support, 2019, www.macmillan.org.uk/assets/health-inequalities-paper-april-2019.pdf.
26. Tania Burchardt et al., 'The Conservatives' record on adult social care', London School of Economics Centre for Analysis of Social Exclusion, 3 November 2020, https://sticerd.lse.ac.uk/CASE/_NEW/PUBLICATIONS/abstract/?index=7525.
27. Age UK, 'The number of older people with some unmet need for care now stands at 1.5 million', 9 November 2019, www.ageuk.org.uk/latest-press/articles/2019/november/the-number-of-older-people-with-some-unmet-need-for-care-now-stands-at-1.5-million/.
28. The Health Foundation, *Social care for adults aged 18-64*, April 2020, www.health.org.uk/publications/reports/social-care-for-adults-aged-18-64.
29. Independent Age, *Free personal care: How to eliminate catastrophic care costs*, 2021, independent-age-assets.s3.eu-west-1.amazonaws.com/s3fs-public/2019-04/Final%20Report_Web_0.pdf.
30. Ibid.
31. Excluding certain costs, like hotel costs.
32. Chris Thomas, 'The "make do and mend" health service: Solving the NHS's Capital Crisis', *Institute of Public Policy Research*, 2019, www.ippr.org/research/publications/the-make-do-and-mend-health-service.
33. Ibid.
34. National Audit Office, *Investigation into the collapse of Carillion*, 2019, www.nao.org.uk/report/investigation-into-the-governments-handling-of-the-collapse-of-carillion/.

35. See Graeme Wearden, 'The rise and fall of Southern Cross', *Guardian*, 1 June 2011, www.theguardian.com/business/2011/jun/01/rise-and-fall-of-southern-cross.
36. Nina Lakhani, 'Southern Cross collapse leaves elderly care in Limbo', *The Independent*, 23 October 2011, www.independent.co.uk/life-style/health-and-families/health-news/southern-cross-collapse-leaves-elderly-care-in-limbo-2312648.html.
37. Care Quality Commission, *Whitchurch care home: Inspection report*, 15 February 2019, https://api.cqc.org.uk/public/v1/reports/64ce2d22-5d83-441e-9a64-1ed6f83adb47?20200129163919.
38. Ibid., p. 3.
39. Name changed to protect anonymity.
40. Skills for Care, *The state of the adult social care sector and workforce in England*, 2020, www.skillsforcare.org.uk/adult-social-care-workforce-data/Workforce-intelligence/documents/State-of-the-adult-social-care-sector/The-state-of-the-adult-social-care-sector-and-workforce-2020.pdf.
41. Ibid.
42. Ibid.
43. Stewart Sutherland, *With respect to old age: A report by the Royal Commission on Long Term Care*, HM Government, 1999, www.scie-socialcareonline.org.uk/with-respect-to-old-age-a-report-by-the-royal-commission-on-long-term-care/r/a11G00000017xMvIAI.
44. Despite the fact they were tripling the NHS's budget, and the health of social care is key to how effectively the NHS can use its own funding.
45. Mark Ivory, *Tony Blair: The Social Care Legacy*, Community Care, 2007, www.communitycare.co.uk/2007/05/09/tony-blair-the-social-care-legacy/.
46. Luke Clements, Louisa Harding Edgar and Allyson M Pollock, 'Why we need a National Care Service – and how to Build One', *Tribune*, 22 June 2020, tribunemag.co.uk/2020/06/towards-a-national-care-service-2.
47. GMB Union, *5 Key Asks: Social Care*, 2021, www.gmb.org.uk/sites/default/files/CARE_GOPUBLIC_2020.pdf.
48. We Own It, *Why we need a National Care Service*, 2021, weownit.org.uk/blog/why-we-need-national-care-service.
49. The Scottish context is slightly different. They did introduce free personal care in 2002. Scottish National Party, 'How is the SNP improving social care services?', 2021, www.snp.org/policies/pb-how-is-the-snp-improving-social-care-services/.
50. Dean Hochlaf and Harry Quilter-Pinner, *Social care: Free at the point of need*, Institute of Public Policy Research, May 2019, www.ippr.org/files/2019-05/social-care-free-at-the-point-of-need-may-19.pdf.
51. Independent Age, 'Three-quarters of adults in England back free personal care for over-65s, says charity', press release, 6 September 2018, www.independentage.org/news-media/press-releases/three-quarters-of-adults-england-back-free-personal-care-for-over-65s.

52. Richard Sloggett and Harry Quilter-Pinner, *Care after Coronavirus: An emerging consensus*, Institute of Public Policy Research, Policy Exchange, May 2020, www.ippr.org/blog/ippr-policy-exchange-social-care-polling.
53. Patel et al., *State of health and care.*
54. CHPI, *Plugging the Leaks in the UK care home industry*, 2019, chpi.org.uk/wp-content/uploads/2019/11/CHPI-PluggingTheLeaks-Nov19-FINAL.pdf.
55. Particularly, disabled people – with the debate often unbalanced in focusing on people over 65 years old.
56. Michel Foucault, *Madness and civilization: A history of insanity in the Age of Reason*, Pantheon Books, 1965.
57. Leonard Cheshire, *Adult Social Care April 2018*, 2018, www.leonardcheshire.org/sites/default/files/2019-03/social_care_briefing_-_leonard_cheshire.docx.
58. Christina R. Victor, 'Loneliness in care homes: a neglected area of research?', *Ageing Health*, 8/6 (2012), www.futuremedicine.com/doi/abs/10.2217/ahe.12.65.
59. NHS England, '1.18i – Social Isolation: percentage of adult social care users who have as much social contact as they would like', survey data, 2016, data.england.nhs.uk/dataset/phe-indicator-90280/resource/9d56da41-f81c-4262-97b3-edaa1abab411.
60. Independent Age, *Free personal care: Insights from Scotland*, 2020, independent-age-assets.s3.eu-west-1.amazonaws.com/s3fs-public/2020-10/Report_Final.pdf?5pNaZ4vjvNoJt4YGK2Bctq4DOnK9_fp2=.
61. Ibid.
62. Social Care Future, *Talking about a brighter social care future*, 2019, socialcarefuture.files.wordpress.com/2019/10/ic-scf-report-2019-h-web-final-111119.pdf.
63. See Reclaiming the Future Alliance, *Independent living for the future*, 2019, dpac.uk.net/2019/04/a-national-independent-living-support-service/.
64. Jack Hunter and Martina Orvolic. *End of life care in England*, Institute of Public Policy Research, May 2018, www.ippr.org/files/2018-05/end-of-life-care-in-england-may18.pdf.
65. Chris Thomas, *Community first social care*, Institute for Public Policy Research, 23 August 2021, www.ippr.org/research/publications/community-first-social-care.
66. Care Quality Commission, *The state of adult social care services: 2014 to 2017.*
67. Maurice McLeod and Chris Thomas, 'Racism is not a culture war toy – it's a fact of life in Tory Britain', *Tribune*, 22 June 2021, tribunemag.co.uk/2021/06/racism-is-not-a-culture-war-toy-its-a-fact-of-life-in-britain.
68. CADTH, 'Dementia Villages: Innovative Residential Care for People with Dementia', 2019, 178, www.cadth.ca/sites/default/files/hs-eh/eh0071-dementia-villages.pdf, pp. 2–28.
69. See Centre for Policy on Ageing, *Foresight Future of An Ageing Population – International Case Studies*, 2016, www.cpa.org.uk/information/reviews/

CPA-International-Case-Study-4-Housing-and-Dementia-Care-in-the-Netherlands.pdf.

70. CADTH, 'Dementia Villages'.
71. Sandy Ransom, *Eden alternative: The Texas Project*, IQILTHC Series Report 2000–2004, 2000, digital.library.txstate.edu/bitstream/handle/10877/4087/fulltext.pdf?sequence=1&isAllowed=y.
72. Daniel Button and Sarah Bedford, *Ownership in social care: Why it matters and what can be done*, New Economics Foundation, 2020, neweconomics.org/2020/01/ownership-in-social-care.
73. Simon Borkin, *Platform co-operatives – solving the capital conundrum*, Nesta, 2019, media.nesta.org.uk/documents/Nesta_Platform_Report_FINAL-WEB_b1qZGj7.pdf.
74. Institute of Public Policy Research, *Prosperity and justice*, www.ippr.org/files/2018-08/1535639099_prosperity-and-justice-ippr-2018.pdf.
75. Ibid.
76. Harry Quilter-Pinner, *Ethical Care: A bold reform agenda for adult social care*, Institute of Public Policy Research, 2019, www.ippr.org/research/publications/ethical-care.
77. Skills for Care, *The state of the adult social care sector and workforce in England*.
78. Marcus Johns, *10 Years of Austerity: Eroding Resilience in the North*, Institute of Public Policy Research, 2020, www.ippr.org/research/publications/10-years-of-austerity.
79. Quilter-Pinner, *Ethical care*.
80. ONS, *Household satellite account, UK: 2015 and 2016*, 2018, www.ons.gov.uk/economy/nationalaccounts/satelliteaccounts/articles/householdsatelliteaccounts/2015and2016estimates.
81. Carers UK, *Juggling work and unpaid care: A growing issue*, 2019, www.carersuk.org/images/News_and_campaigns/Juggling_work_and_unpaid_care_report_final_0119_WEB.pdf.
82. Ibid.
83. New Policy Institute, *Informal carers and poverty in the UK: An analysis of the Family Resources Survey*, 2019, npi.org.uk/files/2114/6411/1359/Carers_and_poverty_in_the_UK_-_full_report.pdf.
84. Raphael Wittenberg, 'Projects of demand for and costs of social care for older people in England, 2010 to 2030, under current and alternative funding systems', *PSSRU*, December 2011, http://eprints.lse.ac.uk/40720/1/2811-2.pdf.
85. Andrea Budnick et al., 'Informal caregivers during the COVID-19 pandemic perceive additional burden', *BMC Health Services Research*, 21/353 (2021), bmchealthservres.biomedcentral.com/articles/10.1186/s12913-021-06359-7.
86. Carers UK, *Caring behind closed doors: Six months on*, 2020, www.carersuk.org/images/News_and_campaigns/Caring_Behind_Closed_Doors_Oct20.pdf.

87. Author's analysis of Office for National Statistics, *Household Satellite Account, UK*, dataset, 2018, www.ons.gov.uk/economy/nationalaccounts/satelliteaccounts/compendium/householdsatelliteaccounts/2005to2014/relateddata.
88. See Chris Thomas, *The state of end of life care*, Institute of Public Policy Research, 2021, www.ippr.org/files/2021-04/end-of-life-care-april21.pdf.
89. Carers UK, *The case for care leave*, 2013, www.carersuk.org/ for-professionals/policy/policy-library?task=download&file=policy_file&id=218.

CHAPTER 5

1. Global Health Security Index, *Global Health Security Index*, 2019, www.ghsindex.org/wp-content/uploads/2019/10/2019-Global-Health-Security-Index.pdf.
2. Johns Hopkins University, *Coronavirus mortality*, dataset, 2021, coronavirus.jhu.edu/data/mortality.
3. In the time between writing and publication, the ability of affluent countries to monopolise the Covid-19 vaccine could change this picture significantly.
4. The sudden name change suggesting a very deliberate change in emphasis (i.e., from protection to security).
5. Matt Hancock, *Reinvigorating our system for international health*, speech, 26 January 2021, www.gov.uk/government/speeches/reinvigorating-our-system-for-international-health.
6. Kate Jones et al., 'Trends in emerging infectious diseases', *Nature*, 451 (2008): 990–993, www.nature.com/articles/nature06536/.
7. Ibid.
8. For instance, HM Government, *Life sciences vision*, 2021, assets.publishing.service.gov.uk/government/uploads/system/uploads/attachment_data/file/1000030/life-sciences-vision.pdf.
9. Katherine Smith et al., 'Global rise in human infectious disease outbreaks', *Journal of the Royal Society*, 6 December 2014, royalsocietypublishing.org/doi/full/10.1098/rsif.2014.0950.
10. World Economic Forum, 'Outbreak Readiness and Business Impact: Protecting Lives and Livelihoods across the Global Economy', white paper, 2019, www.3.weforum.org/docs/WEF%20HGHI_Outbreak_Readiness_Business_Impact.pdf.
11. Ibid.
12. Ibid, p. 5.
13. World Health Organisation, *Crimean Congo Haemorrhagic Fever*, 2013, www.who.int/en/news-room/fact-sheets/detail/crimean-congo-haemorrhagic-fever.
14. World Health Organisation, *Ebola Virus Disease*, 2021, www.who.int/news-room/fact-sheets/detail/ebola-virus-disease,
15. World Health Organisation, *Marburg Virus Disease*, 2018, www.who.int/news-room/fact-sheets/detail/marburg-virus-disease.

16. World Health Organisation, *Lassa Fever*, www.who.int/health-topics/lassa-fever#tab=tab_1.
17. World Health organisation, *Nipah Virus*, 2019, www.who.int/news-room/fact-sheets/detail/nipah-virus.
18. Ibid.
19. Various, 'The threat of pandemic influenza: are we ready? Workshop summary', *The National Archives*, www.ncbi.nlm.nih.gov/books/NBK22148/.
20. Jenny Hope, 'How £300 million was squandered on swine flu jabs that we didn't need', *The Daily Mail*, 7 April 2010, www.dailymail.co.uk/news/article-1263975/Swine-flu--300m-squandered-jabs-didnt-need.html.
21. Gabriele Neumann and Yoshihiro Kawaoka, 'Predicting the Next Influenza Pandemics', *Journal of Infectious Diseases*, 219/S1 (2019): S14–20, www.ncbi.nlm.nih.gov/pmc/articles/PMC6452319/.
22. Zhi-Ping Zhong et al., 'Glacier ice archives fifteen-thousand-year-old viruses', *BioRxiv*, 7 January 2020, www.biorxiv.org/content/10.1101/2020.01.03.894675v1.
23. Elisa Stella et al., 'Permaforst dynamics and the risk of anthrax transmission: a modelling study', *Scientific Reports*, 10, /16460 (2020), www.nature.com/articles/s41598-020-72440-6; www.biorxiv.org/content/10.1101/486290v1.
24. Enoch J. Abbey et al., 'The Global Health Security Index is not predictive of coronavirus pandemic response among Organisation for Economic Cooperation and Development countries', *Plos One*, 7 October 2020, https://doi.org/10.1371/journal.pone.0239398.
25. And vitally, understood outside of the excellent but rather isolated world of global health academia.
26. Stockholm Resilience Centre, *The nine planetary boundaries*, www.stockholmresilience.org/research/planetary-boundaries/planetary-boundaries/about-the-research/the-nine-planetary-boundaries.html.
27. Jonathan Patz, 'A human disease indicator for the effects of recent global climate change', *PNAS*, 99/20 (2020): 12506-12508, www.pnas.org/content/99/20/12506; Jonathan Patz et al., *Climate change and infectious diseases*, World Health Organisation, 2013, www.who.int/globalchange/publications/climatechangechap6.pdf; Jonathan Patz et al., 'Global climate change and emerging infectious diseases', *JAMA*, 275/3 (1996): 217–223, pubmed.ncbi.nlm.nih.gov/8604175/.
28. S. W. Lindsay and W. J. Martens, W.J., 'Malaria in the African highlands: Past, present and future', *Bulletin of the World Health Organization*, 76/1 (1998): 33–45.
29. J. Hartman et al., 'Climate suitability for stable malaria transmission in Zimbabwe under different climate change scenarios', *Global Change & Human Health*, 3/1 (2002): 42–53.
30. Jeffrey Shaman et al., 'Using a dynamic hydrology model to predict mosquito abundances in flood and swamp water', *Emerging Infectious Diseases*, 8/1 (2002): 6–13.

31. Paul R. Epstein, 'West Nile virus and the climate', *Journal of Urban Health*, 78/2 (20021): 367–371.
32. William Checkley et al., 'Effect of El Niño and ambient temperature on hospital admissions for diarrhoeal diseases in Peruvian children', *The Lancet*, 355/9202 (2000): 442–450.
33. Dana A. Fockset al., 'A simulation model of the epidemiology of urban dengue fever: Literature analysis, model development, preliminary validation, and samples of simulation results', *American Journal of Tropical Medicine and Hygiene*, 53/5 (1995): 489–506.
34. Joachim Rocklöv and Robert Dubrow, 'Climate change: an enduring challenge for vector-borne disease prevention and control', *Nature Immunology*, 21/5 (2020): 479-483, www.nature.com/articles/s41590-020-0648-y.
35. United Nations, *Climate change key findings*, www.un.org/en/climatechange/science/key-findings.
36. World Meteorological Organization, *WMO Greenhouse Gas Bulletin*, 14 (November 2018), library.wmo.int/doc_num.php?explnum_id=5455.
37. Jeff Tollefson, 'COVID curbed carbon emissions in 2020 – but not by much', *Nature*, 15 January 2021, www.nature.com/articles/d41586-021-00090-3.
38. William J. Ripple et al., 'World scientists' warning to humanity: a second notice', *BioScience*, 67/12 (2017): 1026–1028, academic.oup.com/bioscience/article/67/12/1026/4605229.
39. Daniel W. O'Neill et al., 'A Good life for all within planetary boundaries', *Nature Sustainability*, 1 (2018): 88–95, www.nature.com/articles/s41893-018-0021-4; Laurie Laybourn-Langton et al., *This is a crisis*, Institute of Public Policy Research, 2019, www.ippr.org/research/publications/age-of-environmental-breakdown.
40. Toph Allen, 'Global Hotspots and correlates of emerging zoonotic diseases', *Nature Communications*, 8/1124 (2017), www.nature.com/articles/s41467-017-00923-8; R. T. Corlett, *State of the tropics*, James Cooks University, www.jcu.edu.au/state-of-the-tropics/publications/2014-state-of-the-tropics-report; S. Nazrul Islam and John Winkel, 'Climate Change and Social Inequality', *DESA Working Paper*, 152 (2017), www.un.org/esa/desa/papers/2017/wp152_2017.pdf; United States Environmental Protection Agency, *International Climate Impacts*, 19 January 2017, 19january2017snapshot.epa.gov/climate-impacts/international-climate-impacts_.html.
41. P. Daszak et al., 'Anthropogenic environmental change and the emergence of infectious diseases in wildlife', *Act Tropica*, 78 (2001): 103-116.
42. Daszak, 'Anthropogenic environmental change and the emergence of infectious diseases in wildlife'.
43. Jason Rohr et al., 'Emerging human infectious diseases and the links to global food production', *Nature Sustainability*, 2 (2019): 445–456, www.nature.com/articles/s41893-019-0293-3.
44. Bryony A. Jones et al., 'Zoonosis emergence linked to agricultural intensification and environmental change', *Proceedings of the National*

Academy of Sciences USA, 110/21 (2013): 8399–8404, www.ncbi.nlm.nih.gov/pmc/articles/PMC3666729/.

45. Jones, 'Zoonosis emergence linked to agricultural intensification and environmental change'.
46. Hannah Ritchie and Max Roser, *Land use*, Our World in Data, 2019, ourworldindata.org/land-use.
47. Ibid.
48. Max Roser, 'Two centuries of rapid global population growth will come to an end', *Our World in Data*, 2019, ourworldindata.org/world-population-growth-past-future.
49. Kate E. Jones et al., 'Global Trends in Emerging Infectious Diseases', *Nature*, 451/7181 (2008): 990–993, www.ncbi.nlm.nih.gov/pmc/articles/PMC5960580/.
50. United Nations, *World Urbanization Prospects 2018*, dataset, 2018, population.un.org/wup/.
51. Hannah Ritchie and Max Roser, *Urbanization*, Our World in Data, 2019, ourworldindata.org/urbanization.
52. Creighton Connolly et al 'Extended urbanisation and the spatialities of infectious disease: Demographic change, infrastructure and governance', *Urban Studies*, 58/2 (2021): 245–263, 254, https://journals.sagepub.com/doi/pdf/10.1177/0042098020910873.
53. Mbih Jerome Tosam et al., 'Global Emerging Pathogens, Poverty and Vulnerability: An Ethical Analysis', in Godfrey Tangwa et al. (eds), *Socio-cultural Dimensions of Emerging Infectious Diseases in Africa*, Springer, 2019: 243–253, link.springer.com/chapter/10.1007/978-3-030-17474-3_18.
54. Alumdena M. Saéz et al., 'Investigation the zoonotic origin of the West African Ebola epidemic', *Embo Press*, 7 (2015): 17–23, embopress.org/doi/full/10.15252/emmm.201404792.
55. Ibid.
56. Development Initiatives, *Poverty trends: Global, regional and national*, fact sheet, 2019, devinit.org/resources/poverty-trends-global-regional-and-national/#note-Yr_3XWnJm.
57. Francisco Ferreira and Carolina Páramo-Sánchez, 'A richer array of international poverty lines', *World Bank*, 13 October 2017, blogs.worldbank.org/developmenttalk/richer-array-international-poverty-lines?CID=POV_TT_Poverty_EN_EXT.
58. Development Initiatives, *Poverty trends*.
59. The World Bank, 'COVID-19 to add as many as 150 million extreme poor by 2021', press release, 2020, www.worldbank.org/en/news/press-release/2020/10/07/covid-19-to-add-as-many-as-150-million-extreme-poor-by-2021.
60. Katherine R. Dean et al., 'Human ectoparasites and the spread of plague in Europe during the second pandemic', *PNAS*, 115/6 (2017): 1304–1309, www.pnas.org/content/115/6/1304.
61. C. J. Duncan and S. Scott, 'What caused the Black Death', *Postgraduate Medical Journal*, 81/955 (2004): 315–320, pmj.bmj.com/content/81/955/315.

62. Molly Blackall, 'Global tourism hits record highs – but who goes where on holiday?', *Guardian*, 1 July 2019, www.theguardian.com/news/2019/jul/01/global-tourism-hits-record-highs-but-who-goes-where-on-holiday.
63. Cited in Esteban Ortiz-Ospina and Diana Beltekian, *Trade and globalisation*, Our World in Data, 2018, ourworldindata.org/trade-and-globalization.
64. Ibid.
65. Kierra Box, 'The climate crisis – how international trade must change', *Friends of the Earth*, 8 February 2021, Policy.friendsoftheearth.uk/opinion/climate-crisis-how-international-trade-must-change.
66. S. M. Soto, 'Human migration and infectious diseases', *Clinical Microbiology and Infection*, 15/S1 (2009): 26–28, www.sciencedirect.com/science/article/pii/S1198743X14604080.
67. Michelle Gayer, 'Conflict and Emerging Infectious Diseases', *Emerging Infectious Diseases*, 13/11 (2007): 1625–1631, www.ncbi.nlm.nih.gov/pmc/articles/PMC3375795/.
68. Ibid.
69. Ibid
70. Mark Wheelis, 'Biological Warfare at the 1346 Siege of Caffa', *Emerging Infectious Diseases*, 8/9 (2002), wwwnc.cdc.gov/eid/article/8/9/01-0536_article.
71. Institute for Economics and Peace, *Ecological Threat Register 2020*, 2020 reliefweb.int/sites/reliefweb.int/files/resources/ETR_2020_web-1.pdf.
72. UNHCR, *UNHCR Projected Global Resettlement Needs 2021*, 2020, reliefweb.int/report/world/unhcr-projected-global-resettlement-needs-2021#:~:text=While%20in%202019%20over%2063%2C000,31%20countries%20will%20be%20met.
73. Laurie Laybourn-Langton, *Inheriting the earth? The unprecedented challenge of environmental breakdown for younger generations*, Institute of Public Policy Research, 2019, www.ippr.org/research/publications/inheriting-the-earth.
74. As per Michael Marmot et al., *The Marmot Review: 10 Years On*, Institute for Health Equity, 2020, www.instituteofhealthequity.org/resources-reports/marmot-review-10-years-on.
75. HM Government, *Contained and controlled: The UK's 20-year vision for antimicrobial resistance*, 2019, assets.publishing.service.gov.uk/government/uploads/system/uploads/attachment_data/file/773065/uk-20-year-vision-for-antimicrobial-resistance.pdf.
76. Ibid.
77. Nick Allen, 'Humans could live to 100, say scientists', *The Telegraph*, 27 May 2021, www.telegraph.co.uk/news/2021/05/27/humans-could-live-150-say-scientists/.
78. Kyree Leary, 'Ageing expert: The first person to live to 1,000 has already been born', *Futurism*, 1 December 2017, futurism.com/aging-expert-person-1000-born; Rob Waugh, 'The first immortals have already been born, Cambridge scientist claims', *Metro*, 14 April 2015, metro.co.uk/2015/04/14/

the-first-immortals-have-already-been-born-cambridge-scientist-claims-5149788/.

79. Laurie Laybourn-Langton et al., *Inheriting the earth?*, Institute for Public Policy Research, September 2019, www.ippr.org/files/2019-11/inheriting-the-earth-july19.pdf.
80. See Stian Westlake, *The end of the Treasury: Breaking up the UK's finance ministry*, Nesta, 3 September 2014, www.nesta.org.uk/report/the-end-of-the-treasury-breaking-up-the-uks-finance-ministry/.
81. For example, the Treasury Select Committee generated very unfavourable headlines for David Cameron, over his involvement in Lobbying for Greensill Capital to be given government contracts. Treasury Select Committee, 'Treasury Committee reports on 'lessons from Greensill Capital'', 2021, https://committees.parliament.uk/committee/158/treasury-committee/news/156684/treasury-committee-reports-on-lessons-from-greensill-capital/.
82. RAN, *The dirty dozen*, 2020, www.ran.org/bankingonclimatechaos2021/#data-panel.
83. Facing Finance, *Dirty profits 7*, 2019, www.facing-finance.org/files/2019/05/ff_dp7_ONLINE_v02.pdf.
84. Banks have long funded these activities. See also: War on Want, *Deadly investments*, 2017, waronwant.org/sites/default/files/Final%20Web%20version%20Deadly%20Investments.pdf.
85. Green New Deal Group, *A Green New Deal*, 2008, https://neweconomics.org/uploads/files/8f737ea195fe56db2f_xbm6ihwb1.pdf.
86. Luke Murphy et al., *Fairness and opportunity*, Institute of Public Policy Research, 2021, Ippr.org/research/publications/fairness-and-opportunity.
87. OECD, *Main science and technology indicators*, dataset, 2021, data.oecd.org/rd/gross-domestic-spending-on-r-d.htm.
88. Shreya Nanda and Chris Thomas, *The science-based economy: The role of health research*, Institute of Public Policy Research, 2020, www.ippr.org/files/2020-07/1594310646_the-science-based-economy-july20.pdf.
89. The exception to this are infectious diseases, where there is a very good case for disproportionate investment on a humanitarian basis.
90. Overall impact in this case means 'Quality-Adjusted Life Years' – a measure that accounts both for quality and length of life.
91. Shreya Nanda and Chris Thomas, *The science-based economy.*
92. Jon Sussex et al., 'Quantifying the economic impact of government and charity funding of medical research on private research and development funding in the United Kingdom', *BMC*, 14/32 (2016), bmcmedicine.biomedcentral.com/articles/10.1186/s12916-016-0564-z
93. For more on the worth of public and charity spending on medical research see King's College London, RAND Europe, Brunel University London, *Medical research: What's it worth?*, research briefing, 2018, acmedsci.ac.uk/file-download/54792223.

CHAPTER 6

1. Douglas Holtz-Eakin et al., *The Green New Deal: Scope, scale and implications*, American Action Forum, 2019, www.americanactionforum.org/research/the-green-new-deal-scope-scale-and-implications/.
2. Lord Ara Darzi et al., *Better health and care for all*, Institute for Public Policy Research, 2018, www.ippr.org/research/publications/better-health-and-care-for-all.
3. Chris Thomas, *Resilient health and care*, Institute of Public Policy Research, 2020, ippr.org/research/publications/resilient-health-and-care. This estimate uses latest estimates available, meaning it does not account for how similar nations healthcare spending will settle following the pandemic.
4. Chris Thomas et al., *The innovation lottery: Upgrading the spread of innovation in the NHS*, Institute of Public Policy Research, 2020, www.ippr.org/research/publications/the-innovation-lottery.
5. As per Chris Thomas, *Resilient health and care*.
6. As costed by Labour Party, *Care for carers*, 2020, labour.org.uk/wp-content/uploads/2020/06/CFCfinal-v2.pdf.
7. As concluded by Luke Raikes et al., *Divided and connected: Regional inequalities in the north, the UK and the developed world – State of the North 2019*, Institute of Public Policy Research North, 2019, www.ippr.org/research/publications/state-of-the-north-2019.
8. Dean Hochlaf and Chris Thomas, *The whole society approach*, Institute of Public Policy Research, 2020, www.ippr.org/research/publications/the-whole-society-approach.
9. UCL Institute for Global Prosperity, *Social prosperity for the future: A proposal for Universal Basic Services*, 2017, www.ucl.ac.uk/bartlett/igp/sites/bartlett/files/universal_basic_services_-_the_institute_for_global_prosperity_.pdf.
10. Department for Education, *Consolidated annual report and accounts*, HM Government, 2020, https://assets.publishing.service.gov.uk/government/uploads/system/uploads/attachment_data/file/932898/DfE_consolidated_annual_report_and_accounts_2019_to_2020__web_version_.pdf.
11. Labour for UBS, *Universal Basic Services manifesto*, 2020, labourforubs.files.wordpress.com/2020/09/labourforubs-policy-consultation-document.pdf.
12. Patel et al., *The state of health and care*.
13. Alfie Stirling and Joe Dromey, *Uncapped potential*, Institute of Public Policy Research, 2017, www.ippr.org/files/2017-11/uncapped-potential-november2017.pdf; London Economics, *The net exchequer impact of increasing pay for agenda for change staff*, 2021, londoneconomics.co.uk/blog/publication/the-net-exchequer-impact-of-increasing-pay-for-agenda-for-change-staff/.
14. OECD, *Environmental taxation*, data, 2020, www.oecd.org/env/tools-evaluation/environmentaltaxation.htm.
15. National Food Strategy, *The plan*, 2021, www.nationalfoodstrategy.org/.
16. Westminster could not introduce public health net zero across all four nations, as powers to this end are devolved.

17. Patel et al., *State of Health and Care.*
18. Chris Thomas, *Community first social care.*
19. Harry Quilter-Pinner, *Social care: Free at the point of need*, Institute of Public Policy Research, 2019, www.ippr.org/files/2019-05/social-care-free-at-the-point-of-need-may-19.pdf.
20. I.e., the £5 billion a year should not count towards government targets on R&D reaching certain per centages of UK GDP.
21. Harry Quilter-Pinner and Dean Hochlaf, *Austerity: There is an alternative and the UK can afford to deliver it*, Institute of Public Policy Research, 2019, www.ippr.org/blog/austerity-there-is-an-alternative-and-the-uk-can-afford-to-deliver-it.
22. Alfie Stirling and Sarah Arnold, 'Austerity by stealth?', *New Economics Foundation*, 17 September 2018, neweconomics.org/2018/09/austerity-by-stealth.
23. Carsten Jung et al., *Boost it like Biden*, Institute of Public Policy Research, 2021, www.ippr.org/research/publications/boost-it-like-biden.
24. Often thought to require its own £50–£100 billion investment programme – as per Tax Research UK, *Funding the UK Green New Deal*, 2019, www.taxresearch.org.uk/Blog/2019/09/16/funding-the-uk-green-new-deal/, or the commitment of £100 billion investment in the 2019 Green Party manifesto or Green New Deal Group, *A national plan for the UK*, 2013, greennewdealgroup.org/wp-content/uploads/2013/09/Green-New-Deal-5th-Anniversary.pdf.
25. Ibid.
26. I.e., Chris Thomas, *The 'make do and mend' health service: Solving the NHS's capital crisis*, Institute of Public Policy Research, 2019, www.ippr.org/files/2019-09/1568215446_the-make-do-and-mend-health-service-summary.pdf.
27. British Medical Association, *Workload undermines PCNs, finds survey*, 2020, www.bma.org.uk/news-and-opinion/workload-undermines-pcns-finds-survey.
28. Nicola Merrifield, 'GPs working average 11-hour day, major survey finds', *Pulse Today*, 1 April 2021, www.pulsetoday.co.uk/news/workload/gps-working-average-11-hour-day-major-survey-reveals/.
29. Chris Thomas and Harry Quilter-Pinner, *Realising the neighbourhood NHS*, Institute of Public Policy Research, 2020, www.ippr.org/research/publications/realising-the-neighbourhood-nhs.
30. NHS Digital, *General practice workforce*, data collection, 2019, digital.nhs.uk/data-and-information/publications/statistical/general-and-personal-medical-services.
31. Though not unheard off. It was called for by the then Labour MP and GP Paul Williams in 2019. Paul Williams, 'As a doctor, I say it's time to nationalise GP surgeries', *Guardian*, 8 May 2019, www.theguardian.com/commentisfree/2019/may/08/gp-nationalise-surgeries-profit-patients-nhs.
32. Probably because their policy is designed by councils of senior doctors who benefit from that status quo.

33. Recreated from Chris Thomas et al., *Levelling-up health for prosperity*, Institute of Public Policy Research/IPPR North, 2020, www.ippr.org/research/publications/levelling-up-health-for-prosperity; Data from Public Health England, *Public health profiles*, data, 2020, fingertips.phe.org.uk/.
34. Chris Thomas et al., *Levelling-up health for prosperity*.
35. The King's Fund, *A Citizen-led approach to health and care: lessons from the Wigan Deal*, 2019, www.kingsfund.org.uk/sites/default/files/2019-06/A_citizen-led_approach_to_health_and_care_lessons_from_the_Wigan_Deal_summary.pdf.
36. Ibid.
37. As per Centre for Public Impact, *The Wigan deal*, 2020, www.centreforpublicimpact.org/case-study/the-wigan-deal.
38. Preston City Council, CLES, *How we built community wealth in Preston*, 2019, www.preston.gov.uk/media/1792/How-we-built-community-wealth-in-Preston/pdf/CLES_Preston_Document_WEB_AW.pdf?m=636994067328930000.
39. Paul Dennett, 'Sensible socialism: The Salford Model', *Tribune*, 19 January 2021, tribunemag.co.uk/2021/01/sensible-socialism-the-salford-model.
40. Ipsos Mori, 'Britains are most likely to think their health care system is overstretched', press release, 1 December 2020, www.ipsos.com/ipsos-mori/en-uk/britons-are-most-likely-think-their-healthcare-system-overstretched.
41. Health Foundation, *What does the public think about NHS and Social Care Services?*, 2017, www.health.org.uk/publications/reports/what-does-the-public-think-about-nhs-and-social-care-services.

EPILOGUE

1. See Richard Sloggett, 'Why both sides want to play political football with the NHS', *Times Red Box*, 19 November 2019, www.thetimes.co.uk/article/why-both-sides-want-to-play-political-football-with-the-nhs-2m6xxjq9d.
2. Ipsos Mori, *Issues Index December 2020*, 2020, www.ipsos.com/sites/default/files/ct/news/documents/2020-01/issues_index_december2019_v1_public.pdf; Ipsos Mori, *Issues index archives*, dataset, www.ipsos.com/ipsos-mori/en-uk/issues-index-2007-onwards; Ipsos Mori, *Best party on key issues: Health care*, www.ipsos.com/ipsos-mori/en-uk/best-party-key-issues-health-care.
3. 1992 has the bearings of a Labour election victory, both in health and beyond. This is true for other issues too – the Conservatives won despite a long period in governance, rising unemployment rates, a long recession, high interest rates and house price collapse. In this case, Labour faced a perfect storm of structural problems – from unpopular leadership, to the media environment, to socio-demographic change – that help explain this outlier.

4. Indicatively, see Tom Harwood, 'Labour's NHS Scaremongering is Based on Shameless Lies', *Free Market Conservatives*, 2019, freemarketconservatives.org/labours-nhs-scaremongering-is-based-on-shameless-lies/.
5. Jack Peat, 'Two in five healthcare workers voted conservative in local elections', *The London Economic*, 2021, www.thelondoneconomic.com/politics/two-in-five-healthcare-workers-voted-conservative-in-local-elections-268479/.

Index

PLUTO PRESS PATREON SUPPORTER | SOLIDARITY LEVEL 4: PLUTONIAN